# Exam Ref AZ-104 Microsoft Azure Administrator

## Second Edition

Charles Pluta

# Exam Ref AZ-104 Microsoft Azure Administrator, Second Edition

Published with the authorization of Microsoft Corporation by: Pearson Education, Inc.

Copyright © 2025 by Pearson Education, Inc.
Hoboken, New Jersey

ISBN-13: 978-0-13-834593-8
ISBN-10: 0-13-834593-7

Library of Congress Control Number: 2024935895

3 2024

## TRADEMARKS

Microsoft and the trademarks listed at *http://www.microsoft.com* on the "Trademarks" webpage are trademarks of the Microsoft group of companies. All other marks are property of their respective owners.

## WARNING AND DISCLAIMER

## SPECIAL SALES

For information about buying this title in bulk quantities, or for special sales opportunities (which may include electronic versions; custom cover designs; and content particular to your business, training goals, marketing focus, or branding interests), please contact our corporate sales department at corpsales@pearsoned.com or (800) 382-3419.

For government sales inquiries, please contact governmentsales@pearsoned.com.

For questions about sales outside the U.S., please contact intlcs@pearson.com.

## CREDITS

EDITOR-IN-CHIEF
Brett Bartow

EXECUTIVE EDITOR
Loretta Yates

ASSOCIATE EDITOR
Shourav Bose

DEVELOPMENT EDITOR
Songlin Qiu

MANAGING EDITOR
Sandra Schroeder

SENIOR PROJECT EDITOR
Tracey Croom

COPY EDITOR
Brie Gyncild

INDEXER
Timothy Wright

PROOFREADER
Charlotte Kughen

TECHNICAL EDITOR
Jim Cheshire

EDITORIAL ASSISTANT
Cindy Teeters

COVER DESIGNER
Twist Creative, Seattle

COMPOSITOR
codeMantra

GRAPHICS
codeMantra

# Contents at a glance

# Contents

# Acknowledgments

I would like to acknowledge my wife, Jennifer, who has supported the unusual hours for projects such as this for over a decade now. I would also like to acknowledge my best friends and colleagues who allow me to bounce ideas off them, provide guidance to them, and share laughs with them: Elias Mereb, Joshua Waddell, Ed Gale, and Aaron Lines. Finally, I have to thank my manager, Julia Nathan, who has been an exemplary coach and role model and continues to support my work on projects such as this book.

# About the Author

**CHARLES PLUTA** is a technical consultant and Microsoft Certified Trainer (MCT) who has authored several certification exams, lab guides, and learner guides for various technology vendors. As a technical consultant, Charles has assisted small, medium, and large organizations by deploying and maintaining their IT infrastructure. He is also a speaker, a staff member, or a trainer at several large annual industry conferences. Charles has a degree in Computer Networking, and holds over 15 industry certifications. He makes a point to leave the United States to travel to a different country every year. When not working or traveling, he plays pool in Augusta, Georgia.

# Introduction

Some books take a very low-level approach, teaching you how to use individual classes and accomplish fine-grained tasks. Like the Microsoft AZ-104 certification exam, this book takes a high-level approach, building on your foundational knowledge of Microsoft Azure and common administrative actions to take in an Azure environment. We provide walk-throughs using the Azure portal; however, the exam might also include questions that use PowerShell or the Azure Command Line Interface (CLI) to perform the same task. You might encounter questions on the exam focused on these additional areas that are not specifically included in this *Exam Ref*.

This book covers every major topic area found on the exam, but it does not cover every exam question. Only the Microsoft exam team has access to the exam questions, and Microsoft regularly adds new questions to the exam, making it impossible to cover specific questions. You should consider this book a supplement to your relevant real-world experience and other study materials. If you encounter a topic in this book that you do not feel completely comfortable with, use the "Need more review?" links you'll find in the text to find more information and take the time to research and study the topic.

## Organization of this book

This book is organized by the "Skills measured" list published for the exam. The "Skills measured" list is available for each exam on the Microsoft Learn website: *microsoft.com/learn*. Each chapter in this book corresponds to a major topic area in the list, and the technical tasks in each topic area determine a chapter's organization. If an exam covers six major topic areas, for example, the book will contain six chapters.

## Preparing for the exam

Microsoft certification exams are a great way to build your resume and let the world know about your level of expertise. Certification exams validate your on-the-job experience and product knowledge. Although there is no substitute for on-the-job experience, preparation through study and hands-on practice can help you prepare for the exam. This book is *not* designed to teach you new skills.

We recommend that you augment your exam preparation plan by using a combination of available study materials and courses. For example, you might use the *Exam Ref* and another study guide for your at-home preparation and take a Microsoft Official Curriculum course for the classroom experience. Choose the combination that you think works best for you. Learn more about available classroom training, online courses, and live events at *microsoft.com/learn*.

Note that this *Exam Ref* is based on publicly available information about the exam and the author's experience. To safeguard the integrity of the exam, authors do not have access to the live exam.

## Microsoft certifications

Microsoft certifications distinguish you by proving your command of a broad set of skills and experience with current Microsoft products and technologies. The exams and corresponding certifications are developed to validate your mastery of critical competencies as you design and develop, or implement and support, solutions with Microsoft products and technologies both on-premises and in the cloud. Certification brings a variety of benefits to the individual and to employers and organizations.

> **MORE INFO** **ALL MICROSOFT CERTIFICATIONS**
>
> For information about Microsoft certifications, including a full list of available certifications, go to *microsoft.com/learn*.

## Access the exam updates chapter and online references

The final chapter of this book, "AZ-104 Azure Administrator exam updates," will be used to provide information about new content per new exam topics, content that has been removed from the exam objectives, and revised mapping of exam objectives to chapter content. The chapter will be made available from the link at the end of this section as exam updates are released.

Throughout this book are addresses to webpages that the author has recommended you visit for more information. We've compiled them into a single list that readers of the print edition can refer to while they read.

The URLs are organized by chapter and heading. Every time you come across a URL in the book, find the hyperlink in the list to go directly to the webpage.

Download the exam updates chapter and the URL list at *MicrosoftPressStore.com/ERAZ1042e/downloads*.

# Errata, updates & book support

We've made every effort to ensure the accuracy of this book and its companion content. You can access updates to this book—in the form of a list of submitted errata and their related corrections—at

*MicrosoftPressStore.com/ERAZ1042e/errata*

If you discover an error that is not already listed, please submit it to us at the same page.

For additional book support and information, please visit *MicrosoftPressStore.com/Support*.

Please note that product support for Microsoft software and hardware is not offered through the previous addresses. For help with Microsoft software or hardware, go to *support. microsoft.com*.

# Stay in touch

Let's keep the conversation going! We're on X/Twitter: *twitter.com/MicrosoftPress*.

# CHAPTER 1

# Manage Azure identities and governance

Microsoft has long been a leader in the identity space. This leadership goes back to the introduction of Active Directory (AD) with Windows 2000 before the cloud even existed. Microsoft moved into cloud identity with the introduction of Azure Active Directory (Azure AD), now Microsoft Entra ID, which is used by more than 5 million companies around the world. The adoption of Microsoft 365 led to this extended use of Entra ID. These two technologies, however, have very different purposes, with AD primarily used on-premises and Entra ID primarily used for the cloud.

Microsoft has poured resources into making on-premises AD and Entra ID work together. The concept is to extend the identity that lives on-premises to the cloud by synchronizing the identities. This ability is provided by Microsoft Entra Connect and Microsoft Entra Connect Sync. Microsoft has also invested in extending those identities to enable scenarios such as single sign-on by using Active Directory Federation Services (ADFS), which is deployed in many large enterprises. (Note that Entra Connect and Entra Connect Sync are not covered on the AZ-104 exam.)

Microsoft has continued pushing forward by developing options for developers to leverage Entra ID for their applications. Microsoft provides the ability for developers to extend a company's identity provider to users outside of the organization. The first option is known as Microsoft Entra External ID. This allows customers to sign in to applications using their social media accounts, such as a Facebook ID. A complementary technology—Entra ID B2B (Business to Business)—extends Entra ID to business partners.

This area of the AZ-104 exam is focused on the management of identities using Entra ID.

In the latter part of this chapter, you will also learn how to manage role-based access control (RBAC) for Azure resources, including the following topics:

- Understand how RBAC works
- Create a custom role assignment
- Provide access to Azure resources using different roles
- Interpret access assignment
- Manage multiple directories

Finally, you will learn how to manage Azure subscriptions and other resources. This includes how to

- Configure Azure Policy to ensure your Azure environment is governed in an effective way while maintaining the agility of the cloud
- Apply governance to Azure resource groups and their child resources through Azure Policy
- Create and manage resource locks
- Apply tags to Azure resources
- Manage the lifecycle of the resources that reside in resource groups
- Manage Azure subscriptions
- Configure management groups
- Govern cost management through quotas and resource tags

By understanding the controls that are available in Azure for subscription and resource management, you enable your organization for success across your Azure estate.

## Skills covered in this chapter:

- Skill 1.1: Manage Microsoft Entra users and groups
- Skill 1.2: Manage access to Azure resources
- Skill 1.3: Manage Azure subscriptions and governance

# Skill 1.1: Manage Microsoft Entra users and groups

In a Microsoft Entra tenant, there are users, groups, and devices that are controlled through the features of Entra discussed in this section. This section focuses on managing users and groups throughout their lifecycles, how to manage device settings, how to perform bulk updates to users using automation tooling such as PowerShell, and how to manage guest accounts.

The latter part of this section discusses how to manage Entra joined devices and how to configure user experience controls, such as self-service password reset (SSPR).

**This skill covers how to:**

- Create users and groups
- Manage user and group properties
- Manage licenses in Microsoft Entra ID
- Manage external users
- Configure Microsoft Entra ID Join
- Configure self-service password reset

# Create users and groups

There are primarily two types of users in Entra ID—cloud-only users and users synchronized from an on-premises directory. Cloud-only users are created and managed exclusively in Entra ID, and their attributes can be updated directly in Entra ID.

You can create cloud-only users through the Azure portal, Azure PowerShell, Azure command-line interface (CLI), or the Microsoft Entra Admin Center or by using the Microsoft Graph. When creating new users, you must be assigned to the Global Administrator or User Administrator role. See Skill 1.2 for more details about various roles and their assignments.

To create users from the Azure portal, type **Microsoft Entra ID** in the search box, or browse to All Azure Services and select Microsoft Entra ID as a user with rights to create users, click Users to open the Users blade, click New User, and click Create A New User. An example of this blade is shown in Figure 1-1. Note that you can also invite users (guest users) to your directory through the Azure portal.

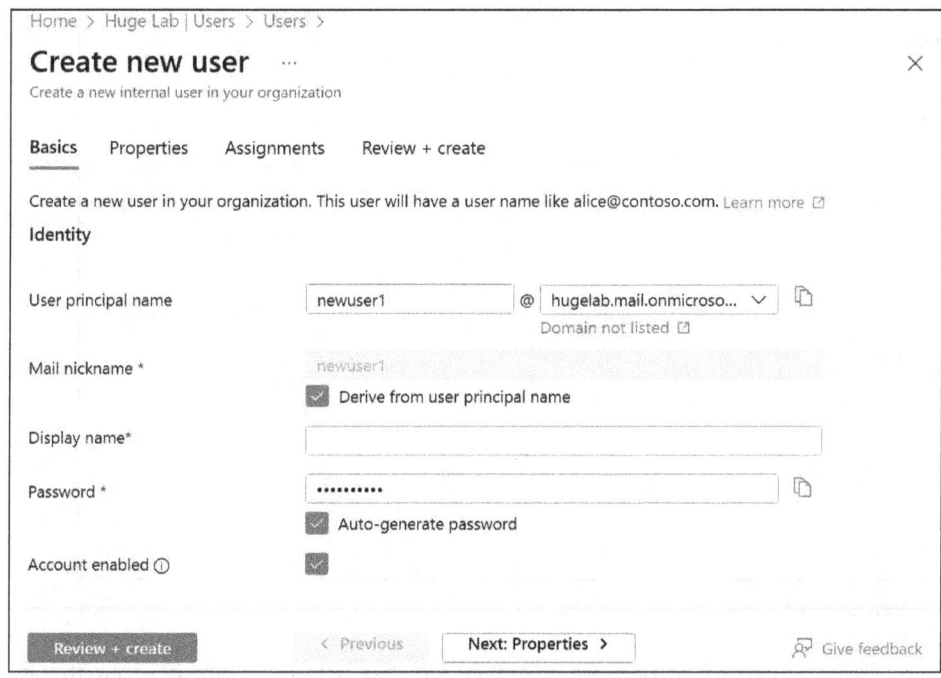

**FIGURE 1-1** Create New User blade in the Azure portal

When creating a new user, the User Principal Name (username), Display Name (the user's given name and surname), and Password fields are mandatory. You can configure additional settings, such as assigning specific groups and roles, blocking sign-ins from a specific location, and so on.

Groups are groups of objects that make role assignments and access permissions easier to manage. A group can contain groups, users, devices, or service principals. When using groups, you eliminate the need to individually assign roles or permissions. Creating groups is a similar experience to creating user accounts and can be performed from the Azure portal, Azure PowerShell, the Azure CLI, Microsoft Entra Admin Center, and Microsoft Graph. To create a group in the Azure portal, type **Microsoft Entra ID** in the Search field or browse to All Azure Services, select Microsoft Entra ID, click Groups to open the Groups blade, and click New Group. The New Group blade is shown in Figure 1-2.

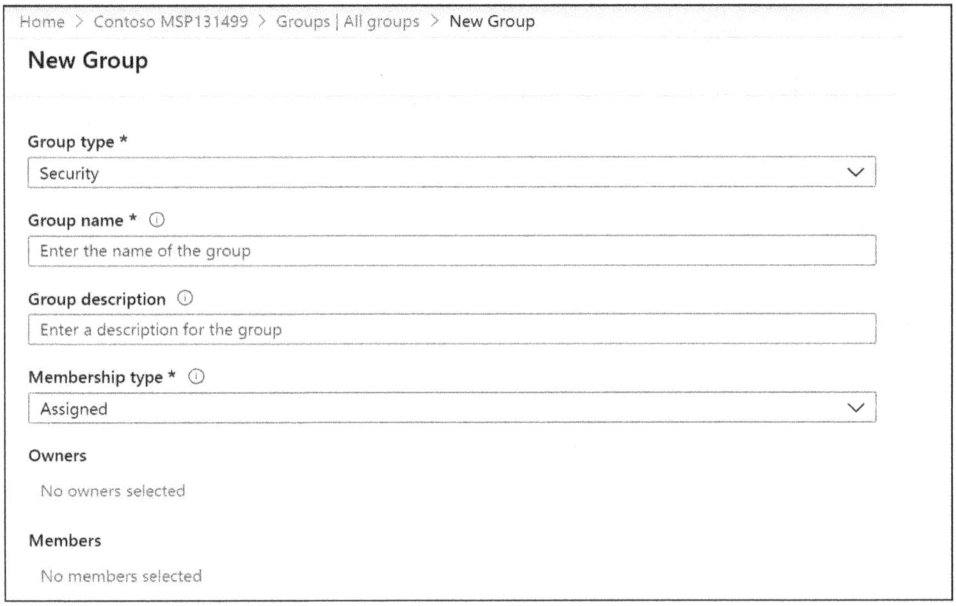

Home > Contoso MSP131499 > Groups | All groups > New Group

**New Group**

Group type *

    Security                                                                  ⌄

Group name * ⓘ

    Enter the name of the group

Group description ⓘ

    Enter a description for the group

Membership type * ⓘ

    Assigned                                                                  ⌄

Owners

    No owners selected

Members

    No members selected

**FIGURE 1-2**    New Group blade in the Azure portal

When you create a new group, there are several factors that dictate the type of group that is created and how that group behaves in Entra and associated workloads, such as Microsoft 365.

First, you must select the type of group you are creating. You have two options: Security and Microsoft 365. Security groups allow you to share Azure resources access to a group of users, devices, or service principals. A Microsoft 365 group allows access to a shared mailbox, calendar, SharePoint site, and so on. Note that even if you are creating groups in an Entra tenant that is not associated with a Microsoft 365 subscription, you will still see the option to create a Microsoft 365 group.

Also, Group Name is a required field. While filling in a Group Description is not required, it is recommended that you include a group description to make it easier to find and identify the purpose of a group later.

The Membership Type drop-down menu provides three options:

- **Assigned**   Use this option to select one or more users and add them to the group. Adding and removing users is performed manually.
- **Dynamic User**   Select this option to use dynamic group rules to automatically add and remove members.
- **Dynamic Device**   Select this option to use dynamic group rules to automatically add and remove devices.

> **IMPORTANT   DYNAMIC GROUP REQUIREMENT**
>
> You can create a dynamic group only if you have a Microsoft Entra ID P1 or P2 license. Otherwise, the Membership Type option is unavailable and is set to Assigned.

For both dynamic user and dynamic device-based groups, the rules associated with the group are evaluated on an ongoing basis. If a user or device has an attribute that matches the rule, that user or device is added to the group. If an attribute changes and the user or device no longer matches the criteria for group membership, the entity will be removed. Membership processing is not immediate. If an error occurs while processing a membership rule, an error is surfaced on the Group blade in the Azure portal. You can always view the current processing status from the Group blade.

It is important to note that you can create a dynamic group for users or devices, but you cannot create both at the same time. You also cannot use user attributes in a device-based rule. It is possible to change the membership type of a group after it has been created, which provides an opportunity to transition from a static (or assigned) membership model to a dynamic membership model or vice-versa.

When creating dynamic groups, rules can be edited in the simple rule format, where you will build the query and conditions in the rule builder, where you can build complex rules with conditional logic. In the example shown in Figure 1-3, a dynamic user group is being created, which will automatically update its membership based on the department attribute and its value in Entra ID.

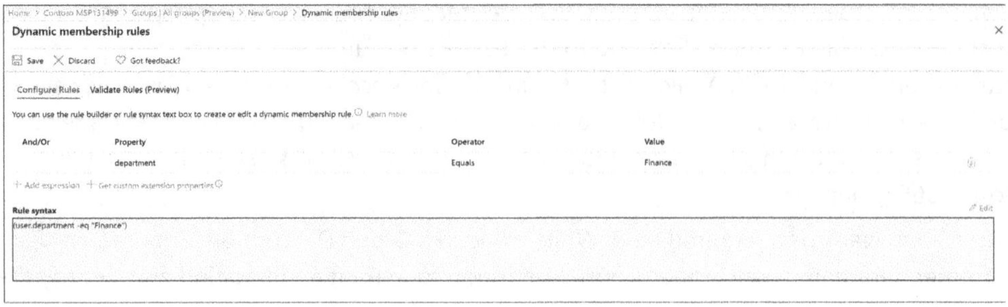

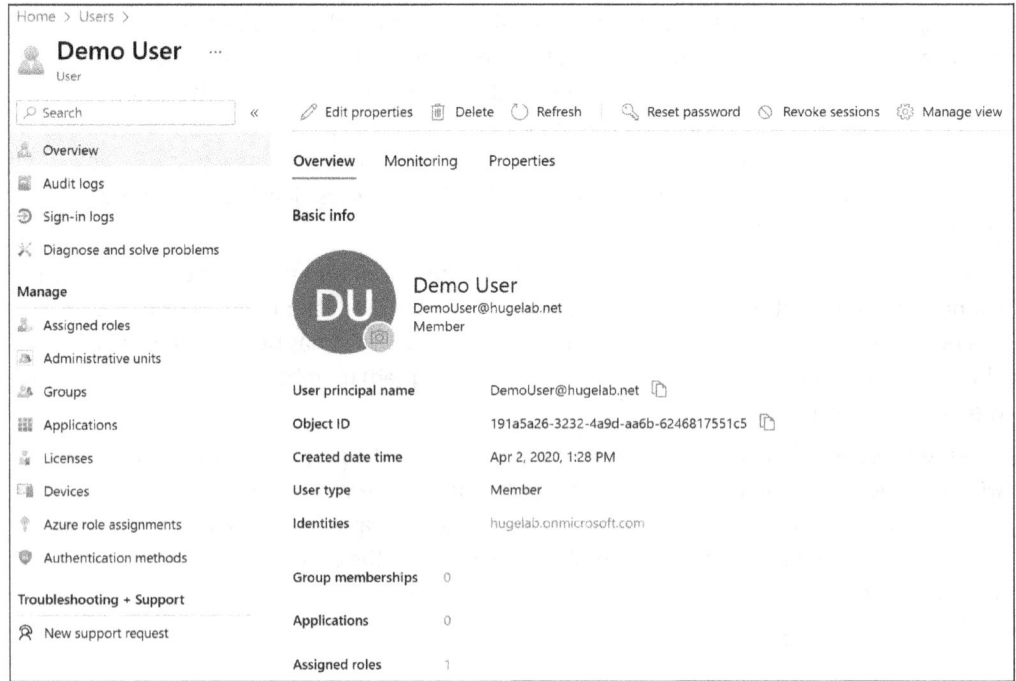

**FIGURE 1-3** Dynamic membership rules

Dynamic groups require an Entra ID Premium P1 or Premium P2 license.

# Manage user and group properties

As users and groups are used, they might need updates to their attributes (or properties). For example, you might need to change a user's job title, or you might need to add or remove members from an existing group.

Users and groups can be updated using management tools such as the Azure portal, Azure PowerShell, Azure CLI, and Microsoft Graph. Figure 1-4 shows an example of the user profile in the Azure portal that can be accessed by browsing to your Entra tenant, selecting Users, choosing a user, and clicking Edit Properties.

**FIGURE 1-4** A user profile in the Azure portal

Groups can be managed through the Azure portal by browsing to your Entra tenant, selecting Groups, choosing a specific group, and then clicking Properties, Members, or Owners, depending on the type of update you want to make. When editing a group, you will not be able to change the Group Type (such as changing a Security group to a Microsoft 365 group), but you will be able to update the Group Name, Group Description, and the Membership Type, as shown in Figure 1-5. Changing a static group to dynamic group will remove all the members from the static group and apply dynamic membership rules. This change will also affect the access to the resources if the static group has any previously assigned access for its members.

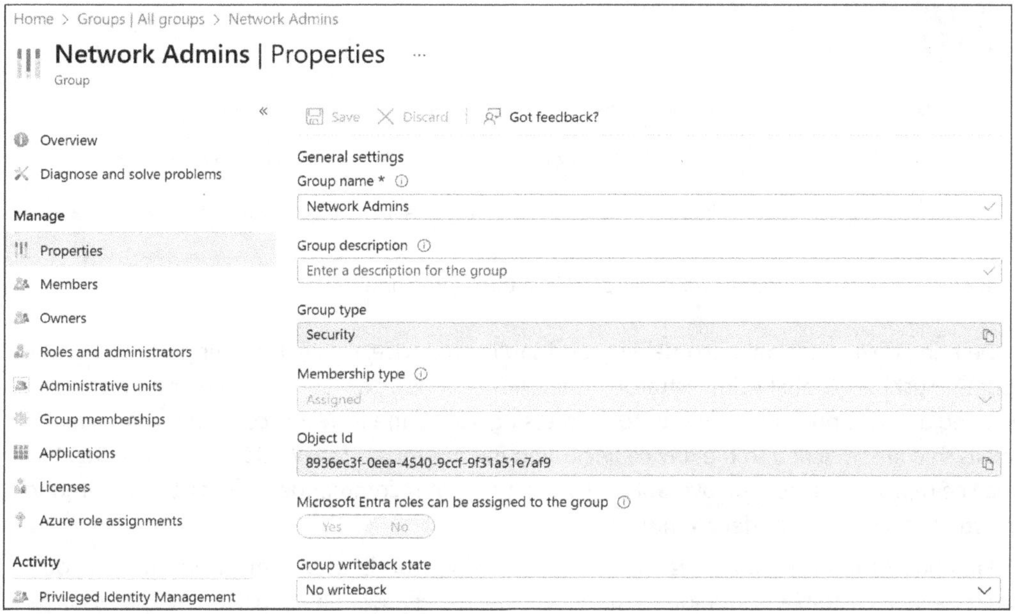

**FIGURE 1-5** Group properties in the Azure portal

Registered and joined devices in Entra ID can be managed in two areas in the Azure portal:

- Browse to your Entra tenant in the Azure portal, and select Devices. Overview is the default view, but you can also choose other views, such as All Devices, Device Settings, BitLocker Keys, and so on.
- Open the Devices blade for an individual user.

With either option, you will be able to search for devices using the device name as a filter, view a detailed overview of any registered and joined devices, and perform common device-management tasks.

To enable and disable devices, you must be a Global Administrator, Intune Administrator, or Cloud Device Administrator. Disabling a device prevents it from accessing Entra ID resources. Note that this does not prevent the user from accessing resources in general; it only prevents the user from accessing resources from that disabled device. Figure 1-6 shows the Disable option.

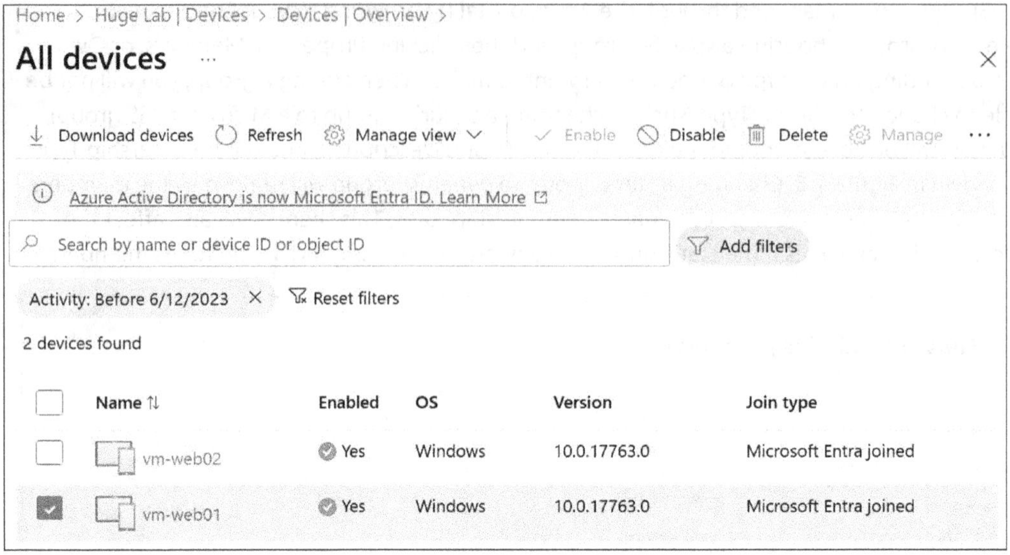

**FIGURE 1-6** Disable option in the All Devices blade in the Azure portal

Deleting devices is similar to enabling or disabling a device. Again, the user performing the update must be a Global Administrator, Intune Administrator, or Cloud Device Administrator. Deleting a device prevents a device from accessing your Entra ID resources and removes all details that are attached to the device (including BitLocker keys for Windows devices). Deleting a device represents a non-recoverable activity and is not recommended unless it is required for an activity such as device decommissioning.

Previously, the Azure portal was only helpful for single updates to users, which meant you had to rely on custom automation solutions (mostly using PowerShell) for updating users in bulk. Because of recent updates, you can now perform bulk operations (such as creating, inviting, and deleting users in batches) using the Azure portal as well as the Entra admin center at *https://entra.microsoft.com*.

You can access this functionality by navigating to your Entra tenant in the Azure portal and then clicking Users. You will see these options at the top of the blade, as shown in Figure 1-7.

**FIGURE 1-7** Bulk update options in the Users blade in the Azure portal

Clicking Bulk Create opens the Bulk Create User blade, which is shown in Figure 1-8.

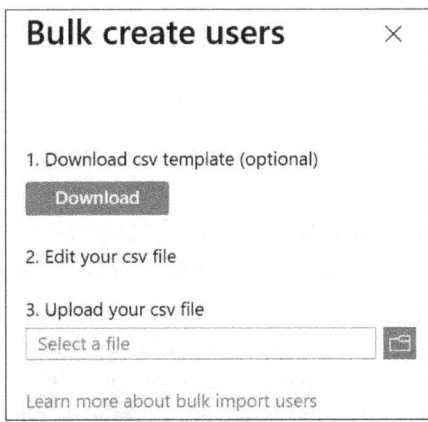

**FIGURE 1-8** Bulk Create Users blade in the Azure portal

Bulk user creation is a three-step process:

1. Click Download on the Bulk Create User blade to download a CSV (comma-separated values or comma-delimited) template (UserCreateTemplate.csv). This is a standard template with mandatory attributes, such as Name, User Name, Initial Password, and Block Sign In. You can also specify optional attributes such as First Name, Last Name, Job Title, and so on.

2. Edit the CSV file with bulk update values. You just need to update appropriate values and save the changes. The sample mandatory values are already included in the template for reference.

3. Upload the updated CSV file and submit the operation.

After submitting the operation, you can check the status of the bulk operation by navigating to Bulk Operation Results under the Activity section of the Users blade (see Figure 1-9).

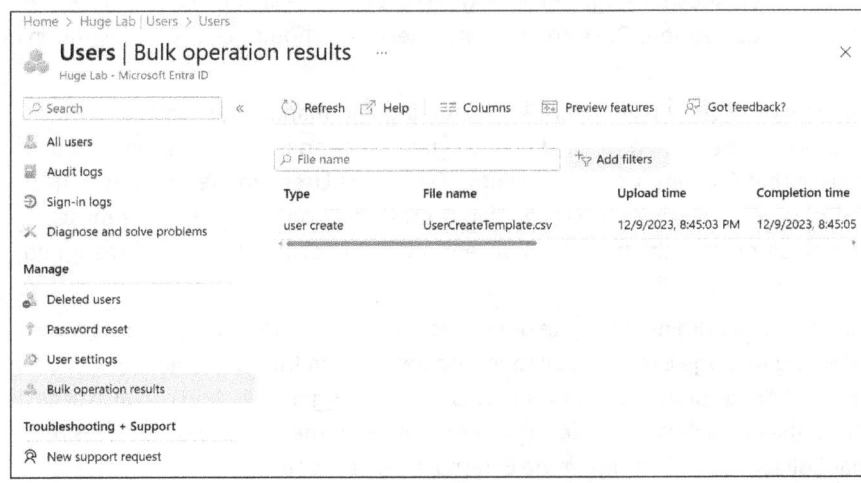

**FIGURE 1-9** Bulk Operation Results blade in the Azure portal

## Manage licenses in Microsoft Entra ID

There are a few different license types available with Entra ID:

- Microsoft Entra ID Free
- Microsoft Entra ID Premium P1
- Microsoft Entra ID Premium P2
- Microsoft Entra ID Governance

Note that either P1 or P2 licenses are included with other bundles and suites of licenses, such as the Enterprise Mobility + Security suite. To be able to assign a license to a user account, two things must first occur.

First, the license must be purchased and associated with the tenant. For small and medium businesses, you might be able to do this by using the Entra or Microsoft 365 admin portals. For Cloud Solution Providers and Enterprise Agreement organizations, you most likely need to speak with an account representative to get the licenses added to your contract.

Second, the user account that you plan to assign a license to must have their Usage Location property configured. The Usage Location property defines the primary country or region that the user resides or works in and can determine if certain features of a license can actually be used.

After purchasing the licenses and ensuring that your user accounts have their Usage Location defined, you can associate the licenses with users, or you can assign a license based on group membership. Using Dynamic Groups is a great way to automate license management based on user properties.

## Manage external users

To create guest users from the Azure portal, browse to your Entra tenant as a user with rights to create users, select the Users blade, choose New User, and then select Invite External User. An example of this blade is shown in Figure 1-10. A guest user can be anyone who is invited to collaborate with your organization. Once created, the guest user should receive an invitation in their mailbox.

Creating and managing guest users is similar to creating and managing normal user accounts. Guest users can be invited to the directory, group, or application. As soon as you invite the guest user, that account is created in Entra ID with the User Type set to Guest. The guest user will receive an email invitation immediately after creation. The guest user must accept the invitation along with the first-time consent process in order to access the assigned resources.

By default, all users and admins can invite guests. You can restrict the way guest users can be invited by selecting Manage External Collaboration Settings on the Users blade under User Settings. The External Collaboration Settings blade is shown in Figure 1-11. You can also access these settings from the Entra tenant by clicking User Settings on the left, and then choosing Manage External Collaboration Settings in the External Users section.

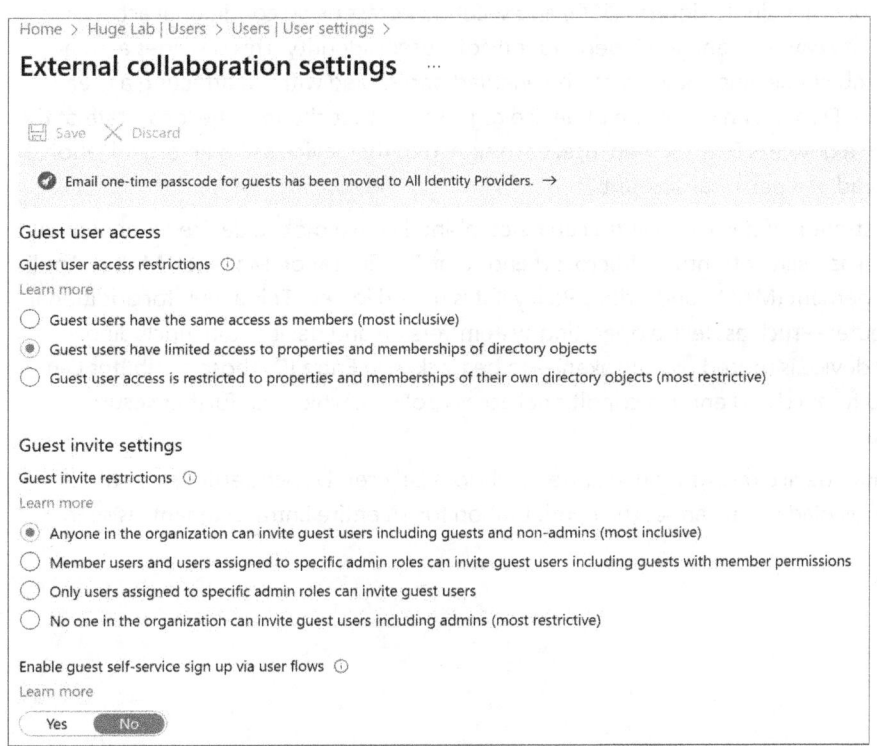

Home > Huge Lab | Users > Users >

## Invite external user ...
Invite an external user to collaborate with your organization

Basics    Properties    Assignments    Review + invite

Invite a new guest user to collaborate with your organization. The user will be emailed an invitation they can accept in order to begin collaborating.
Learn more ☑

**Identity**

Email ⓘ*               [                                        ]

Display name          [                                        ]

**Invitation message**

Send invite message   ☑

Message               [                                        ]

Cc recipient          [                                        ]

Review + invite    < Previous    Next: Properties >                          ⚲ Give feedback

**FIGURE 1-10**    Invite External User blade in the Azure portal

Home > Huge Lab | Users > Users | User settings >

## External collaboration settings ...

💾 Save    ✕ Discard

  ✅ Email one-time passcode for guests has been moved to All Identity Providers. →

**Guest user access**

Guest user access restrictions ⓘ
Learn more

  ○ Guest users have the same access as members (most inclusive)
  ◉ Guest users have limited access to properties and memberships of directory objects
  ○ Guest user access is restricted to properties and memberships of their own directory objects (most restrictive)

**Guest invite settings**

Guest invite restrictions ⓘ
Learn more

  ◉ Anyone in the organization can invite guest users including guests and non-admins (most inclusive)
  ○ Member users and users assigned to specific admin roles can invite guest users including guests with member permissions
  ○ Only users assigned to specific admin roles can invite guest users
  ○ No one in the organization can invite guest users including admins (most restrictive)

Enable guest self-service sign up via user flows ⓘ
Learn more
  ( Yes    No )

**FIGURE 1-11**    External Collaboration Settings blade in the Azure portal

When a guest user is added, the Consent Status for the guest user (viewable in PowerShell) is PendingAcceptance. This value will be changed to Accepted immediately after the guest user accepts the invitation. The guest user will appear as an "invited user" in the Azure portal until the user accepts the invitation.

## Configure Microsoft Entra Join

Microsoft Entra includes the ability to manage device identity, which enables single sign-on to devices and the applications and services managed through Entra that are accessed from that device. Managed devices include both enterprise and bring-your-own-device (BYOD) scenarios. This allows users to work from any device, including personal devices, all while protecting corporate intellectual property with the necessary regulatory and compliance controls.

Using Entra ID Join, you can control these devices, the applications installed and accessed from them, and how those applications interact with your corporate data.

When associating devices with Entra, you have three options: Register A Device, Join A Device, and Use Hybrid Joined. Registering devices would be appropriate for personal devices, while joining devices is useful for corporate-owned devices. Hybrid joined devices are joined to your on-premises Active Directory and are registered with your Entra ID tenant.

When you associate a device with Entra ID, you can manage a device's identity by implementing features like single sign-on (SSO) and securing access using conditional access. Note that this identity can be managed independently of a user's identity. This provides a great degree of flexibility because devices can be enabled or disabled without affecting a user account. Entra ID Join is an extension of device registration that changes the local state of the device. When a device is Entra-joined, users can sign in to the device using an organizational account instead of a personal account.

Also, registration of devices in Entra can be combined with a mobile device management solution, such as Microsoft Intune, Microsoft Endpoint Configuration Manager, Mobile Application Management (MAM), and Group Policy if it is hybrid joined. This allows for additional device attributes—such as device operating system version and device state (including whether the device is rooted or jailbroken)—to be tracked in Entra ID. Those attributes can then be used to build and enforce conditional access policies, which can further secure corporate data.

To configure device registration in Entra ID, choose Devices, Device Settings. On the Device Settings blade, you can set the configuration for an entire Entra ID tenant, as seen in Figure 1-12.

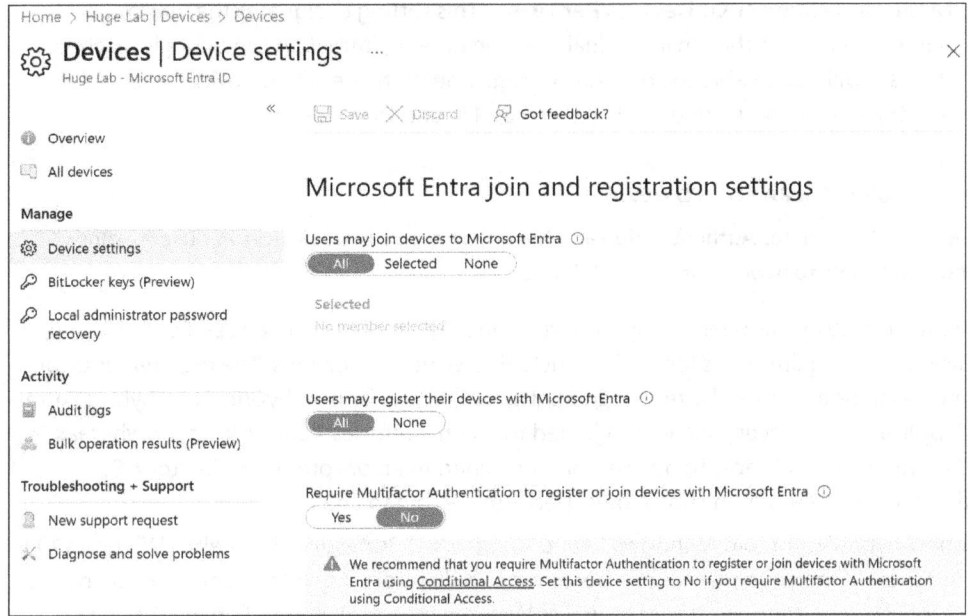

**FIGURE 1-12** Configure device registration settings

On this blade, you can configure the following settings:

- **Users May Join Devices To Microsoft Entra**   Use this setting to select the users and groups that can join devices to Entra. This setting only applies to Entra Join on Windows 10 or Windows 11 devices. The default value is All and can be changed to Selected or None.

- **Additional Local Administrators On Microsoft Entra Joined Devices**   With Entra ID Premium or with the Enterprise Mobility + Security suite, you can choose which users are granted Local Administrator rights to the device. Global Administrators and the device owner are granted Local Administrator rights by default. The default value is None and can be changed to Selected. If the value is set to Selected, any users added here are also added to the Device Administrators role in Entra ID.

- **Users May Register Their Devices with Microsoft Entra**   Allow users to register their devices with Microsoft Entra (Workplace Join). Enrollment with Microsoft Intune or Mobile Device Management for Office 365 requires device registration. If you have configured either of these services, All will be selected, and the button associated with the setting will be disabled.

- **Require Multifactor Auth To Join Devices**   Multifactor authentication (MFA) is recommended when adding devices to Entra. When set to Yes, users who are adding devices from the internet must first use a second method of authentication. Prior to enabling this setting, you must ensure that multifactor authentication is configured for the users who are able to register devices and that those users have set up MFA.

- **Maximum Number Of Devices Per User**  This setting designates the maximum number of devices that an individual user can have in Entra ID. If the quota is reached, the user will not be able to add a device until one of their existing devices is removed. Valid values for this setting are 5, 10, 20, 50, 100, and Unlimited.

> *NOTE*  **HYBRID ENTRA JOINED DEVICES**
>
> The Require Multifactor Authentication and Maximum Number Of Devices Per User settings are not applicable to hybrid Entra joined devices.

After the directory has been configured, you can begin registering devices. For Entra Join, there are several requirements for devices, including Windows versions. The requirements for Windows versions are driven by the type of Entra Join: hybrid or non-hybrid. Non-hybrid Entra Join is applicable to devices that are not joined to an on-premises Active Directory, whereas hybrid Entra Join is applicable to devices that are joined to an on-premises directory. For hybrid Entra Join, an IT administrator must perform the join to Entra ID.

For non-hybrid Entra Join, Windows 10 and Windows 11 Professional as well as Windows 10 and Windows 11 Enterprise devices can be joined to a directory. For hybrid Entra Join scenarios, you can join current Windows devices, such as Windows 11 and Windows Server 2016. Also, there is support for a hybrid join with down-level devices, including Windows 7, Windows 8.1, Windows Server 2008, Windows Server 2008 R2, Windows Server 2012, and Windows Server 2012 R2.

## Configure self-service password reset

The password reset is one of the highest cost-incurring activities for many organizations, and many organizations have dedicated front-line help desks to handle such requests. Self-service password reset (SSPR) allows users to reset their own passwords in Microsoft Entra ID, including the ability to optionally write the password back to an on-premises environment when properly licensed and configured by using password writeback and Entra Connect or Entra Connect Sync. SSPR allows users to change their passwords, reset their passwords when they cannot sign in, and unlock their accounts, all without the intervention of an IT department.

Each scenario above addresses both cloud-only and hybrid users. Also, licensing requirements vary. Table 1-1 details each scenario, the type of user it applies to, and any required licenses.

**TABLE 1-1** Self-service password reset license requirements

| Scenario | User Type | License Requirements |
|---|---|---|
| Password Change | Cloud-only user | Included in all license types of Entra ID |
| Password Reset | Cloud-only user | Microsoft 365 Business Standard, Microsoft 365 Business Premium, Entra ID P1, Entra ID P2 |
| Password Change/Unlock/Reset | Hybrid user | Microsoft 365 Business Premium, Entra ID P1, Entra ID P2 |

SSPR can be enabled through the Azure portal by browsing to your Entra tenant and selecting Password Reset. When enabling SSPR, you can scope the functionality to a group, which will allow you to roll out the feature in waves as users are onboarded into the service. As a part of configuration, you will also select the Authentication Methods for SSPR: Mobile App Notification, Mobile App Code, Email, Mobile Phone, Office Phone, and/or Security Questions (as shown in Figure 1-13). Finally, using the Registration blade, you will configure registration options such as whether registration is required to use SSPR and the number of days for reconfirmation.

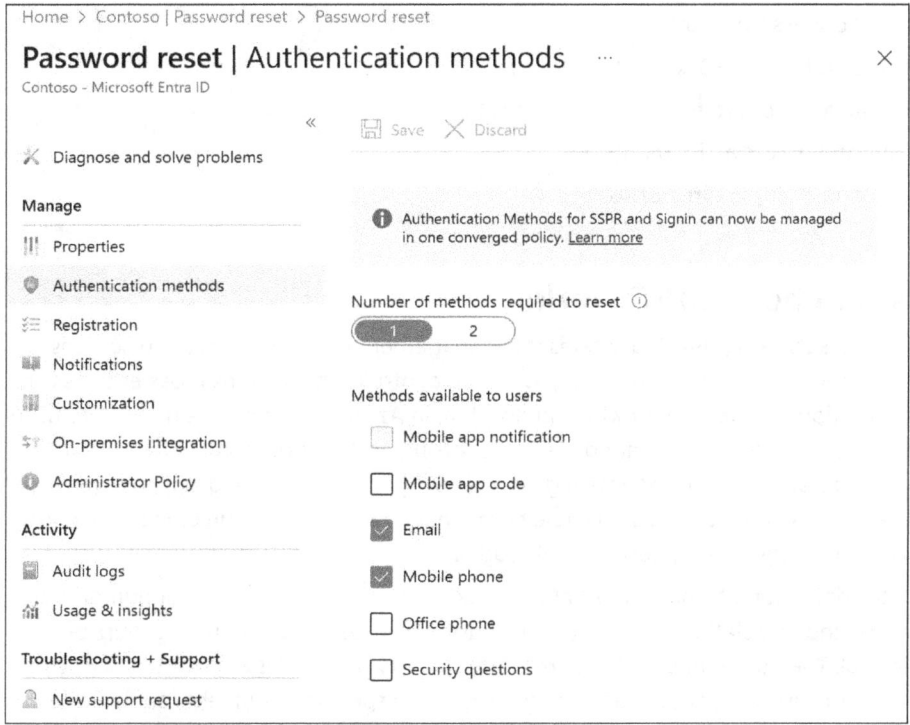

**FIGURE 1-13** Configure SSPR authentication methods

Additionally, you can also control how notifications are triggered to users and admins using the Notifications blade. There is an option available to customize a helpdesk link to notify the administrator directly, which can be configured using the Customization blade. If on-premises integration is enabled, you can also control writeback passwords to your on-premises directory and allow users to unlock accounts without resetting their passwords using the On-Premises Integration blade.

> *NEED MORE REVIEW?* **SELF-SERVICE PASSWORD RESET WRITEBACK**
>
> You can find details on self-service password reset writeback using the following link:
> *https://learn.microsoft.com/en-us/entra/identity/authentication/concept-sspr-writeback.*

# Skill 1.2: Manage access to Azure resources

Access control in Microsoft Azure is an important part of an organization's security and compliance requirements. Implementing role-based access control (RBAC) defines access rights at a very granular level, based on each user's assigned tasks or the day-to-day activities those users need to perform in their roles. This ensures that each person can perform the task they need to accomplish.

---

**This skill covers how to:**

- Understand how RBAC works
- Create a custom role
- Interpret access assignments
- Manage multiple directories

---

## Understand how RBAC works

Role-based access control (RBAC) facilitates the management of access to Azure resources by entities referred to as security principals, as well as controls what actions those entities can perform. In addition to determining who can do what, in Azure, access can be granted to users, groups, service principals, and managed identities through role assignments, which are then applied at a scope, such as a management group, subscription, a resource group, or even an individual resource. Azure RBAC is applicable to the management of resources created in the Azure Resource Manager (ARM) deployment model.

A role is the definition of what actions are allowed and/or denied. RBAC is configured by selecting a role and associating the role with a security principal, such as a user, group, or service principal. Then this combination of role and security principal is applied to a scope of a management group, subscription, a resource group, or a specific resource through a role assignment.

Azure RBAC also includes role inheritance, where child resources inherit the role assignments of any parents. For example, if a user is granted Reader access to a subscription, that user will have Reader access to all the resource groups and resources in that subscription. If a managed identity is granted Contributor rights for a single resource group, that security principal can only interact with that resource group and its child resources, but it cannot create new resource groups or access resources in other resource groups unless an explicit role assignment is made.

Azure RBAC uses the additive model. As you begin to apply roles to security principals in Azure, it is not uncommon to have overlapping assignments where a security principal is assigned a different role assignment at both a parent and a child scope. For example, if a user is granted Contributor rights at the management group scope and then is granted Reader rights in a subscription, the user will still have Contributor rights across the subscription along with

Contributor rights to any other subscriptions under the management group. Another way to think of this is that the most privileged access right takes precedence.

Before a security principal such as a user or group can interact with Azure resources, they must be granted access at a scope through a role assignment. Once a security principal has been granted access, it can perform any action that it has rights to perform. It is always recommended to provide the minimum privileges to an object or user to perform actions as needed. Figure 1-14 shows a suggested access pattern that adheres to the principles of least privilege. In this example, a security group in Entra ID, called IT Audit, is granted Reader access rights at the subscription scope, granting members of the group Reader access to all resource groups and resources in the subscription. A security group called Application Admins is granted Contributor access rights to only selected resource groups. Another security group called Application Owners is granted Owner access rights to selected resource groups as well. By using multiple security groups and role assignments at the proper scope, access can be granted in the future just by updating the security group membership in Entra ID.

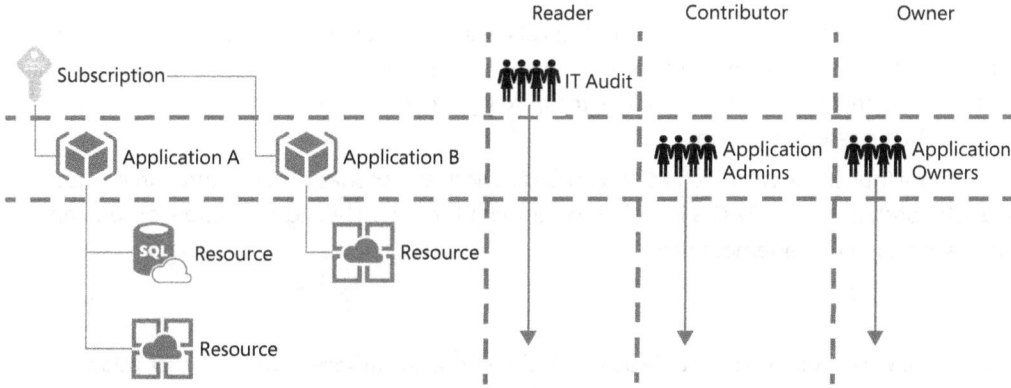

**FIGURE 1-14** Azure RBAC role assignments

> **IMPORTANT  USING GROUPS WITH AZURE RBAC**
>
> RBAC role definitions are associated with a user, group, service principal, or managed identity via a role assignment. When assigning roles to a group, all users in the group will inherit the assigned role. You can assign roles to a group for easier management and greater flexibility when applying RBAC at scale.

The specific permissions that are applied to a resource with RBAC are defined in a role definition. A role definition contains the list of permissions—or declared permissions—and those permissions define what actions can or cannot be performed against a type of resource, such as read, write, or delete.

Role definitions, or roles, can be either built-in or custom. There are a number of built-in role definitions in Azure. An example of a built-in role is the Owner role, which includes permissions to manage resources, security, and the application of role assignments. Also, there are built-in roles with limited permission sets, such as a Storage Blob Data Reader, which allows the assigned security principal to only read and list containers and blobs.

There are many built-in roles in Azure, which can be found at *https://docs.microsoft.com/azure/role-based-access-control/built-in-roles*. Microsoft adds new built-in roles as services evolve or as new services are introduced.

> **IMPORTANT  AZURE ROLES AND ENTRA ID ROLES**
>
> RBAC roles are different from the Entra ID administrative roles. RBAC roles are used to manage access and allow or restrict users to Azure resources, while Entra ID administrative roles are used to allow or restrict admins to perform identity tasks, such as creating new users, resetting the users' passwords, and so on. For example, a user who is granted Global Administrator rights in Entra ID does not have permissions to create resources in Azure, but they can perform all the identity tasks for an Entra tenant.

The access rights are controlled with a logical boundary known as scope. For example, to grant a user Contributor rights to all the resources in a resource group, the Contributor role can be assigned to the group at the resource group scope where it is then inherited by all of the resources in the resource group.

There are four scopes at which RBAC can be applied, and scopes are structured in a parent-child relationship where RBAC is inherited by any child scopes. The highest scope, or top-most parent scope, is a management group.

> **EXAM TIP**
>
> Management groups are not applicable in all scenarios, and in some cases, a subscription will be the highest scope you will work with when applying role assignments. This will be determined by your organization's Azure landing zone deployment stamp.

Under the management group are more management groups and/or subscriptions; under subscriptions are resource groups; and under resource groups are resources. Figure 1-15 shows a sample hierarchy with a parent management group and two subscriptions, each with a resource group and child resources. Note that you can also create another management group under a root management group. An Entra ID tenant can support up to 10,000 management groups.

> **IMPORTANT  RBAC INHERITANCE**
>
> The concept of RBAC inheritance is critical. Granting a user access to the Owner role at the management group scope will grant that user Owner rights to all the resources (management groups, subscriptions, resource groups, and resources) under the management group that is inclusive of all the resource groups and resources within them.

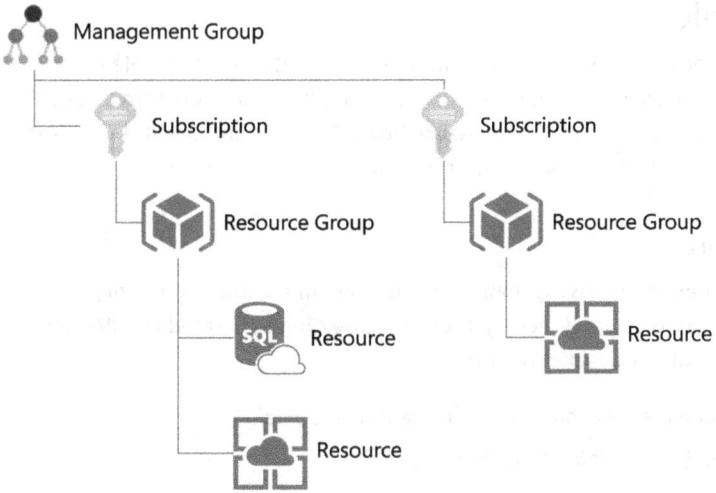

**FIGURE 1-15** Scope hierarchy

After you have identified the role, security principal, and scope at which the role will be assigned, you can make the assignment. Remember, security principals do not have access to Azure resources until a role assignment is made, and that access can be revoked by removing a role assignment.

> **IMPORTANT    ROLE ASSIGNMENT LIMITS**
>
> You can have up to 4,000 role assignments in each subscription, and you can have up to 500 role assignments per management group. These limits are independent of each other and have been adjusted over time. Check the latest quotas and limits at *https://learn.microsoft. com/en-us/azure/azure-resource-manager/management/azure-subscription-service-limits*.

To create and remove role assignments, you must have Microsoft.Authorization/role-Assignments/* permission at the necessary scope. This permission is granted through the Owner or User Access Administrator built-in roles, or it can be included in custom roles.

> **NOTE    AZURE ROLE ASSIGNMENTS**
>
> With Azure role assignments, there is no way to revoke access rights at a child scope through the application of a more restrictive role assignment because the role assignment is inherited from the parent. It is, however, possible to apply a deny assignment at a scope when using Azure Blueprints, Deployment Stacks, and resource locks. Deny assignments are evaluated before role assignments and can be used to exclude service principals from accessing child scopes. For more information, see *https://learn.microsoft.com/en-us/azure/governance/ blueprints/tutorials/protect-new-resources*.

# Create a custom role

In addition to built-in roles available in Azure, you might need to create a custom role to provide a set of permissions that are not available in any of the built-in roles. Custom roles can be created and assigned through the Azure portal, Azure PowerShell, Azure CLI, and REST API. This chapter primarily covers how to create a custom role using the Azure portal.

> **IMPORTANT** **CUSTOM ROLES**
>
> Custom roles can be shared between subscriptions that trust the same Entra ID directory. There is a limit of 5,000 custom roles per directory, though Azure China operated by 21Vianet can have up to 2,000 custom roles for each directory.

There are three ways you can create custom roles in the Azure portal:

- Clone from the existing built-in roles available
- Start from scratch
- Start from a JSON file to define the custom permissions

To clone a built-in role, open the Access Control (IAM) blade by accessing a subscription or resource group and then choosing Add, Add Custom Role, as shown in Figure 1-16.

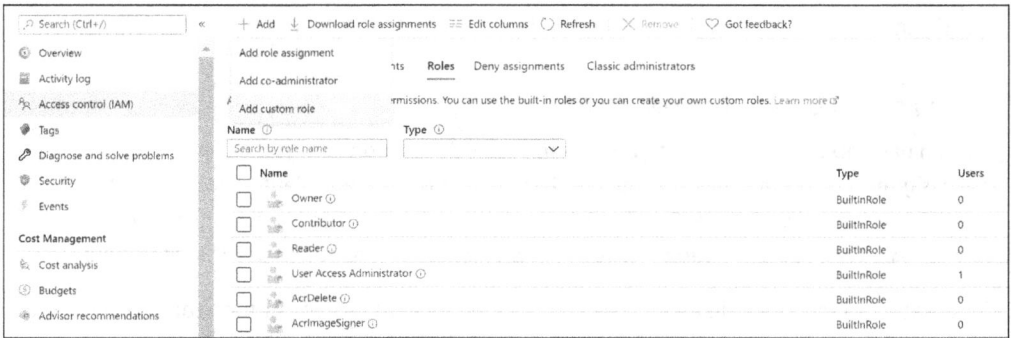

**FIGURE 1-16** Add Custom Role option in the Access Control (IAM) blade

On the Create A Custom Role blade, next to Baseline Permissions, select Clone A Role. Next, from the Role To Clone drop-down menu, select the desired role, such as Virtual Machine Contributor, as shown in Figure 1-17. You can select the role with the nearest identical permissions from the built-in roles.

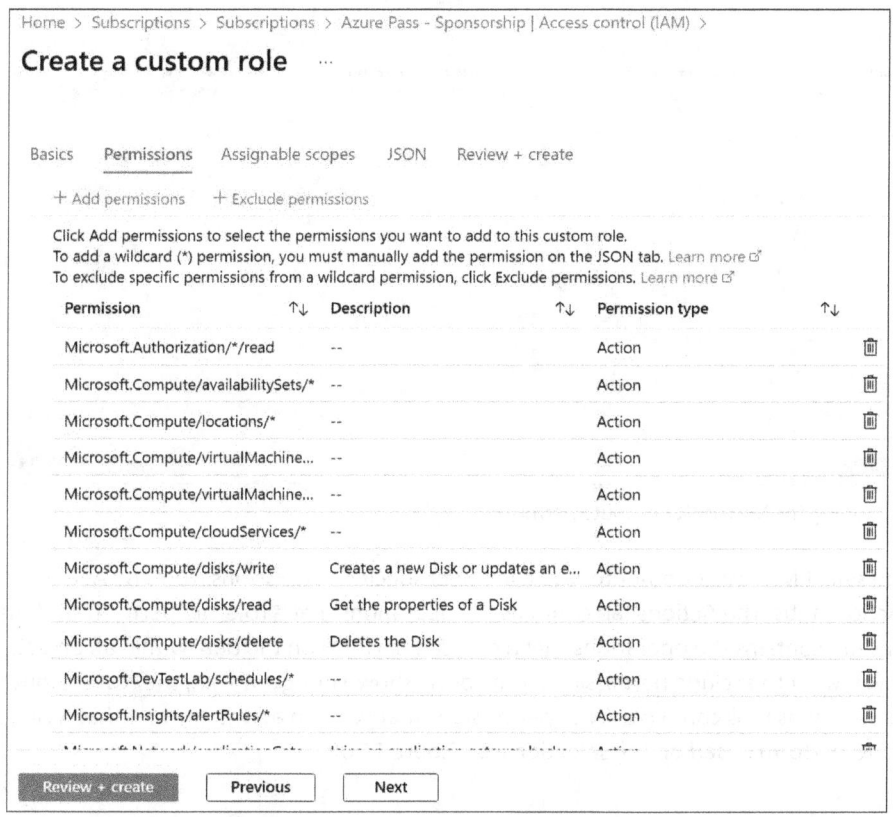

**FIGURE 1-17** Creating a custom role

Click Next to open the Permissions tab, shown in Figure 1-18. This tab displays all the permissions associated with the built-in role you selected in the Basics tab.

Home > Subscriptions > Subscriptions > Azure Pass - Sponsorship | Access control (IAM) >

## Create a custom role ...

Basics    **Permissions**    Assignable scopes    JSON    Review + create

+ Add permissions    + Exclude permissions

Click Add permissions to select the permissions you want to add to this custom role.
To add a wildcard (*) permission, you must manually add the permission on the JSON tab. Learn more ☐
To exclude specific permissions from a wildcard permission, click Exclude permissions. Learn more ☐

| Permission ↑↓ | Description ↑↓ | Permission type ↑↓ | |
|---|---|---|---|
| Microsoft.Authorization/*/read | -- | Action | 🗑 |
| Microsoft.Compute/availabilitySets/* | -- | Action | 🗑 |
| Microsoft.Compute/locations/* | -- | Action | 🗑 |
| Microsoft.Compute/virtualMachine... | -- | Action | 🗑 |
| Microsoft.Compute/virtualMachine... | -- | Action | 🗑 |
| Microsoft.Compute/cloudServices/* | -- | Action | 🗑 |
| Microsoft.Compute/disks/write | Creates a new Disk or updates an e... | Action | 🗑 |
| Microsoft.Compute/disks/read | Get the properties of a Disk | Action | 🗑 |
| Microsoft.Compute/disks/delete | Deletes the Disk | Action | 🗑 |
| Microsoft.DevTestLab/schedules/* | -- | Action | 🗑 |
| Microsoft.Insights/alertRules/* | -- | Action | 🗑 |

[ Review + create ]    [ Previous ]    [ Next ]

**FIGURE 1-18** Add or exclude permissions while creating a custom role

When you click Add Permissions, you can search from all the different permissions available from the catalog. For example, type **virtual machine**, as shown in Figure 1-19. You can select Microsoft Compute to access operations available for this resource provider.

> **IMPORTANT    ARM RESOURCE MANAGER PROVIDER OPERATIONS**
>
> To explore all the operations available for each Azure Resource Manager resource provider, see *https://docs.microsoft.com/en-us/azure/role-based-access-control/resource-provider-operations*.

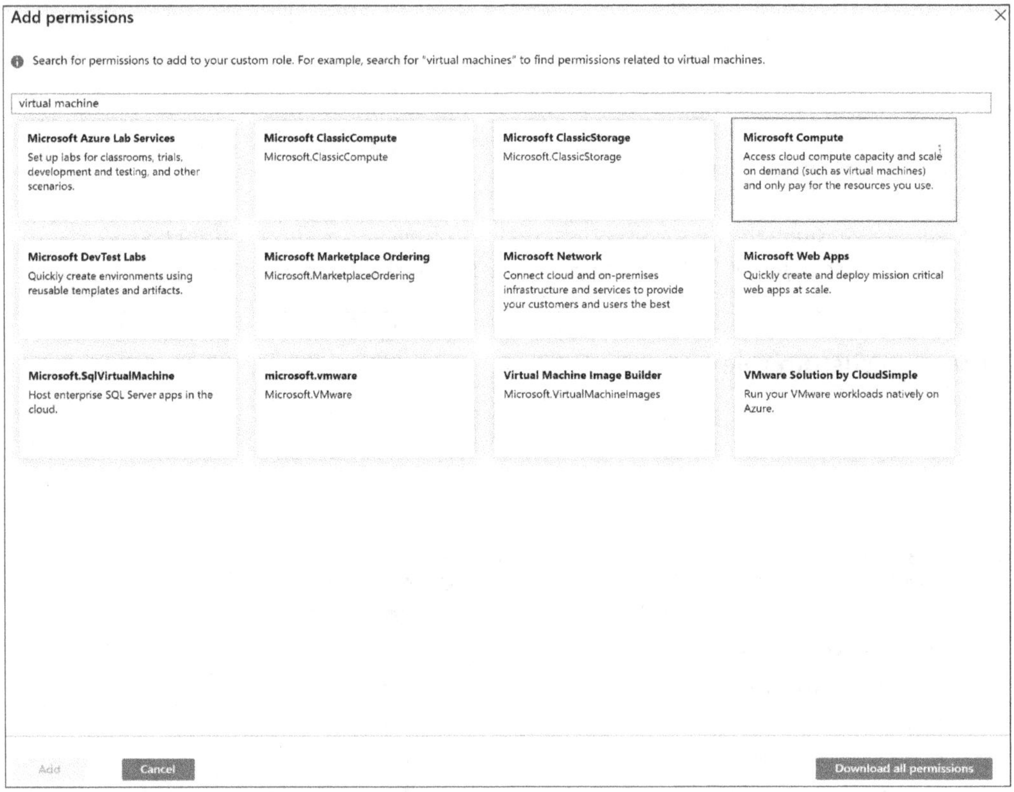

**FIGURE 1-19**   Adding the Microsoft Compute permission

Once you select Microsoft Compute, you can select specific permissions from the Actions and Data Actions tabs. The Actions tab contains the operations that a role can perform, and the Data Actions tab contains the operations that a role can perform on the data within an object. Similarly, if you want to exclude permissions (previously shown in Figure 1-18), the Not Actions and Not Data Actions tabs contain a list of permissions that you can add to the role that it is not allowed to perform based on the selection (see Figure 1-20).

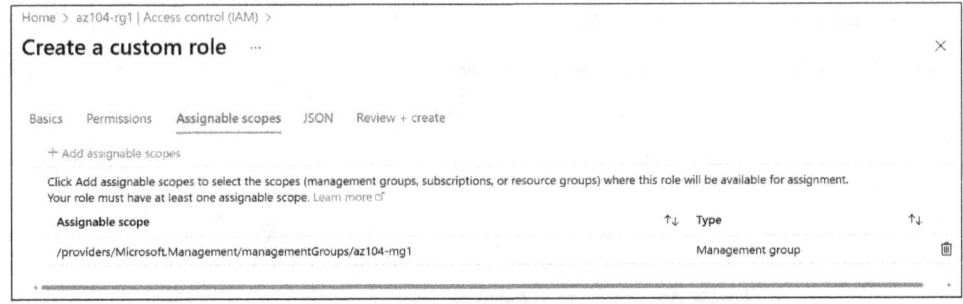

**Microsoft.Compute permissions**

‹ All resource providers

ⓘ Search for permissions to add to your custom role. For example, search for "virtual machines" to find permissions related to virtual machines.

virtual machine

(Actions) Data Actions

☑ Permission | Description

**Microsoft.Compute/availabilitySets/vmSizes**

☐ Read : List Virtual Machine Sizes for Availability Set ⓘ | List available sizes for creating or updating a virtual machine in the availability set

**Microsoft.Compute/locations/vmSizes**

☑ Read : List Available Virtual Machine Sizes in Location ⓘ | Lists available virtual machine sizes in a location

**Microsoft.Compute/locations/vsmOperations**

☐ Read : Get Operation for Virtual Machine Scale Set with the Virtual Machine Runtime Service Extension ⓘ | Gets the status of an asynchronous operation for Virtual Machine Scale Set with the Virtual Machine Runtime Service Extension

**Microsoft.Compute/virtualMachineScaleSets**

☐ Read : Get Virtual Machine Scale Set ⓘ | Get the properties of a Virtual Machine Scale Set

☐ Write : Create or Update Virtual Machine Scale Set ⓘ | Creates a new Virtual Machine Scale Set or updates an existing one

☐ Delete : Delete Virtual Machine Scale Set ⓘ | Deletes the Virtual Machine Scale Set

☐ Other : Delete Virtual Machines in a Virtual Machine Scale Set ⓘ | Deletes the instances of the Virtual Machine Scale Set

☐ Other : Start Virtual Machine Scale Set ⓘ | Starts the instances of the Virtual Machine Scale Set

☐ Other : Power Off Virtual Machine Scale Set ⓘ | Powers off the instances of the Virtual Machine Scale Set

☐ Other : Restart Virtual Machine Scale Set ⓘ | Restarts the instances of the Virtual Machine Scale Set

☐ Other : Deallocate Virtual Machine Scale Set ⓘ | Powers off and releases the compute resources for the instances of the Virtual Machine Scale Set

☐ Other : Manual Upgrade Virtual Machine Scale Set ⓘ | Manually updates instances to latest model of the Virtual Machine Scale Set

☐ Other : Reimage Virtual Machine Scale Set ⓘ | Reimages the instances of the Virtual Machine Scale Set

☐ Other : Reimage all Disks for a Virtual Machine Scale Set ⓘ | Reimages all disks (OS Disk and Data Disks) for the instances of a Virtual Machine Scale Set

Add     Cancel

**FIGURE 1-20** Permission list under the Actions tab

After you select the required permissions, you must select Add Assignable Scopes to define a scope for this custom role. The scope can be defined as a Management Group, Sub-scription, Resource Group, or Resource Level. The custom role must have at least one valid scope assigned (see Figure 1-21).

Home › az104-rg1 | Access control (IAM) ›

**Create a custom role**   ⋯                                              ✕

Basics   Permissions   **Assignable scopes**   JSON   Review + create

+ Add assignable scopes

Click Add assignable scopes to select the scopes (management groups, subscriptions, or resource groups) where this role will be available for assignment. Your role must have at least one assignable scope. Learn more ⬚

**Assignable scope**   ↑↓ | **Type**   ↑↓

/providers/Microsoft.Management/managementGroups/az104-mg1 | Management group   🗑

**FIGURE 1-21** Assignable scopes selection while creating a custom role

Select Next or the JSON tab to display the JSON (JavaScript Object Notation) code based on the selection made on the prior screens. This code can be downloaded as a .json file, or it can be copied to reuse later. You can proceed to the Review + Create tab to create the custom role (see Figure 1-22).

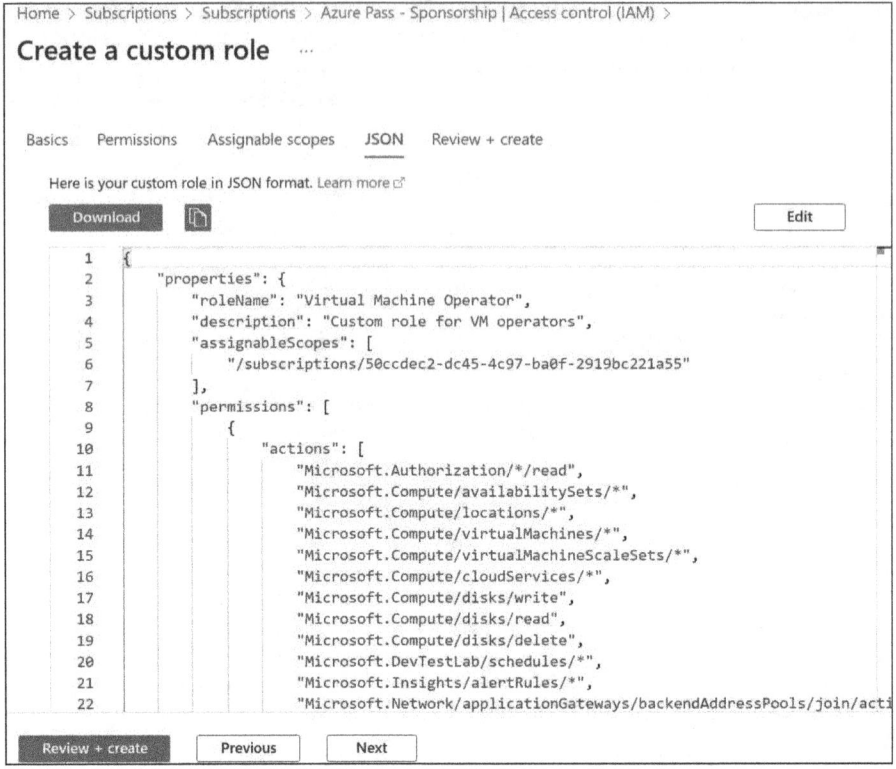

Home > Subscriptions > Subscriptions > Azure Pass - Sponsorship | Access control (IAM) >

## Create a custom role ...

Basics    Permissions    Assignable scopes    **JSON**    Review + create

Here is your custom role in JSON format. Learn more ☐

[Download]  [copy icon]                                                    [Edit]

```
 1  {
 2      "properties": {
 3          "roleName": "Virtual Machine Operator",
 4          "description": "Custom role for VM operators",
 5          "assignableScopes": [
 6              "/subscriptions/50ccdec2-dc45-4c97-ba0f-2919bc221a55"
 7          ],
 8          "permissions": [
 9              {
10                  "actions": [
11                      "Microsoft.Authorization/*/read",
12                      "Microsoft.Compute/availabilitySets/*",
13                      "Microsoft.Compute/locations/*",
14                      "Microsoft.Compute/virtualMachines/*",
15                      "Microsoft.Compute/virtualMachineScaleSets/*",
16                      "Microsoft.Compute/cloudServices/*",
17                      "Microsoft.Compute/disks/write",
18                      "Microsoft.Compute/disks/read",
19                      "Microsoft.Compute/disks/delete",
20                      "Microsoft.DevTestLab/schedules/*",
21                      "Microsoft.Insights/alertRules/*",
22                      "Microsoft.Network/applicationGateways/backendAddressPools/join/acti
```

[Review + create]    [Previous]    [Next]

**FIGURE 1-22**    JSON view while creating a custom role

Newly created custom roles can be accessed from the Roles tab (see Figure 1-23). Custom roles appear in the Azure portal with an orange resource icon.

+ Add    ☰☰ Edit columns    ⟳ Refresh    ✕ Remove    ♡ Got feedback?

Check access    Role assignments    Deny assignments    Classic administrators    **Roles**

A role definition is a collection of permissions. You can use the built-in roles or you can create your own custom roles. Learn more ☐

Name ⓘ                              Type ⓘ
[virtual machine]                   [                        ▽]

Showing 5 of 168 roles

| | Name | Type |
|---|---|---|
| ☐ | Classic Virtual Machine Contributor ⓘ | BuiltInRole |
| ☐ | Virtual Machine Administrator Login ⓘ | BuiltInRole |
| ☐ | Virtual Machine Contributor | BuiltInRole |
| ☑ | Virtual Machine Operator ⓘ | CustomRole |
| ☐ | Virtual Machine User Login ⓘ | BuiltInRole |

Custom role for Virtual Machine Operators

**FIGURE 1-23**    Roles selection while creating a custom role

Alternatively, built-in roles can be cloned by selecting a role from the Roles tab. For example, you could select Virtual Machine Contributor, click the ellipsis (...), and then select Clone (see Figure 1-24).

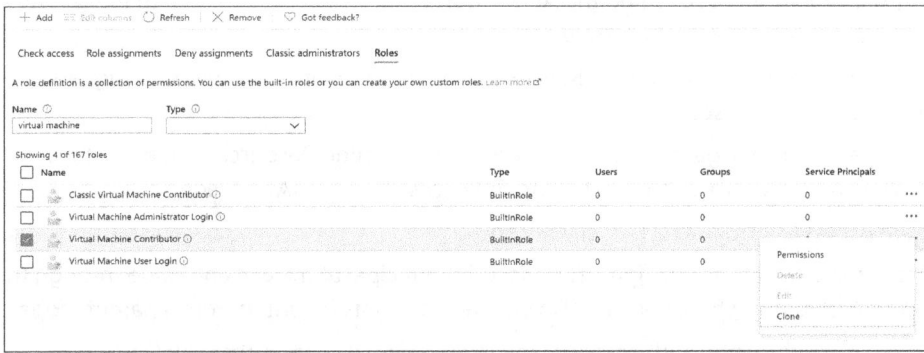

**FIGURE 1-24** Cloning a role

> **IMPORTANT   REQUIRED PERMISSION TO CREATE A CUSTOM ROLE**
>
> To create a custom role, you must have the Microsoft.Authorization/roleDefinitions/write permission on all AssignableScopes.

You can also create a custom role by choosing Start From Scratch from the Baseline Permissions. This option could be time-consuming because you need to select all the permissions, one by one, to create a custom role from scratch.

Similarly, custom roles can be defined using a JSON file by selecting Start From JSON under Baseline Permissions. The JSON file contains the role definitions:

- A name represented by the Name attribute
- An identifier represented by the Id attribute
- A description represented by the Description attribute

A flag that denotes whether the role is custom or built-in is represented by the IsCustom attribute, which is set to false for built-in roles; this should be set to true when authoring custom roles.

The actions that can or cannot be performed within the Azure management plane are represented by the Actions[] and NotActions[] attributes, respectively.

Optionally, the scopes at which the role is available through the AssignableScopes[] attribute.

## Interpret access assignments

To manage access (role) assignments, you can use the Azure portal, the Azure CLI, Azure PowerShell, Azure SDKs, or the Resource Manager REST APIs. The following section describes how to manage role assignments using the Azure portal.

In the Azure portal, the Access Control (IAM) blade is used to manage access to resources, and it is where role assignments are applied or removed. The Access Control (IAM) blade is available at any scope where role assignments can be made (management group, subscription, resource group, and resource). To find the Access Control (IAM) blade, navigate to the resource or service where you want to manage role assignments.

In the following example, the Virtual Machine Contributor built-in role will be assigned to a user at the resource group scope.

In the Azure portal, navigate to a resource group by selecting Resource Groups on the left, selecting a resource group, and then selecting the Access Control (IAM) blade.

From the Access Control (IAM) blade, you can

- Check the effective access rights for a security principal at the current scope through the Check Access tab, where you can also view access rights inherited from a parent scope
- Edit role assignments, both granting and revoking access rights through the Role Assignments tab
- View deny assignments, which are controlled by Microsoft, through the Deny Assignments tab
- View and manage permissions to classic resources through the Classic Administrators tab
- View the available roles, both built-in and custom, through the Roles tab

> **IMPORTANT** **DENY ASSIGNMENTS IN THE IAM BLADES**
>
> The Deny Assignments tab of the Access Control (IAM) blade cannot be used to make or alter deny assignments. Deny assignments are set and controlled by applying a resource lock for resources or by using Deployment Stacks. For more information on Deployment Stacks, visit *https://learn.microsoft.com/en-us/azure/azure-resource-manager/bicep/deployment-stacks*.

To create a role assignment, navigate to the Role Assignments tab and click Add, as shown in Figure 1-25.

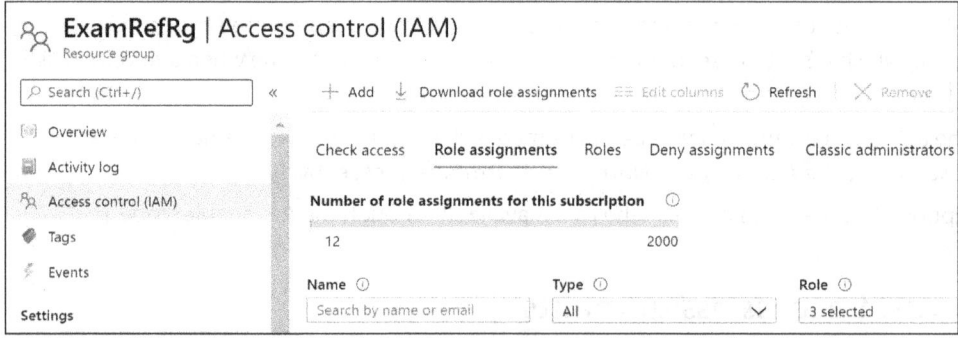

**FIGURE 1-25**   Role Assignments tab on the Access Control (IAM) blade

After clicking Add, select Add Role Assignment, as shown in Figure 1-26.

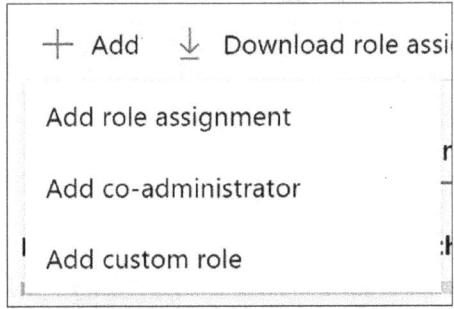

**FIGURE 1-26** Add role assignment

On the Add Role Assignment blade, there are three tabs: Role, Members, and Review + Assign. The Role tab lists roles as job function roles or privileged administrator roles. Select the role from the Role tab that you plan to assign, then on the Members tab, select the user, group, security principal, or managed identity you want to assign the role to. Click Review + Assign when you are done. Figure 1-27 shows an example where the user, DemoUser@hugelab.net, is being granted access to the Virtual Machine Contributor role. In the example directory, two security principals were returned from the filtered list using the search term "User"—(VM User and Demo User). A single principal (Demo User) was selected (displayed under Selected Members) to apply to the Virtual Machine Contributor role assignment.

**FIGURE 1-27** Select members

After clicking Review + Assign, you will see the role assignment on the Role Assignments tab. To remove a role assignment, select one or more security principals and click Remove. An example is shown in Figure 1-28.

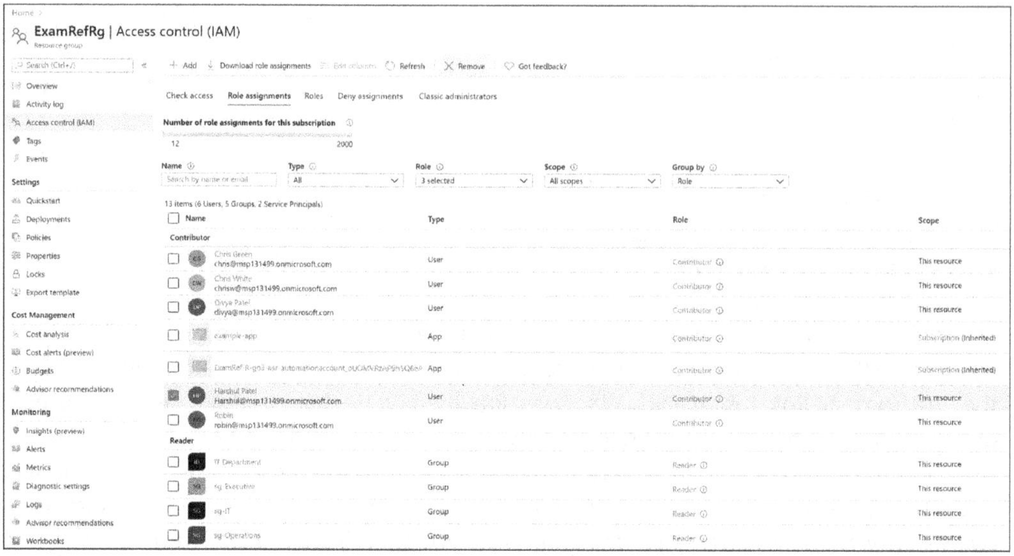

**FIGURE 1-28** Remove a role assignment

## Manage multiple directories

Each Entra ID tenant (or directory) is managed as an independent resource. There is no parent-child relation between directories, although users from one directory can be invited to another directory through Entra External Identities features.

Because each tenant is an independent resource, directories can be created and deleted as needed. This also means that each directory can have independent administrators and role assignments. Deleting an existing directory can affect resources outside the directory. For example, when deleting a directory where external users are present, those users will no longer be able to access any applications or resources that have been shared with them.

Finally, each directory can be synchronized independently as well. This means if you have two domains on-premises that need to be synchronized to two different Entra ID tenants, you have the flexibility you need when implementing hybrid identity with Entra.

> **IMPORTANT  ENTRA ORGANIZATIONS AND AZURE SUBSCRIPTION**
>
> There is no parent-child relationship between Entra ID organizations and an Azure subscription. If your subscription is canceled or no longer valid, you can still access your Entra tenant using the Microsoft 365 admin center or PowerShell. Also, you can add another subscription to the existing organization later.

Managing directories can include deleting directories or even an entire Entra ID tenant. To delete a tenant, Global Administrator rights are required. When a directory is deleted, all the resources or objects within that directory are deleted as well.

There are several prerequisites that must be satisfied prior to deleting a directory:

- There are no existing users or groups except for the single global admin.
- There are no enterprise application registrations in the directory.
- No MFA providers are linked to the directory.
- There are no subscriptions for Azure, Microsoft 365, or other Microsoft SaaS services associated with the directory.

# Skill 1.3: Manage Azure subscriptions and governance

An Azure subscription, which forms the core of an Azure environment, is a foundational component of every Azure implementation. Every resource that you create in Azure resides in an Azure subscription, which is a billing boundary for Azure resources with per-resource, role-based access controls.

As you build and deploy services in Azure, you will create many types of resources. For instance, when creating your first virtual machine, you will also deploy many other resources including

- A disk for the operating system
- A network interface for the VM
- A virtual network and subnet for that network interface to bind to
- A network security group (in a default portal configuration)

It is important to understand that many services in Azure create multiple resources, and how you manage those resources will be driven by organizational policy and the lifecycle of your infrastructure hosted in Azure.

**This skill covers how to:**

- Configure Azure policies
- Configure resource locks
- Apply and manage tags on resources
- Manage resource groups
- Manage Azure subscriptions
- Configure management groups
- Configure cost management

A resource in Azure is a single-service instance, which can be a virtual machine, a virtual network, a storage account, or any other Azure service (see Figure 1-29).

**FIGURE 1-29**   Azure resource

Resource groups are logical groupings of resources or those single-service instances (Figure 1-30).

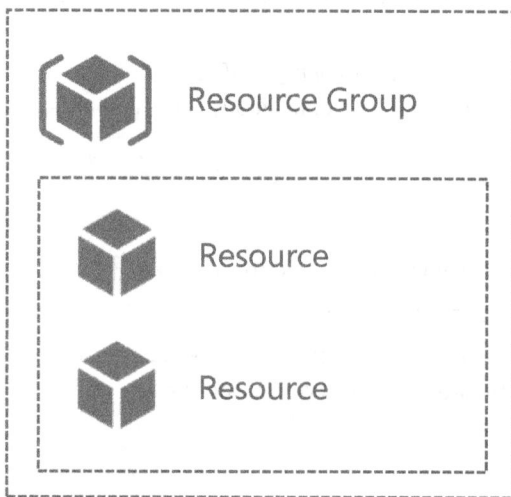

**FIGURE 1-30**   Azure hierarchy

Each resource in Azure can only exist in one resource group, and resource groups cannot be renamed. There are no limitations to the types of resources that can be logically contained within a resource group, and there are no limitations on the regions in which resources must reside when in a resource group.

Figure 1-31 shows this hierarchy within an Azure subscription, multiple resource groups, and the resources that reside within those resource groups.

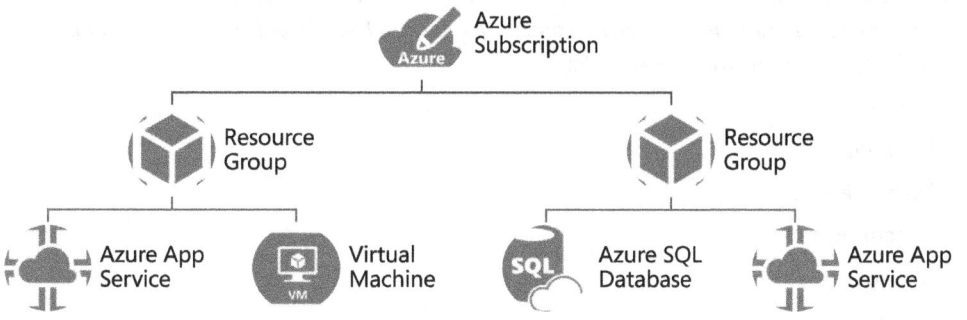

**FIGURE 1-31** Azure hierarchy

## Configure Azure policies

Azure Policy is an Azure service that can be used to create, assign, and manage policies that enforce governance in your Azure environment. This includes the application of rules that allow or deny a given resource type, apply tags automatically, and even enforce data sovereignty. Azure RBAC and Azure Policy are often used in combination. Where Azure RBAC controls individual user access, group access, and rights to your Azure environments at a specific scope, Azure Policy provides a mechanism to express how the environment is governed for all users at a specified scope regardless of any RBAC assignments. Another way to state this is that Azure RBAC is a default deny mechanism with an explicit allow mechanism, whereas Policy is a default allow mechanism with an explicit deny system.

To implement Policy, a policy definition must first be authored. That policy definition is then assigned a specific scope using a policy assignment. Recall that scope refers to what your policy is assigned to with valid scopes, a management group, a subscription, a resource group, or a resource.

Policy definitions can also be packaged using initiative definitions and applied to a scope using initiative assignments. Policy and initiative definitions both support parameter sets, which help simplify the reuse of a policy at multiple scopes.

A policy definition describes your desired behavior for Azure resources at the time resources are created or updated. Through a policy definition, you declare what resources and resource features are considered compliant within your Azure environment and what should happen when a resource is noncompliant. For example, you can create a policy that states that resources can only be created in the East US and West US regions for an entire subscription. If a user attempts to create a resource in East US 2, Azure Policy can deny the creation of the resource because it does not meet the stated compliance goal for allowed regions. In this example, Policy is used to deny the creation of a resource and to enforce organizational standards. As you further explore Policy, you will learn that Policy can be used not just as a deny mechanism but also as an auditing and creation mechanism.

Policy definitions are authored in JSON. The schema for Azure Policy can be downloaded from *https://schema.management.azure.com/schemas/2020-10-01/policyDefinition.json.* A policy definition contains these elements:

- Mode
- Parameters
- Display Name
- Description
- Metadata
- Policy Rule
- Logical Evaluation
- Effect

> **NOTE   POLICY DEFINITION**
>
> While you do not need to memorize the schema, it is worthwhile to understand the elements of a policy definition and how to build your own policies from a blank template when necessary. Microsoft offers a number of built-in policy definitions and maintains a repository of samples at *https://learn.microsoft.com/en-us/azure/governance/policy/samples/* and *https://github.com/Azure/azure-policy/tree/master/samples.*

Policy definitions can be created through the Azure portal by browsing to the Policy service at All Services and then choosing Policy, Definitions. From this blade, you can manage both built-in policies and any custom policies that you create. Figure 1-32 shows a list of the built-in policies for selected subscription.

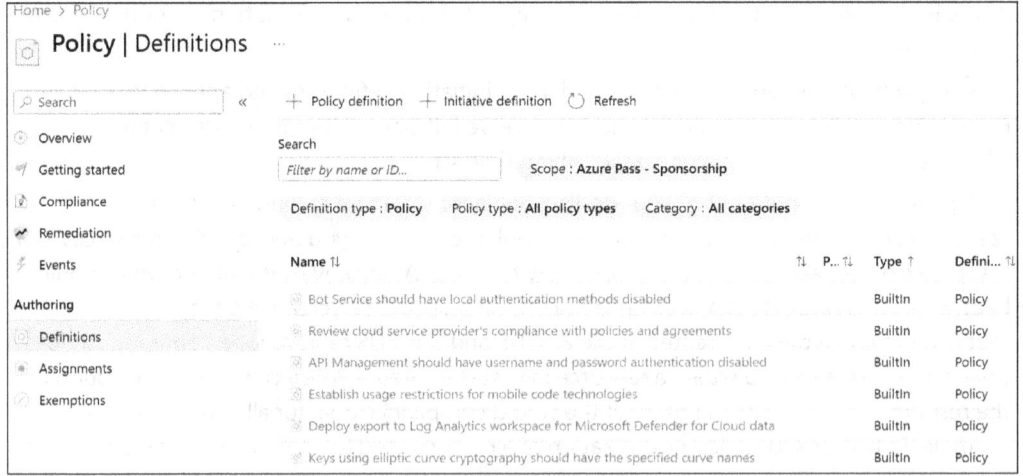

**FIGURE 1-32**   Azure built-in policies

Keep in mind that Policy can also be managed and applied at the management group scope. By associating policies with management groups, policy definitions and policy assignments can be shared across multiple subscriptions. This includes the ability to monitor multiple subscriptions for compliance. It also allows you to secure the management of organization-wide policy at a level above a single subscription.

When managing resource groups—and in many cases the multiple Azure services that reside within them—Azure Policy with policy definitions and policy assignments can be used to govern those resources. Initiative definitions and initiative assignments can be used to govern those same resources, but instead of applying multiple policy definitions and making multiple policy assignments, you can package or group multiple definitions into a single initiative and then assign that initiative to your desired scope.

Controlling resource groups with Azure Policy is done by scoping the assignment of policy and initiatives. Recall that Azure Policy supports multiple scopes:

- **Management group**   Assignments scoped at the management group (either the Tenant Root Group or a child group) apply to all child resources in the management group including child management groups, all subscriptions, resource groups, and resources.
- **Subscription**   Assignments scoped to a subscription apply to all child resources in the subscription resource groups and resources.
- **Resource group**   Assignments scoped to a resource group apply to all child resources in the resource group.

When creating assignments, it is also possible to configure excluded scopes. You can always exclude a subscope. For example, when scoping an assignment to a management group, any subscriptions, resource groups, or even resources that are children of the management group, can be excluded. When scoping an assignment to a subscription, child resource groups and resources can be excluded. When scoping an assignment to a resource group, only child resources can be excluded.

The flexibility of policy scoping is a powerful feature of Azure Policy. This allows you to model your environments with rich declarations in the form of policy definitions that are applied exactly as required by your organization's governance needs.

Imagine you have an environment with the following requirements:

- All resources should be tagged with the tag "Environment" and the value "Dev/Test".
- Only A-Series and D-Series virtual machines can be created, specifically Standard A0, A1, and D2 virtual machines that are not promotional.
- Resources in the rgCoreNetwork resource group are exempt from these policies.

To model this environment with Azure Policy, you can create two policy definitions (or use built-in policy definitions where applicable), as shown in Table 1-2.

**TABLE 1-2** Azure Policy definitions example

| Policy Field | Policy Effect | Description |
|---|---|---|
| Type | Deny | Do not create virtual machines if they are not in the A-Series or D-Series SKU. |
| tags | Append | Append tag name "Environment" and tag value "Dev/Test" to all resources. |

1. In the Azure portal, browse to the Policy service and select the Definitions blade. To reduce administrative overhead, a new initiative definition will be created. Initiative definitions are a collection of policy definitions that are focused on the same goal. They allow for a set of policies to be grouped as a single item.

2. From the Definitions blade, click Initiative Definition, as shown in Figure 1-33.

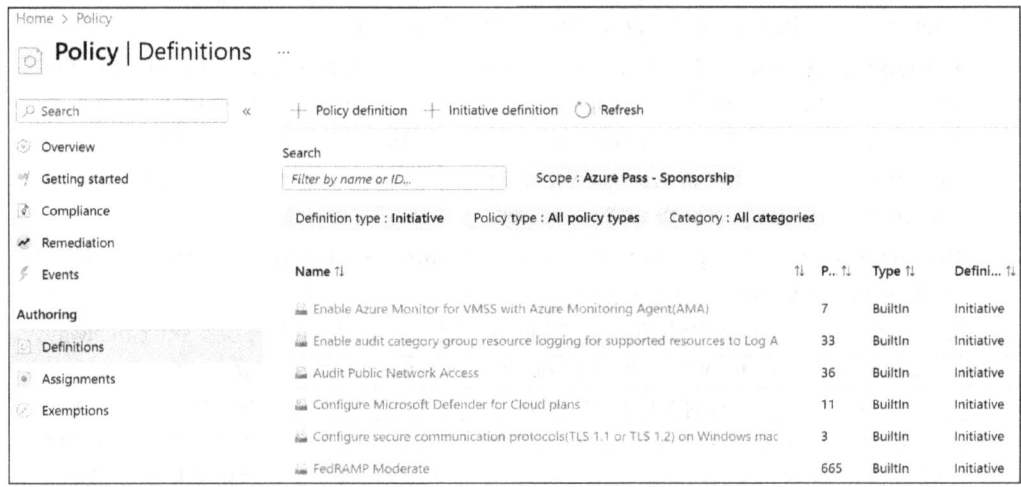

**FIGURE 1-33** Azure Policy Definitions blade

3. Type **Dev/Test Compliance** into the Name field, select the Definition Location, and choose Create New from the Category options. Type **Custom** in the Category field, as shown in Figure 1-34.

4. On the Policies tab, select Add Policy Definition(s) to add built-in policy definitions to the initiative. In this example, add the following built-in policies to the definition and set the values as noted (see Figure 1-35):

   - Require A Tag And Its Value On Resources
   - Allowed Virtual Machine Size SKUs

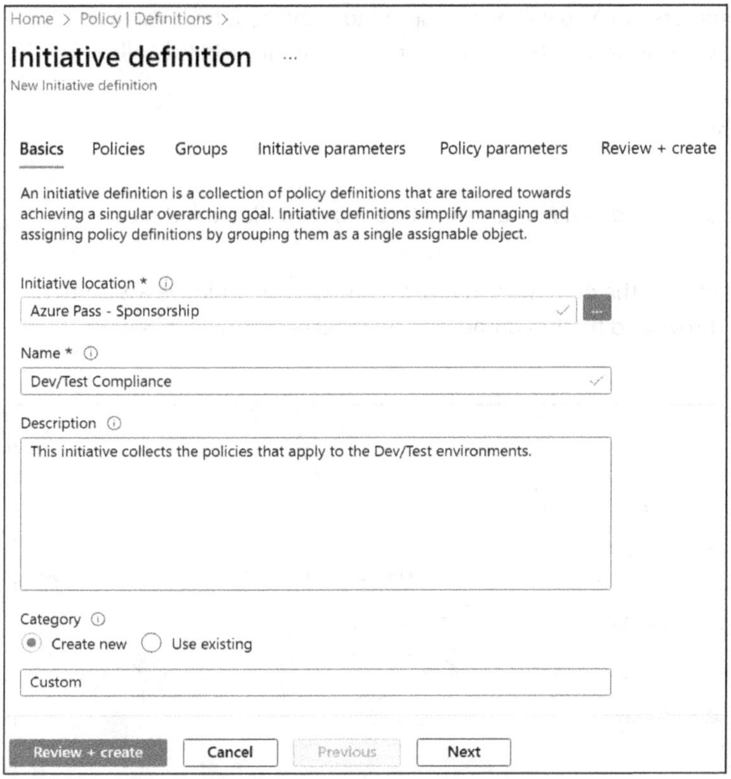

**FIGURE 1-34**  Azure Policy Initiative Definition blade

**FIGURE 1-35**  New initiative definition policies and parameters for Azure Policy

5. Select the Policy Parameters tab. You will be prompted to provide default values for the policy definitions that were added to the initiative. In this scenario, add the following tag name and value:

- Name: Environment
- Value: Dev/Test

6. In the Allowed Size SKUs drop-down menu, select three sizes available for your subscription.

7. Click Review + Create to save the definition so it can be used in an initiative assignment. From the Policy page, browse to the Assignments blade and click Assign Initiative (see Figure 1-36).

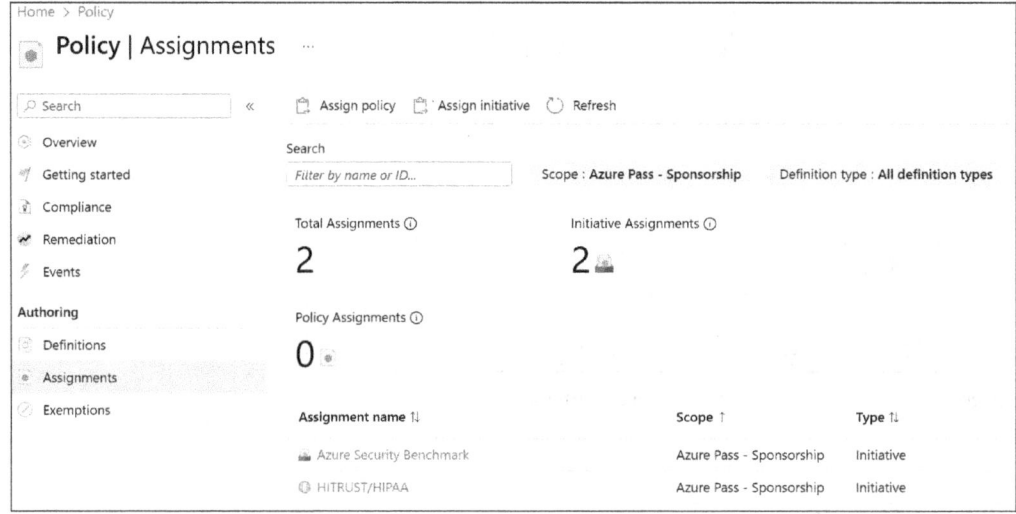

**FIGURE 1-36** Azure Policy Assignments blade

8. To meet the environmental requirements, set the Scope of the assignment to the target subscription and configure the Exclusions to exclude the rgCoreNetwork resource group. Also, set the Initiative Definition to Dev/Test Compliance and set the Assignment Name to Dev/Test Compliance. Lastly, make sure Policy Enforcement is enabled (see Figure 1-37). Then click Review + Create, and click Create.

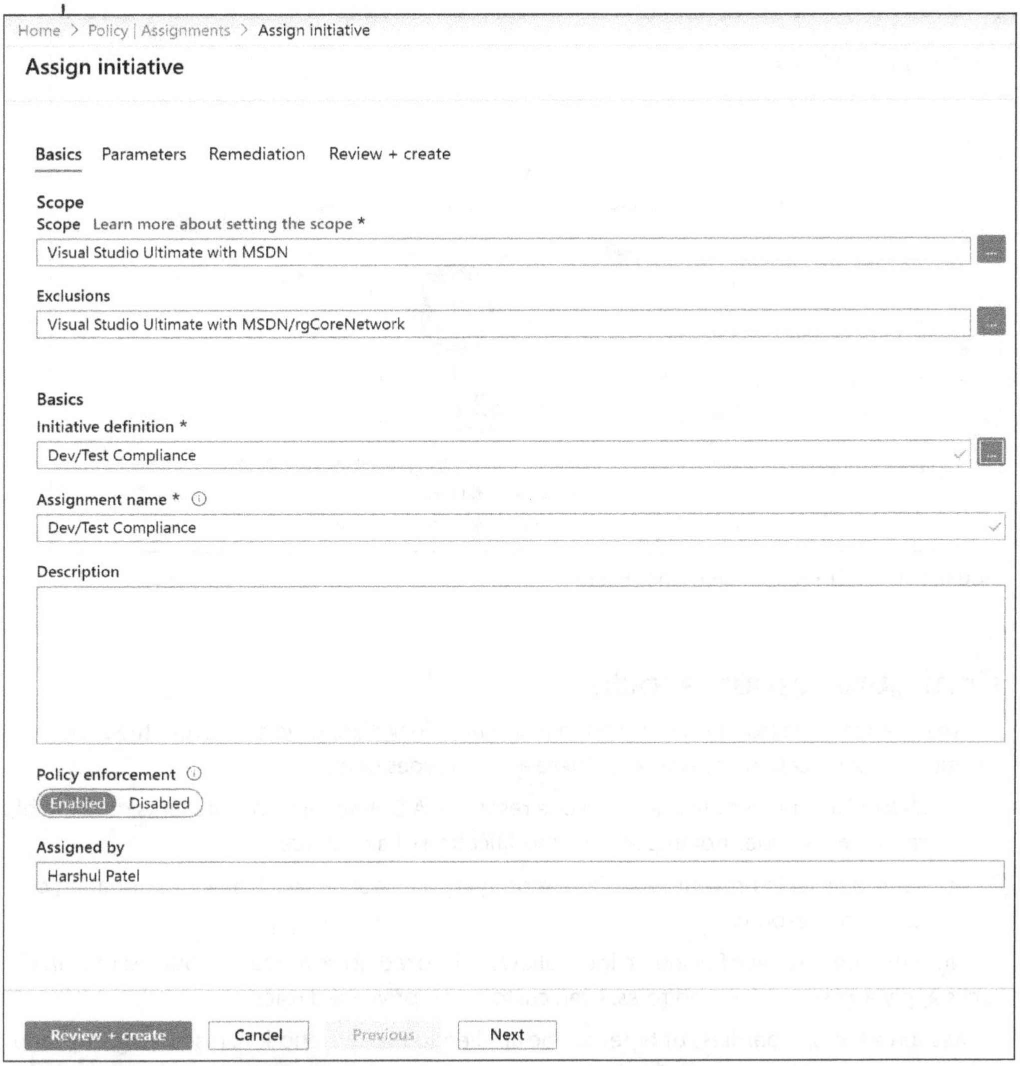

FIGURE 1-37   Azure Policy Assign Initiative blade

After policy definitions have been assigned, either through policy assignments or initiative assignments, the effects of the policy will be immediately applicable. Policy evaluation for compliance happens about once an hour, which means you might not be able to view the compliance state of a new assignment immediately.

Compliance state can be viewed on the Compliance blade of the Azure Policy service. You can delete, edit, and duplicate the policy assignment by right-clicking it on the Compliance blade, as shown in Figure 1-38.

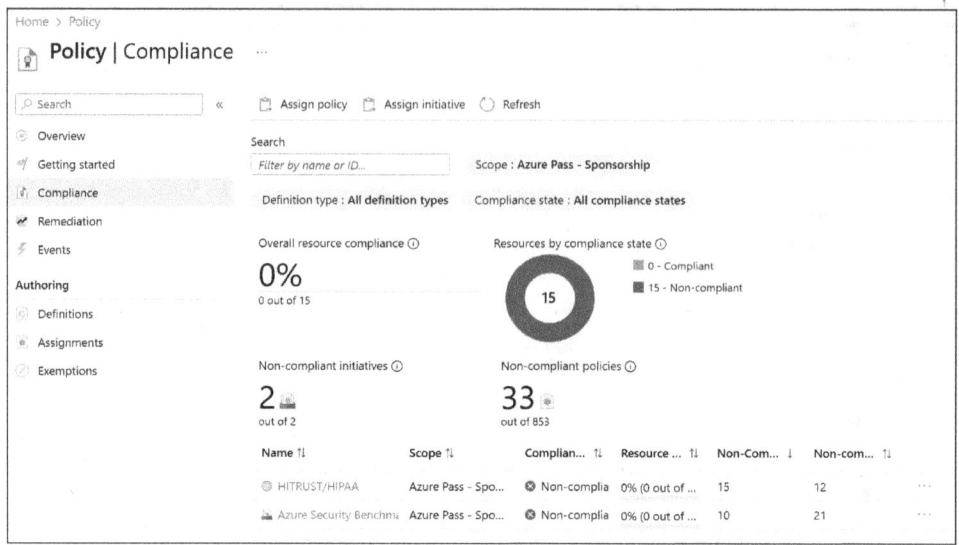

**FIGURE 1-38**   Azure Policy Compliance blade

## Configure resource locks

Azure resource locks (sometimes called management locks) are used to prevent the accidental deletion or modification of resources. There are two types of locks:

- *Delete locks* prevent the deletion of a resource. A Delete lock only prevents deletion of a resource and does not impede the modification of a resource.

- *Read-only locks* prevent users from modifying a resource, which includes updating or deleting a resource.

Note that both types of resource locks allow authorized users to read resources; resource locks apply across all users and roles, even custom and privileged roles.

Resource locks, regardless of type, can be applied to the subscription, resource group, and resource scopes. When you apply a lock to a scope, the resources within that scope inherit the lock. This means that a lock applied to the resource group scope applies to all the resources in the resource group. Resource locks apply to all service instances and resources within a scope.

Lock inheritance applies to the child resources of the scope that you are configuring the lock on. For example, a lock on a resource group applies to all resources in the group. If a Delete lock is applied to one of the resources in the resource group and you attempt to delete that resource group, it will fail. When you try to delete the resource group, the operation tries to delete all the underlying resources first and won't be able to delete the resource with a Delete lock, hence the resource group deletion would also fail.

Note that resource locks get applied to the management plane of Azure. This means resource locks don't affect the resource's own functionality; instead, they restrict the interactions with other Azure resources. For example, a Read-only lock applied to a storage account would prevent users from reading the access keys. If you attempt to read or modify the access keys, the operation will fail with a "Cannot perform write operation because the following scope(s) are locked" error, as shown in Figure 1-39.

**FIGURE 1-39**   Read-only lock applied to a storage account

When creating locks, exercise caution because they can cause unexpected results. Many operations appearing to be read operations require write access within the Azure management plane. For example, the same Read-only lock on a storage account would prevent users from creating new blob containers because the action requires write access.

Once you have determined the type of lock you will apply based on your requirements, you can apply the lock through the Azure portal, Azure PowerShell, the Azure CLI, Resource Manager templates, or the REST API.

To create a lock through the Azure portal, browse to the desired scope and select the Locks blade. From the blade, click Add to create a new lock. Give the lock a Lock Name, select the Lock Type, and describe the lock in the Notes field, as shown in Figure 1-40.

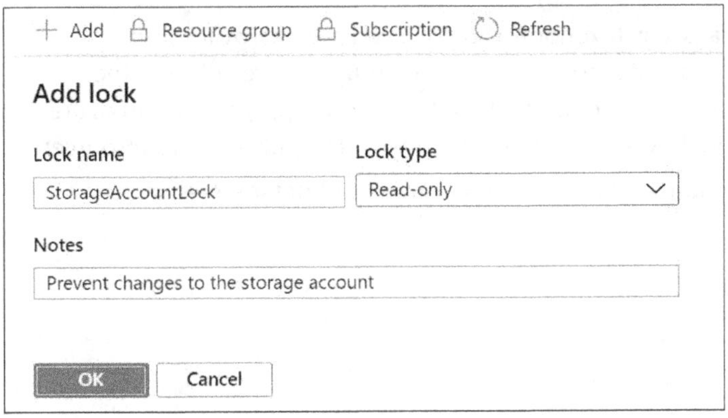

**FIGURE 1-40**   Creating a lock

# Apply and manage tags on resources

Resource tags allow you to apply custom metadata to your Azure resources to logically orga-nize them and to build out custom taxonomies. A tag is a name and a value pair. For example, suppose as you deploy resources in Azure, you want to track the environment the resource is associated with. To do this, you can create a tag called Environment and the value Production for all resources in production. For downstream environments such as development or test environments, you can use the same Environment tag with the Dev/Test value. Common tags include the environment with which a resource is associated, a cost center or billing code, and resource owner.

As tags are applied, you can query the resources in your subscription using your tags, and you can even do this across resource groups. This allows you to understand related resources across resource groups for both billing and management. Tags are also included in the billing data for Azure Cost Management + Billing. Cost Management + Billing gives a clear line of sight for chargeback to understand resource usage and cost. Figure 1-41 shows an example of an export with resource tags from an Azure EA subscription.

| PublisherTy ▾ | ChargeTy ▾ | ServiceName ▾ | ServiceTier ▾ | Meter ▾ | PartNumt ▾ | CostUSD |
|---|---|---|---|---|---|---|
| azure | usage | storage | premium ssd managed | p10 disks | | 10.740285 |
| azure | usage | storage | standard page blob | disk read operations | | 97.188975 |
| azure | usage | storage | standard page blob | disk write operations | | 889.79809 |
| azure | usage | storage | tables | batch write operations | | 9.9925364 |

**FIGURE 1-41** Azure detailed usage export

> **NOTE**  **TAGS AND USAGE REPORTS**
>
> Tags must be applied at the resource scope to be visible in detailed usage exports. Tags applied at the resource group scope are not inherited by child resources. This means that as you are applying tags to your resources in Azure, you should think about applying tags to each resource to have the clearest line of sight into your usage based on your organizational tags.

When planning for resource tags, any taxonomy should include a strategy for both on-demand (or self-service) tagging and automatic tagging through Azure Policy. In the "Configure Azure policies" section, you learned how to automatically apply tags using Azure Policy. In this section, you will learn how to create tags and manually apply them to resources.

As you plan your tagging taxonomy, be mindful of the limitations of tags in Azure, as detailed in Table 1-3.

**TABLE 1-3** Azure tag limitations

| TaG LIMIT | Notes |
|---|---|
| Resource support | ▪ Not all resource types support tags. This means that you will not be able to apply tags to everything in Azure. For example, management groups, network interfaces and generalized VMs don't support tags. Refer to this link: *https://learn.microsoft.com/en-us/azure/azure-resource-manager/ management/tag-support.* |
| Number of tags | ▪ *Most* resources, resource groups, and subscriptions are limited to 50 tags. Each resource can have different tags. Some resources, such as Azure Automation, DNS zones, and Azure CDN, are limited to 15 tags. |
| Tag name | ▪ Tag names cannot exceed 512 characters. For storage accounts, tag names are limited to 128 characters. |
| Tag value | ▪ Tag values cannot exceed 256 characters. |
| Tag inheritance | ▪ Tags are not inherited by child resources. Tags applied to a resource group are not applied to resources in that resource group. |
| Classic resources | ▪ Tags cannot be applied to classic resources and are only available for resources created in the Azure Resource Manager model. |
| Illegal characters | ▪ Tag names cannot contain the following characters: <, >, %, &, \, ?, /. Additionally, some resources such as Azure Front Door also restrict using # or : in the tag name. |

To apply tags to a subscription, resource group, or resource, the user applying the tag must have write access to the resource (Contributor role or higher access).

Tags can be created and applied to Azure resources through

- The Azure portal
- Azure PowerShell
- The Azure CLI
- Resource Manager templates
- Resource Manager REST API

This means tags can be applied both in an imperative manner and declaratively through Resource Manager templates. While this can be done through the Azure portal, PowerShell, the CLI, or Resource Manager, templates or policies are better suited when this is being done as resources are created because you don't want to perform this manually for each resource after deployment.

Tags can be applied at the subscription, resource group, and/or the resource level. Note again that there is no inheritance for tags. If you need a tag to be applied to all resources in a resource group, each resource must be tagged individually.

## Manage resource groups

When creating resource groups, it is important that you consider factors for your resource group design:

- A resource can be a member of only one resource group.
- A resource group cannot be nested in another resource group.

- You can move a resource from one resource group to another.
- A resource group can be used to scope access control.
- A resource group can be used to scope policy.
- A resource in a resource group can interact with resources in another resource group.
- A resource group is created in a location, also known as an Azure region. The location of a resource group specifies where the metadata for the resource group is stored. If you have compliance or geography constraints, this is an important consideration.
- Microsoft recommends that all resources in a resource group share the same lifecycle.
- It is not mandatory to have all Azure resources belong to a resource group.
- Creating a resource group through the Azure portal can be an easier task. You just need region or location details along with a valid resource group name (see Figure 1-42).

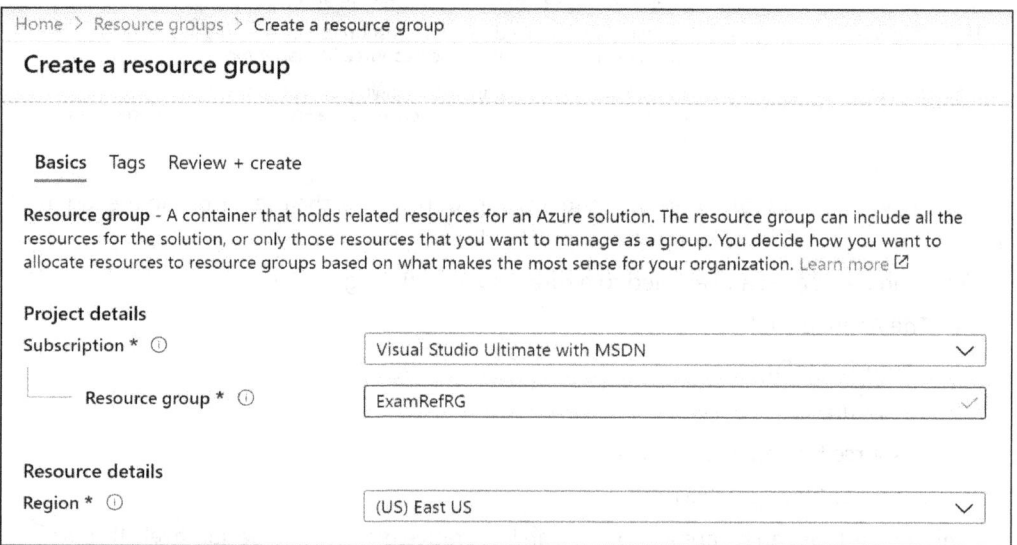

**FIGURE 1-42** Create A Resource Group blade

## Move resources across resource groups

Some resources in Azure can be moved between resource groups and even across subscriptions, but support for move operations varies based on the service. A reference of services that can be moved can be found at *https://learn.microsoft.com/en-us/azure/azure-resource-manager/management/move-support-resources*. In Figure 1-43, the VM in Resource Group 2 can be moved into Resource Group 1, and it can also be moved across subscriptions into the resource group in Subscription 2.

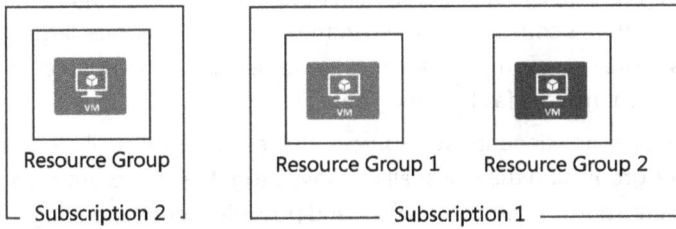

**FIGURE 1-43**  Moving resources diagram

During a move operation, your resources will be locked. Both write and delete operations to the Azure resource will be blocked, but the underlying service will continue to function. For example, if you move a web app in Azure App Service, the app will continue to serve web requests to visitors. It can take up to four hours for a move operation to complete. If the move operation fails within the four-hour window, Resource Manager will reattempt the move operation.

To move resources between subscriptions, both subscriptions must be associated with the same Entra tenant. If the subscriptions do not belong to the same tenant, you can update the target subscription to use the source Entra tenant by transferring ownership of the subscription to another account. Note that this operation can have unexpected effects because the Entra tenant associated with a subscription is used for RBAC to any currently deployed Azure services.

When moving resources between subscriptions, the *resource provider* of the source resource must also be registered in the target subscription. A resource provider is the underlying service that allows that service to function and operate in your subscription. To see the list of resource providers, navigate to your subscription. On the subscription, select the Resource Providers blade. This is not a concern when moving resources within the same subscription because the resource provider will already be registered.

If you are moving resources between subscriptions, you must also be mindful of resource quotas. For example, if you are moving many virtual machines, you will need to make sure that the target subscription has enough vCPUs available, or the move operation will fail. Make sure you validate any quotas prior to moving a resource.

Finally, there are limitations in Azure Resource Manager that affect the number of resources you can move in a single operation. A single move operation in Resource Manager cannot move more than 800 resources. With this constraint, it is recommended that you break large operations into smaller batches. Note that even if you are moving fewer than 800 resources in a single move request, the operation may still fail by timing out.

If the resource you are moving has any dependent resources, the resources must all be located within the same resource group, and they must all be moved together. For example, a virtual machine is dependent on the network interface, disks, and possibly more depending on the deployment.

Once you have met the stated prerequisites to a move operation, you are ready to perform the move operation. You can move the resources with the Azure portal, Azure PowerShell, the Azure CLI, or the REST API. Note that Azure performs basic validation before performing the actual move operation, irrespective of the method being used. Additionally, you can validate the move operation through the REST API with the validateMoveResources method without actually performing the move operation. This API validates whether resources can be moved from one resource group to another resource group. If validation succeeds, an HTTP 204 will be returned, and if it fails, an HTTP 409 with an error message will be returned in the response. This method can be called with a POST request to

```
https://management.azure.com/subscriptions/%7bsubscriptionId%7d/resourceGroups/%7bsource
ResourceGroupName%7d/validateMoveResources?api-version=2021-04-01
```

In a POST request, include a request body with "resources" and "targetResourceGroup" properties:

```
{
  "resources": ["<resource-id-1>", "<resource-id-2>"],
  "targetResourceGroup": "/subscriptions/<subscription-id>/resourceGroups/<target-group>"
}
```

If the request is properly formatted, the operation will return output like the following:

```
Response Code: 202
cache-control: no-cache
pragma: no-cache
expires: -1
location: https://management.azure.com/subscriptions/<subscription-id>/
operationresults/<operation-id>?api-version=2018-02-01
retry-after: 15
...
```

The HTTP 202 response code shows the request was accepted. The location URI can be used in an HTTP GET that you can use to check the status of the long-running operation for the final HTTP 204 or HTTP 409 status code. Figure 1-44 shows the output of an operation to validate a move request for an Azure Automation account associated with a Log Analytics workspace. As expected, the validation operation returned an HTTP 409 because this move request cannot be executed.

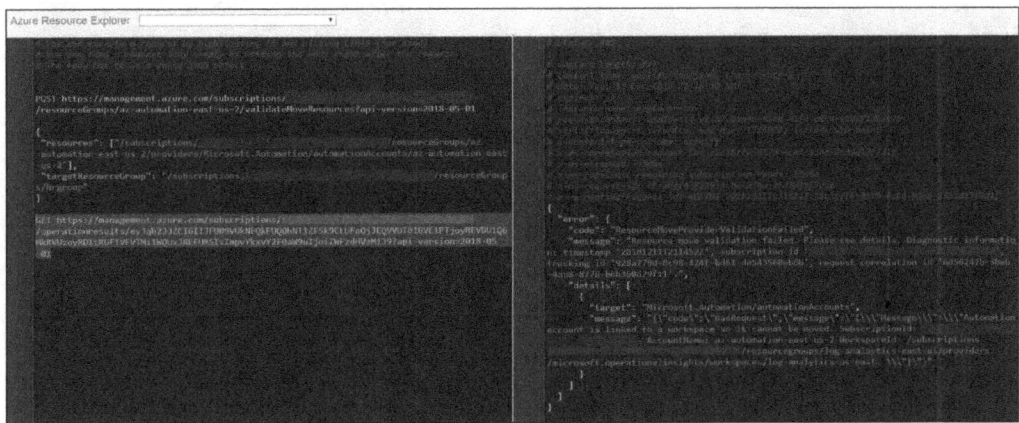

**FIGURE 1-44** ValidateMoveResources API response

To use the Azure portal, browse to the resource group containing the resources, click Move, and choose Move To Another Resource Group or Move To Another Subscription, as shown in Figure 1-45.

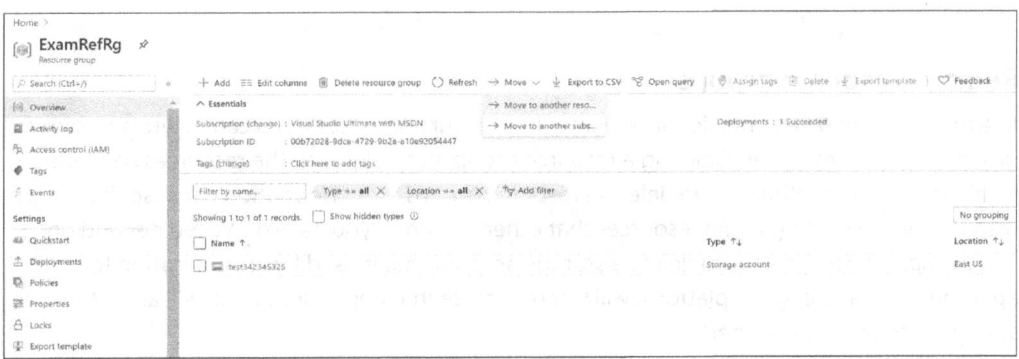

**FIGURE 1-45** Move menu in the Azure portal

You can now select the resources to move and select the destination resource group. Note that you must acknowledge that you might need to update existing tools or scripts to account for the changes in resource IDs (see Figure 1-46).

**FIGURE 1-46** Move Resources blade

## Remove resource groups

In Azure, you can delete individual resources in a resource group, or you can delete a resource group and all its resources. Deleting a resource group also deletes all the resources contained within it in one operation. When deleting resource groups, exercise caution because the resource group might contain resources that other resources you have deployed depend on. For example, if you attempt to delete a storage account that is used by an application to store application data, the Azure platform will not recognize that dependency and will allow the storage account to be deleted.

For resources that have dependent resources, you will not be able to delete the target resource until the dependencies have been cleared. For example, to delete a resource group that contains an App Service plan, you must first remove or disassociate any App Service apps that depend on that plan. An example of attempting to delete an App Service plan with existing App Service app associations is shown in Figure 1-47.

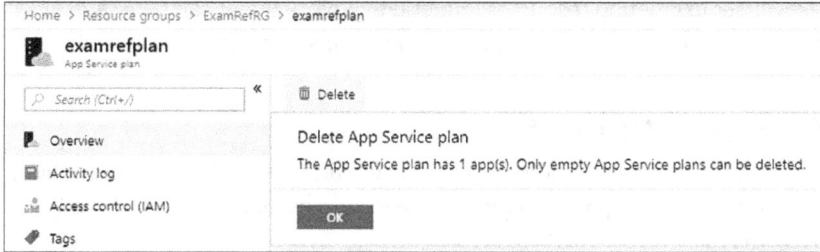

**FIGURE 1-47**  Delete an Azure resource with dependencies

To delete a resource group, you can use the Azure portal, Azure PowerShell, the Azure CLI, or the REST API.

To delete a resource group in the Azure portal, browse to the resource group and click Delete Resource Group (see Figure 1-48).

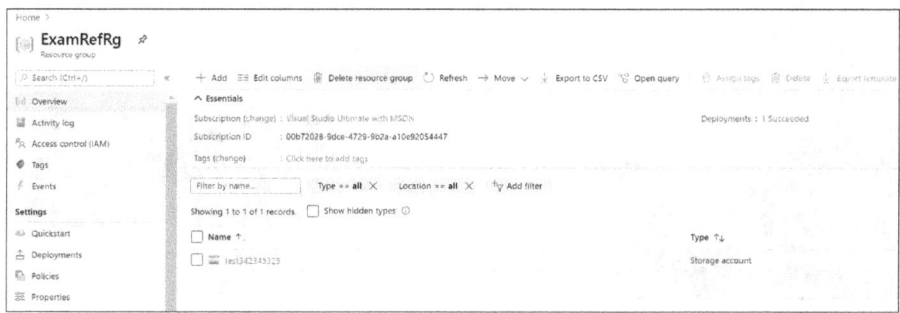

**FIGURE 1-48**  Delete Resource Group option

In the Are You Sure You Want To Delete ["resource group name"]? dialog box that opens, you will need to type the resource group name to confirm that you want to delete it. As shown in Figure 1-49, the blade will also show the affected resources and warn you that the operation is irreversible.

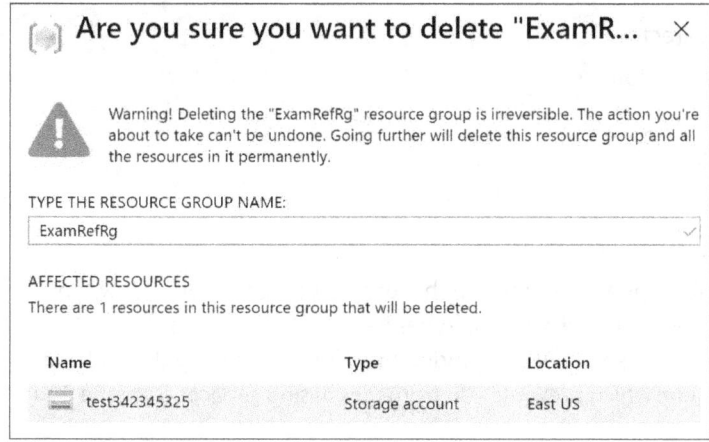

**FIGURE 1-49**  Azure resource group deletion confirmation

Selecting Delete will begin deleting resources immediately. Note that it can take several minutes for a resource group to be deleted because each resource is deleted individually.

## Manage Azure subscriptions

Azure subscriptions include controls that govern access to the resources within a subscription, govern cost through quotas and tagging, and govern the resources that are allowed in an environment with Azure Policy.

As discussed earlier, a subscription is a logical unit of Azure services linked to an Azure account, which is an identity in Entra ID. Entra ID is an identity provider for Azure and provides authentication to resources in an Azure subscription. The resources themselves then have role-based access controls applied to them that provide authorization to the resources (see Figure 1-50).

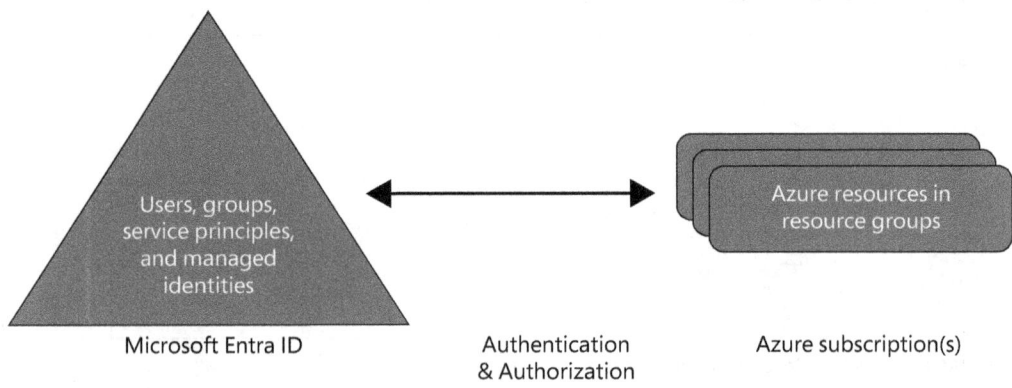

**FIGURE 1-50**   Entra ID and Azure subscription relationship

There are multiple ways to obtain an Azure subscription, and a wide range of subscription types (or offers). Some common types include the following:

- Free trial
- Pay-As-You-Go/Web Direct
- Visual Studio/MSDN subscriptions
- Microsoft Resellers
- Cloud Solution Provider
- Microsoft Open Licensing
- Enterprise Agreements

The capabilities of each subscription are similar in that each subscription type allows you to create and manage resources. Some subscription types have restrictions on supported resource types and locations. For example, Visual Studio subscriptions typically do not have a credit card associated with them, which prevents you from purchasing services from the Azure

Marketplace, such as network virtual appliances. Visual Studio subscriptions for Azure only have access to a limited number of Azure regions. The regional restrictions for each offer can be viewed at *https://azure.microsoft.com/regions/offers/*.

## Assign administrator permissions

Azure has many different roles for managing access to Azure resources. These include classic subscription administrative roles like Account Administrator, Service Administrator, or Co-Administrator, as well as Azure role-based access controls (RBAC) that are available in Azure Resource Manager (ARM). Note that classic roles and resources are scheduled to be deprecated in August 2024. When managing access to Azure subscriptions and resources, it is recommended to use Azure RBAC roles whenever possible.

> **NEED MORE REVIEW?**  **ROLES AND RELATIONSHIPS**
>
> To learn more about the correlation between classic subscription administrator roles, Azure RBAC roles, and Entra roles, see *https://learn.microsoft.com/en-us/azure/role-based-access-control/rbac-and-directory-admin-roles*.

Classic subscription administrators have full access to an Azure subscription. They can manage resources through the Azure portal, Resource Manager APIs (including through PowerShell and the CLI), and the classic deployment model APIs.

By default, the account that is used to sign up for an Azure subscription is automatically set as both the Account Administrator and the Service Administrator. They both are authorized to perform subscription management activities, but creation of new Azure subscriptions and billing changes can be performed only by the Account Administrator. There can be only one Account Administrator per account and one Service Administrator per subscription.

Once the subscription has been created, more Co-Administrators can be added. The Co-Administrator has the same level of access as the Service Administrator but cannot change the association of subscriptions to Azure directories. There can be up to 200 Co-Administrators per subscription.

Users assigned with the Service Administrator and Co-Administrator roles have the same access as a user who is assigned the Azure RBAC Owner role at the subscription scope.

In the Azure portal, you can view the current assignments for the Account Administrator and Service Administrator roles by browsing to a subscription in the Azure portal and selecting the Properties blade, as seen in Figure 1-51.

**FIGURE 1-51** Azure subscription properties

## Built-in Azure RBAC roles

Azure RBAC roles are more flexible than classic administrator roles and allow for more fine-grained access management. Azure RBAC has more than 70 built-in roles, but there are four foundational roles, as shown in Table 1-4.

**TABLE 1-4** Azure RBAC roles

| Azure RBAC role | Permissions | Notes |
|---|---|---|
| Owner | <ul><li>Full access to all resources</li><li>Delegate access to others</li></ul> | <ul><li>The Service Administrator and Co-Administrators are assigned the Owner role at the subscription scope.</li><li>Applies to all resource types.</li></ul> |
| Contributor | <ul><li>Create and manage all types of Azure resources</li><li>Cannot grant access to others</li></ul> | <ul><li>Applies to all resource types.</li></ul> |
| Reader | <ul><li>View Azure resources</li></ul> | <ul><li>Applies to all resource types.</li></ul> |
| User Access Administrator | <ul><li>Manage user access to Azure resources</li></ul> | |

## Configure management groups

Management groups can also be used to apply Azure RBAC to a subscription. Using management groups, you can apply governance consistently across subscriptions, including the application of common RBAC controls and the application of Azure Policy, as discussed later in this chapter.

Within management groups, subscriptions can be organized in a multi level hierarchy, providing a number of tangible benefits:

- **Reduced overhead**  There is no need to apply governance on every subscription.
- **Enforcement**  Company admins can apply governance at the management group level, outside the control of the subscription admin and the controls implemented at the management group can be applied to both existing and new subscriptions. This eliminates inconsistencies in the application of governance as the same controls are applied the same way to the desired subscriptions.
- **Reporting**  Azure Policy provides reports of compliance. With management groups, the reporting can span across multiple or all subscriptions in an organization.

Management groups form a hierarchy that is up to six levels deep, excluding the root and subscription levels. Each group has exactly one parent and can have multiple children. An example hierarchy is shown in Figure 1-52. In such a hierarchy, one common set of Policy could be applied at the root management group, which all child management groups and subscriptions would inherit. Then, as needed, those children can have additional controls applied.

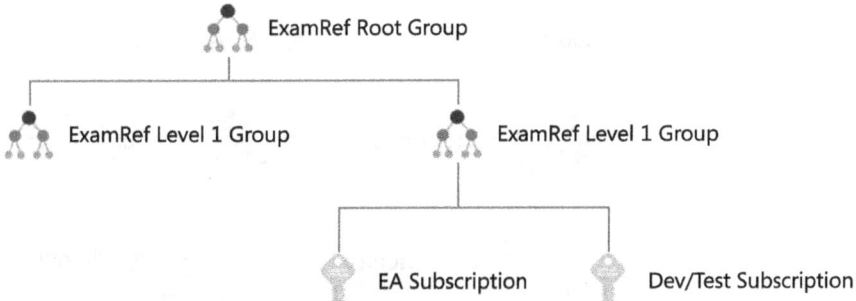

**FIGURE 1-52**  Example management group hierarchy

There is a single root management group at the root of the hierarchy. This management group is associated with the Entra tenant that is then associated with an Azure subscription. It cannot be moved or deleted. Individual subscriptions, including new subscriptions, are added to a management group.

Like RBAC, Azure Policy is also applied at a specific scope. The scope can be a subscription, a resource group, or an individual resource. For example, when a policy is applied at the subscription scope, it is inherited by all the resource groups and resources in the subscription, as shown in Figure 1-53.

Management groups introduce an additional scope above a subscription. When applied at the management group scope, each subscription under the management group inherits the RBAC and policy assignments of the management group as shown in Figure 1-54.

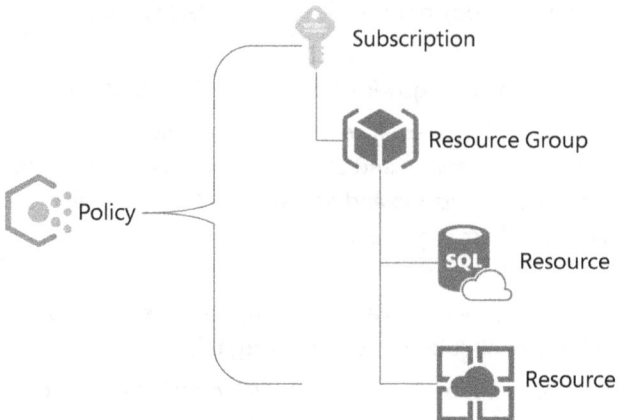

**FIGURE 1-53**  Example policy applied at the subscription scope

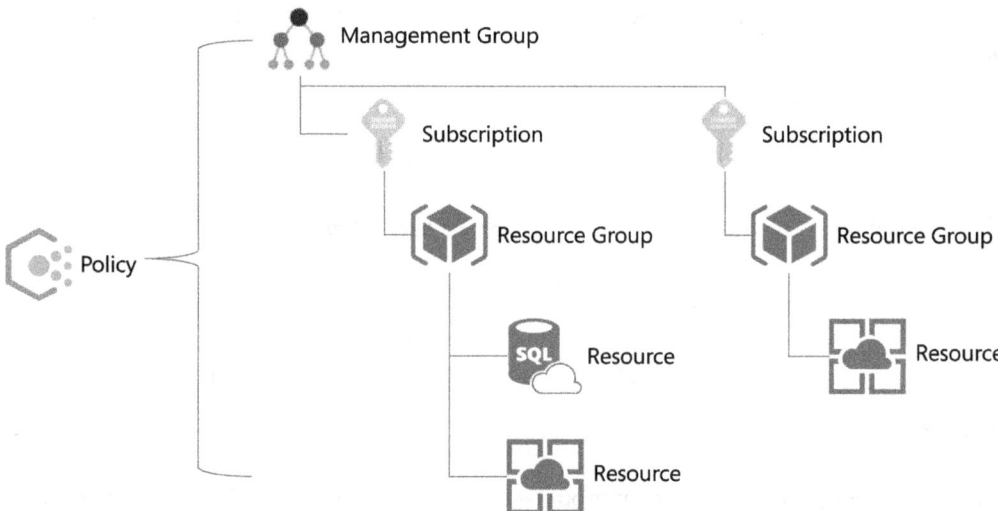

**FIGURE 1-54**  Example policy applied at the management group scope

To add a role assignment to a management group, browse to management groups in the Azure portal. Select a management group and then click Details next to that group's name. Select the Access Control (IAM) blade, click Add, and choose Add Role Assignment, just as you would for an Azure subscription, as shown in Figure 1-55.

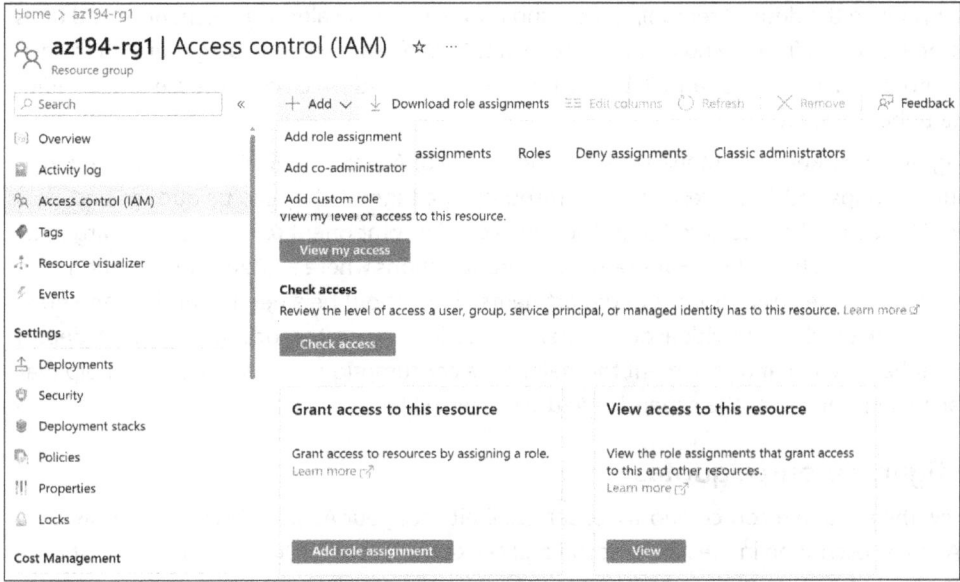

**FIGURE 1-55** Access control (IAM) blade for an Azure management group

> **IMPORTANT    RBAC AND MANAGEMENT GROUPS**
>
> RBAC applied at the management group level is inherited by all the child resources within the scope of the management group (subscriptions, resource groups, and resources). For instance, if you add a user as an Owner at the management group scope, that user will become an Owner in all the subscriptions associated with the management group, and the role is inherited by subscriptions in child management groups too.

## Configure cost management

In Azure, there are several types of quotas that are applicable to subscriptions, including resource quotas and spending quotas. With Azure resource quotas (or limits), Azure administrators can view the current consumption and usage of resources within an Azure subscription and understand how that consumption can be affected by Azure resource limits. Administrators can also request quota increases for certain resource types. For instance, in most subscription types, the number of cores available for virtual machines is limited to 20 per region by default. This limit can be increased by submitting a request to Microsoft support. Some quota requests are automatically approved in the Azure portal, and other requests require a support ticket with justification and manual approval.

You can also configure spending limits for your Azure subscription. Spending quotas allow administrators to set alerts or budgets within an Azure subscription to inform the business when their Azure spending has hit a certain threshold. While a resource limit can stop resources from being created (for example, there are not enough cores available to the subscription in the desired region), a spending quota acts as an alerting mechanism and does not stop resources from being created or consumed. While an alert can be generated from a spending quota, resources can still be created and consumed which could cause the spending quota to be exceeded.

Tags in Azure Resource Manager allow consumers of Azure to logically categorize Azure resource groups and Azure resources. As resources are tagged, they can be queried and tracked based on the associated tags. Tags are a crucial component to implement chargeback within an Azure subscription. For example, in organizations where an Azure subscription is shared by multiple business units or departments, there might be a need to understand how resources are used for individual departments and show the cost associated with each department, either to bill that department for their Azure consumption (chargeback) or to help that department understand their spend in Azure (showback).

## Configure resource quotas

To view the existing resource quotas (or service limits) for your Azure subscription, browse to the Azure subscription in the Azure portal and select the Usage + Quotas blade. From this blade, you can view existing quotas by service, resource provider, and location. You also filter the list by resource types you have deployed.

To increase a quota, click New Quota Request, as shown in Figure 1-56.

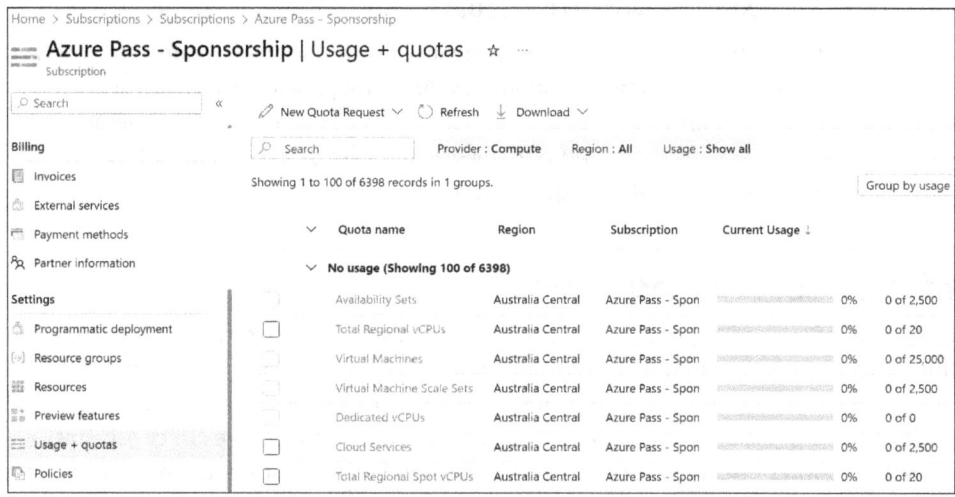

**FIGURE 1-56**  Azure subscription resource quotas

Clicking Request Increase begins the process to open a new support request. As a part of the request, you must select the quota type (for example, Compute/VM cores or Machine Learning service) and provide a description of your request.

The consumption of resources within a subscription against a resource quota can also be viewed with PowerShell. There are multiple cmdlets available in the Az (formerly AzureRm) PowerShell module for querying per-service quota usage. For example, to view the current usage of vCPU quotas, use Get-AzVMUsage, and to view the current resource usage for the storage service, use Get-AzStorageUsage.

**EXAM TIP**

In this chapter and throughout the remaining reference, PowerShell cmdlets are referenced using the Az module. See *https://learn.microsoft.com/en-us/powershell/module/az.accounts/ enable-azurermalias?view=azps-11.2.0&viewFallbackFrom=azps-10.2.0* for more detail.

## Configure cost center quotas

One of the key factors in managing an Azure subscription is being able to plan for and drive organizational accountability for Azure spend. One of the best ways to drive accountability is to make sure that the consumers of Azure resources understand their cost, including current usage and forecasting future spend based on current resource consumption.

Budgets in Azure Cost Management provide Azure customers subscriptions under many offer types with the ability to proactively manage cost and monitor Azure spend over time at a subscription level.

**EXAM TIP**

The full list of supported accounts and offers for Azure Cost Management can be found at *https://learn.microsoft.com/en-us/azure/cost-management-billing/costs/understand-cost-mgt-data*.

Budgets are a monitoring mechanism only with set thresholds and notification rules. When a budget threshold is exceeded, a notification is triggered but resources continue to run.

To use budgets with an Azure subscription, that subscription must be a supported offer type as previously stated. Users must have at least read access (Reader rights) to a subscription to view budgets and must have Contributor (or higher) rights to create and manage budgets. There are also specialized roles that can be used to grant principals access to Cost Management data including Cost Management Contributor and Cost Management Reader.

To create a budget in the Azure portal, navigate to Cost Management + Billing, click Cost Management, and then click Budgets.

Click Add, and in the Create Budget blade, enter a budget name and budget amount. You can also change your desired scope by clicking Change Scope. Choose the Reset Period (monthly, quarterly, or annual) and an Expiration Date. Budgets require at least one Cost Threshold (percent of budget) and an email address for the alert recipient. Figure 1-57 shows an example for a monthly budget for $10,000.

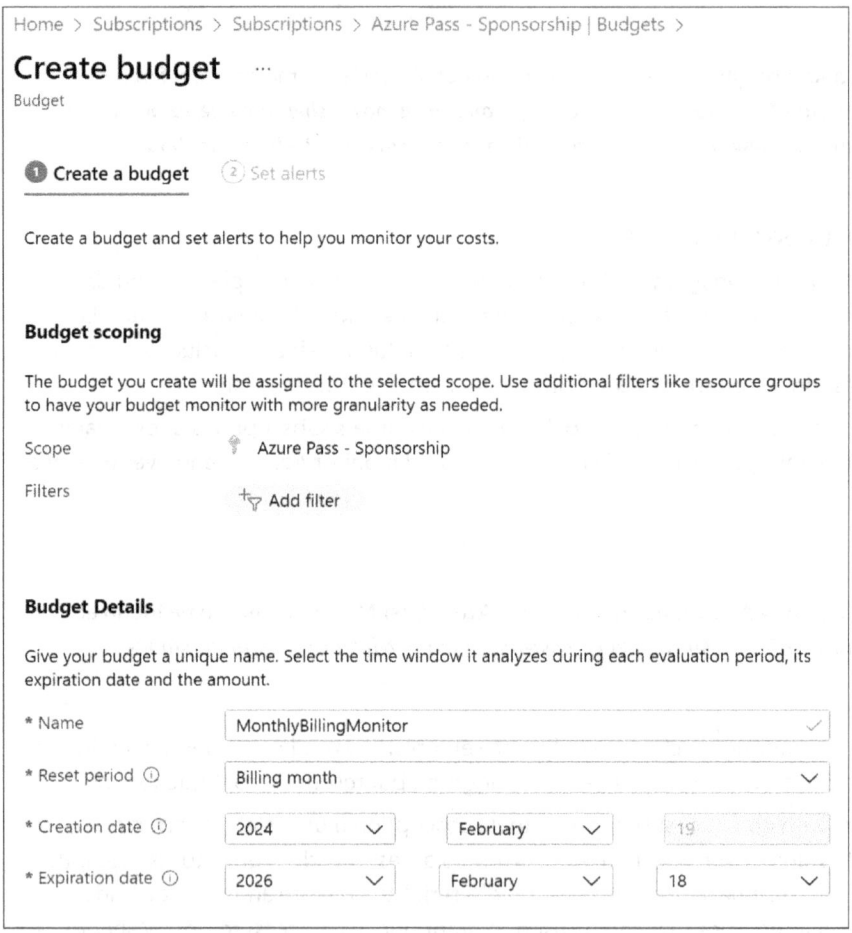

**FIGURE 1-57**  Azure budgets

Figure 1-58 shows a threshold set at 90 percent of the budget ($9,000).

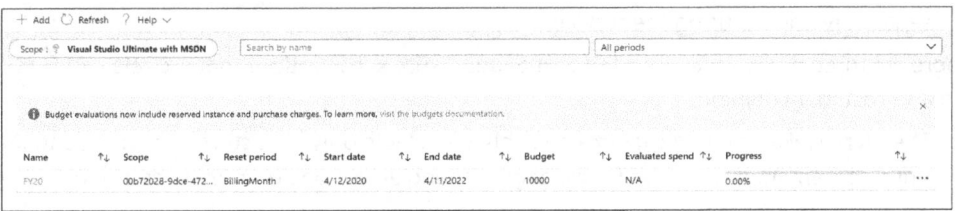

FIGURE 1-58   Azure budget alerts

After your budgets have been created, they can be viewed through the Budgets blade. When viewing the subscription scope, you will see the budgets for both the subscription and any resource group scoped budgets in a single view, as shown in Figure 1-59.

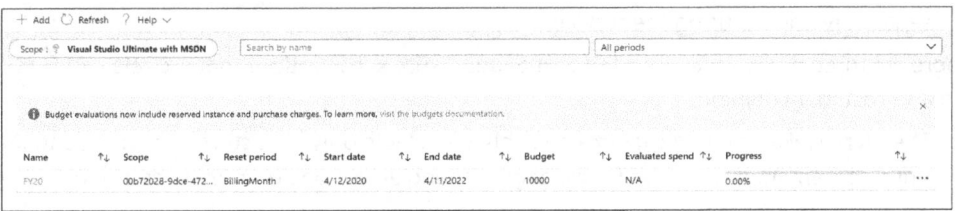

FIGURE 1-59   Azure budgets

## Monitor and report spend

While Azure Advisor and its cost recommendations provide one method for monitoring spend and unused resources, Azure has many other tools that can help you monitor the cost of your resources and report on that cost.

There are several considerations that you must account for when reporting on the cost associated with your Azure resources:

- Azure services are available to customers in over 140 countries worldwide.

- Billing is supported across over 20 major currencies.

- Azure subscriptions are billed monthly. If you are paying by credit card, note that prepaid cards and virtual credit cards are not accepted.

- You can also pay for Azure by monthly invoice. To apply for invoice payment, raise an appropriate billing support ticket from the Azure management portal. Processing the request takes five to seven days, depending on the time required for the necessary credit checks. Invoice payment is only available to business customers, and once a subscription has been moved to invoice payment, it cannot be moved back to credit card payment. If you choose invoice payment, you will get an invoice, and you will pay with a wire transfer or check.

- Customers on an Enterprise Agreement (EA) can add up-front commitments to Azure and then create multiple subscriptions under the agreement, which draw from the monetary commitment.

  - EA commitments are billed immediately, and then consumed throughout the year against the Azure resources consumed.

  - If the committed spend is exceeded, the extra spend, or "overage," is billed at the same discounted EA rate. Billing for overage is annual if the overspend is under 50 percent of the commitment, or quarterly if over 50 percent.

- Azure Marketplace third-party services are billed separately with a potentially different billing period, separate invoice, and separate credit card charge. Each service has its own billing model, which will be described in the Azure portal at the time of purchase. These range from pay-as-you-go per-minute billing to fixed monthly charges. Some services also offer a "bring your own license" model, which must provide a license purchased separately prior to using the service.

There are three portals that are used to manage Azure subscriptions that are relevant for billing and cost management:

- The EA portal at *https://ea.azure.com*. This is available only to customers with an Enterprise Agreement and is used for managing spend across one or more subscriptions.

- The Azure portal at *https://portal.azure.com*. This is available for all subscriptions and includes Azure Cost Management.

- The Azure Sponsorships portal at *https://www.microsoftazuresponsorships.com*. This is available only to those who have a sponsorship subscription from Microsoft.

The EA portal can be used to monitor spend across multiple subscriptions with the ability to view costs by the entire organization or by the business unit. Organizations can view historical spending, broken out by commitment, and overage or third-party Azure Marketplace consumption (see Figure 1-60). They can also download their current price sheet to see their EA discount rates, which often differ from the public pricing shown in the Azure portal and in the pricing calculator.

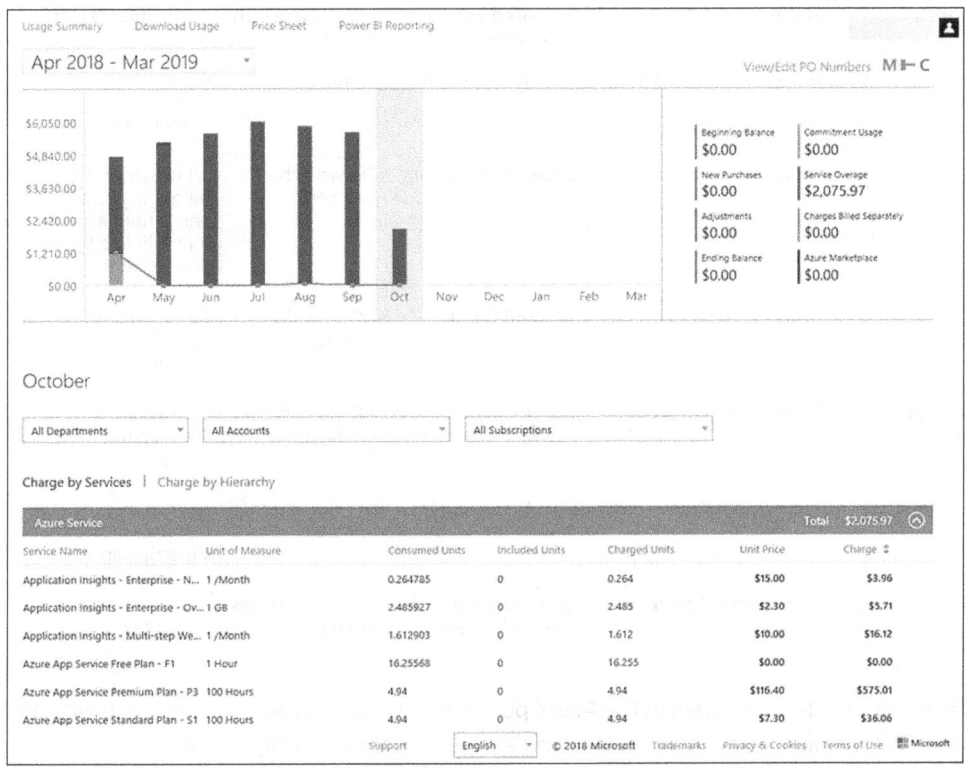

**FIGURE 1-60** Azure EA portal usage summary

EA customers can create spending quotas and set notification thresholds through the EA portal. This is in addition to the budget alerts available through the Cost Management and Billing tools of your Azure subscription. An advantage of using the EA portal to configure spending notifications is that a quota alert can be triggered based on aggregate spending across all the subscriptions within a department. Cost centers can be assigned to the departments that accounts and subscriptions roll up to for EA customers, making it easier to track cost by business unit and operate a showback or chargeback model.

Within the Azure portal, EA customers can also use Azure Cost Management for tracking cost for individual subscriptions. Cost Management includes features for performing cost analysis, setting per-subscription budgets and alerts, setting recommendations for optimization, and exporting cost management data to perform deeper analysis.

Access to the Cost Management service is dictated by scopes and might vary depending on the type of agreement or subscription that you have with Microsoft. A user must have at least read access to one of the following scopes shown in Table 1-5 to view data in Cost Management.

**TABLE 1-5** Cost Management access scopes

| Scope | Defined at | Required access to view data | Prerequisite EA setting | Consolidates data to |
|---|---|---|---|---|
| Billing account | *https://ea.azure.com* | Enterprise Admin | None | All subscriptions from the enterprise agreement |
| Department | *https://ea.azure.com* | Department Admin | DA view charges enabled | All subscriptions belonging to an enrollment account that is linked to the department |
| Enrollment account | *https://ea.azure.com* | Account Owner | AO view charges enabled | All subscriptions from the enrollment account |
| Management group | *https://portal.azure.com* | Cost Management Reader (or Reader) | AO view charges enabled | All subscriptions below the management group |
| Subscription | *https://portal.azure.com* | Cost Management Reader (or Reader) | AO view charges enabled | All resources/ resource groups in the subscription |
| Resource group | *https://portal.azure.com* | Cost Management Reader (or Reader) | AO view charges enabled | All resources in the resource group |

To access Cost Management in the Azure portal, browse to Cost Management + Billing and choose Cost Management. Finally, select Cost Analysis, as shown in Figure 1-61.

If you have access to more than one scope, you can filter by scope and begin interacting with the data. From the Cost Analysis blade, you can view the total costs for the current month, view the budget (if available), set the granularity (Accumulated, Daily, or Monthly), and apply the filters. You can filter by Location, Meter, Meter Category, Meter Subcategory, Resource, Resource Group Name, Resource Type, Service Name, Service Tier, Subscription ID, Subscription Name, and Tag.

The data in a view can be downloaded from the Cost Analysis blade as a CSV file. Any filtering that you have applied, including groupings, are applied to the file.

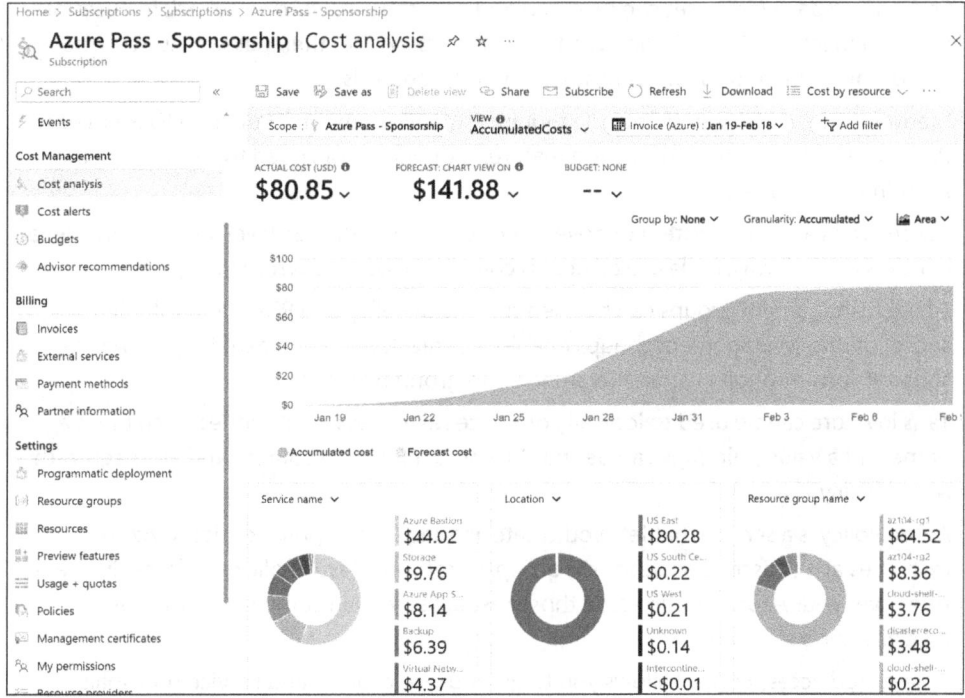

**FIGURE 1-61** Azure Cost Management Cost Analysis

# Chapter summary

Here are some of the key takeaways from this chapter:

- Windows 10 can be added to Entra ID as a device to be managed, enabling BYOD or corporate cloud only deployments with Entra Join.

- Entra Join enables administrators to manage device identity independently of users. For example, dynamic security groups can be created based on device attributes and then conditional access policies could be applied to those groups.

- Downstream Windows clients can be managed through Entra ID using Entra hybrid join.

- Conditional access is a feature of Entra ID which allows administrators to control access to cloud applications through additional checks such as user location, the device the user is accessing the cloud app from, and more.

- Multiple Entra tenants can be created and managed through Azure. This includes creating new directories and deleting existing directories.

- Users and groups can be created through the Azure portal, Azure PowerShell, the Azure CLI, and the Graph API.

- Users and groups can be managed in bulk with tools like PowerShell.

- Self-service password reset can be combined with the password writeback features of Entra Connect and Entra Cloud Sync to allow users to reset their passwords from the cloud while adhering to on-premises password standards.

- Many advanced features of Entra ID require Entra ID Premium P1 or Entra ID Premium P2 licenses. When considering Entra features, administrators need to be aware of the licensing boundaries.

- Azure offers a rich ecosystem of governance controls with user-level and platform-level controls in the form of role-based access control (RBAC) and Azure Policy.

- Azure management groups can be used to control Policy and RBAC for multiple subscriptions. Management groups enable organizational alignment for your Azure subscriptions through custom hierarchies and groupings.

- Tags in Azure can be used to logically organize resources by categories. Each tag is a name and a value pair. Tags can be shared across multiple resources and enforced with Azure Policy.

- Azure Policy is a service that lets you create, manage, and apply policies to Azure resources at a subscription, resource group, or resource level. Policies enforce different rules over your Azure resources, so those resources remain compliant with your organization's standards.

- Role-based access control allows you to grant users, groups, and service principals access to Azure resources at the subscription, resource group, or resource scopes with RBAC inheritance. The three core roles are Owner, Contributor, and Reader.

- You can create resources from the portal, PowerShell, the CLI tools, and Azure Resource Manager templates. You should understand when to use which tool and how to configure the resource during provisioning and after provisioning.

- A resource is simply a single service instance in Azure. Most services in Azure can be represented as a resource. For example, a web app instance is a resource. An App Service plan is also a resource. Even a SQL Database instance is a resource.

- A resource group is a logical grouping of resources. For example, a resource group where you deploy a VM compute instance may be composed of a network interface card (NIC), a virtual machine, a virtual network, and a public IP address.

- An ARM template is a JSON file that allows you to declaratively describe a set of resources. These resources can then be added to a new or existing resource group. For example, a template can contain the configuration necessary to create two API app instances, a mobile app instance, and an Azure SQL Database instance.

- A template can simplify orchestration because you only need to deploy the template to deploy all your resources.

- With a template, you can configure multiple resources simultaneously and use variables/parameters/functions to create dependencies between resources.

# Thought experiment

In this thought experiment, apply what you have learned. You can find answers to these questions in the next section.

You are responsible for creating and tracking resources in Azure for two business units within your organization: HR and Marketing. Your organization has an Enterprise Agreement (EA). Each business unit needs to deploy its own resources. Your Finance department needs to be able to understand the consumption of resources for each business unit for chargeback purposes. Finance would also like to be able to receive a notification when a defined monetary threshold is reached for each business unit.

The resources that each business unit will deploy are from a known set of resources and users should be prevented from creating unapproved resources. There will be resources within a subscription that are not billed back directly to the business units, but will be billed to IT. These resources must be differentiated for Finance.

1. How will you ensure that users can only create approved resources in Azure?

2. How will you grant access to create resources and restrict each business unit's users from impacting the other business units?

3. How will Finance access billing data for Azure and how will they be able to tell where each cost is coming from?

4. How will Finance be notified when each business unit is nearing their spending threshold?

# Thought experiment answers

This section contains the solution to the thought experiment for the chapter.

For each business unit, HR and Marketing, a separate subscription can be created. This will allow for the separation of resources by business unit and allow for segregated and aggregated cost reporting and monitoring for Finance through the EA portal.

1. To ensure users can only create approved resources, policies should be defined that can be assigned to each subscription. The policies will deny the creation of any unapproved resources and compliance can be monitored through Azure Policy as well.

2. Each business unit will be placed into its own subscription. Within a subscription, resource groups will be created, and users will be granted appropriate rights at the resource group level. As RBAC is inherited by child resources, with the appropriate rights granted, users will be able to create and manage resources as needed without affecting others in the subscription. This will be layered with Azure Policy to ensure that only allowed resources can be created. This can be extended further by creating Azure Resource Manager templates, which can be used by business unit users to deploy their resources with well-known configurations.

Alternatively, you can also use management groups to segregate the business units. You can still use RBAC to inherit the access subscription and child resources from a management group.

3. Users in the Finance department can be granted access to the EA portal and/or Azure Cost Management by configuring access through the required scopes. To make sure that they can tell where each resource cost is coming from, tags should be applied to all resources using a taxonomy defined by Finance. For example, "BusinessUnit" can be a tag with the allowed values "HR," "Marketing," and "IT." That taxonomy should be governed through Azure Policy to ensure that all resources are tagged with required and valid tags.

4. To manage thresholds, Department quotas can be configured in the EA portal. In addition, Budgets can be created in Cost Management. Budgets in Cost Management can provide more flexibility as multiple notification thresholds can be set and each notification can have a different receiver. This would allow a single budget to send notifications to both business unit owners and Finance.

# CHAPTER 2

# Implement and manage storage

Implementing and managing storage is one of the most important aspects of building or deploying a new solution using Azure. There are several services and features available for use, and each has its own place. Azure Storage is the underlying storage for most of the services in Azure. It provides service for the storage and retrieval of blobs and files, and it has services that are available for storing large volumes of data through tables. Azure Storage includes a fast and reliable messaging service for application developers with queues. This chapter reviews how to implement and manage storage with an emphasis on Azure storage accounts.

## Skills covered in this chapter:

- Skill 2.1 Configure access to storage
- Skill 2.2: Configure and manage storage accounts
- Skill 2.3: Configure Azure Files and Azure Blob Storage

> **NOTE  MICROSOFT EXAM OBJECTIVES**
>
> The sections in this chapter align with the objectives that are listed in the AZ-104 study guide from Microsoft. However, the sections are presented in an order that is designed to help you learn and do not directly match the order that is presented in the study guide. On the exam, questions will appear from different sections in a random order. For the full list of objectives, visit *https://learn.microsoft.com/en-us/credentials/certifications/resources/study-guides/az-104*.

## Skill 2.1: Configure access to storage

An Azure storage account is a resource that you create that is used to store data objects such as blobs, files, queues, tables, and disks. Data in an Azure storage account is durable and highly available, secure, massively scalable, and accessible from anywhere in the world over HTTP or HTTPS.

# Create and configure storage accounts

Azure storage accounts provide a cloud-based storage service that is highly scalable, available, performant, and durable. Within each storage account, a number of separate storage services are provided:

- **Blobs** Provides a highly scalable service for storing arbitrary data objects such as text or binary data.
- **Tables** Provides a NoSQL-style store for storing structured data. Unlike a relational database, tables in Azure Storage do not require a fixed schema, so different entries in the same table can have different fields.
- **Queues** Provides reliable message queueing between application components.
- **Files** Provides managed file shares that can be used by Azure VMs or on-premises servers.
- **Disks** Provides a persistent storage volume for Azure VM that can be attached as a virtual hard disk.

There are three types of storage blobs: block blobs, append blobs, and page blobs. Page blobs are generally used to store VHD files when deploying unmanaged disks. (Unmanaged disks are an older disk storage technology for Azure virtual machines. Managed disks are recommended for new deployments.)

When creating a storage account, there are several options that must be set: Performance Tier, Account Kind, Replication Option, and Access Tier. There are some interactions between these settings. For example, only the Standard performance tier allows you to choose the access tier. The following sections describe each of these settings. We then describe how to create storage accounts using the Azure portal, PowerShell, and Azure CLI.

## Storage account names

When you name an Azure storage account, you need to remember these points:

- The storage account name must be globally unique across all existing storage account names in Azure.
- The name must be between 3 and 24 characters and can contain only lowercase letters and numbers.

## Performance tiers

When creating a storage account, you must choose between the Standard and Premium performance tiers. This setting cannot be changed later.

- **Standard** This tier supports all storage services: blobs, tables, files, queues, and unmanaged Azure virtual machine disks. It uses magnetic disks to provide cost-efficient and reliable storage.

- **Premium** This tier is designed to support workloads with greater demands on I/O and is backed by high-performance SSD disks. Premium storage accounts support block blobs, page blobs, and file shares.

## Account types

There are three possible storage account types for the Standard tier: StorageV2 (General-Purpose V2), Storage (General-Purpose V1), and BlobStorage. There are four possible storage account types for the Premium tier: StorageV2 (General-Purpose V2), Storage (General-Purpose V1), BlockBlobStorage, and FileStorage. Table 2-1 shows the features for each kind of account. Key points to remember are

- The Blob Storage account is a specialized storage account used to store Block Blobs and Append Blobs. You can't store Page Blobs in these accounts; therefore, you can't use them for unmanaged disks.

- Only General-Purpose V2 and Blob Storage accounts support the Hot, Cool, and Archive access tiers.

General-Purpose V1 and Blob Storage accounts can both be upgraded to a General-Purpose V2 account. This operation is irreversible. No other changes to the account kind are supported.

> **NOTE  LEGACY STORAGE ACCOUNT TYPES**
>
> Standard General-Purpose V1 and standard Blob Storage accounts are considered legacy storage accounts, and they can be deployed but are not recommended by Microsoft. You can find more information about legacy storage account types at *https://learn.microsoft.com/en-us/azure/storage/common/storage-account-overview#legacy-storage-account-types*.

**TABLE 2-1** Storage account types and their supported features

|  | General-Purpose V2 | General-Purpose V1 | Blob Storage | Block Blob Storage | File Storage |
|---|---|---|---|---|---|
| Services supported | Blob, File, Queue, Table | Blob, File, Queue, Table | Blob (Block Blobs and Append Blobs only) | Blob (Block Blobs and Append Blobs only) | File only |
| Unmanaged Disk (Page Blob) support | Yes | Yes | No | No | No |

| | General-Purpose V2 | General-Purpose V1 | Blob Storage | Block Blob Storage | File Storage |
|---|---|---|---|---|---|
| Supported Performance Tiers | Standard Premium | Standard Premium | Standard | Premium | Premium |
| Supported Access Tiers | Hot, Cool, Archive | N/A | Hot, Cool, Archive | N/A | N/A |
| Replication Options | LRS, ZRS, GRS, RA-GRS, GZRS, RA-GZRS | LRS, GRS, RA-GRS | LRS, GRS, RA-GRS | LRS, ZRS | LRS, ZRS |

## Replication options

When you create a storage account, you can also specify how your data will be replicated for redundancy and resistance to failure. There are four options, as described in Table 2-2.

**TABLE 2-2** Storage account replication options

| Replication Type | Description |
|---|---|
| Locally redundant storage (LRS) | Makes three synchronous copies of your data within a single datacenter. Available for General-Purpose or Blob Storage accounts, at both the Standard and Premium Performance tiers. |
| Zone redundant storage (ZRS) | Makes three synchronous copies to three separate availability zones within a single region. Available for General-Purpose V2 storage accounts only, at the Standard Performance tier only. Also available for Block Blob Storage and File Storage accounts. |
| Geographically redundant storage (GRS) | This is the same as LRS (three local synchronous copies), plus three additional asynchronous copies to a second Azure region hundreds of miles away from the primary region. Data replication typically occurs within 15 minutes, although no SLA is provided. Available for General-Purpose or Blob Storage accounts, at the Standard Performance tier only. |
| Read access geographically redundant storage (RA-GRS) | This has the same capabilities as GRS, plus you have read-only access to the data in the secondary data center. Available for General-Purpose or Blob Storage accounts, at the Standard Performance tier only. |
| Geographically zone redundant storage (GZRS) | This is the same as ZRS (three synchronous copies across multiple availability zones in the selected region), plus three additional asynchronous copies to a different Azure region hundreds of miles away from the primary region. Data replication typically occurs within 15 minutes, although no SLA is provided. Available for General-Purpose v2 storage accounts only, at the Standard Performance tier only. |
| Read access geographically zone redundant storage (RA-GZRS) | This has the same capabilities as GZRS, plus you have read-only access to the data in the secondary data center. Available for General-Purpose V2 storage accounts only, at the Standard Performance tier only. |

These replication options control the level of durability and availability of the storage account. When the entire datacenter is unavailable, LRS would incur an outage. If the primary region is unavailable, both the LRS and ZRS options would incur an outage, but the GRS and GZRS options would still provide the secondary region that takes care of the requests during the outage. However, not all the replication options are available in all regions. You can find supported regions with these replication options at *https://learn.microsoft.com/en-us/azure/storage/common/storage-redundancy*.

When creating a storage account via the Azure portal, the replication and performance tier options are specified using separate settings. When creating an account using Azure Power-Shell, the Azure CLI, or via a template, these settings are combined within the SKU setting.

For example, to specify a Standard storage account using locally redundant storage using the Azure CLI, use --sku Standard_LRS.

## Access tiers

Azure Blob Storage supports four access tiers: Hot, Cool, Cold, and Archive. Each represents a trade-off of availability and cost. There is no trade-off on the durability (probability of data loss), which is defined by the SKU and replication, not the access tier.

Access tiers apply to Block Blob Storage only. They do not apply to other storage services, including append or page Blob Storage.

The tiers are as follows:

- **Hot**   This access tier is used to store frequently accessed objects. Relative to other tiers, data access costs are low while storage costs are higher.
- **Cool**   This access tier is used to store large amounts of data that is not accessed frequently and that is stored for at least 30 days. The availability SLA can vary depending on the replication model selected. Relative to the Hot tier, data access costs are higher and storage costs are lower.
- **Cold**   This access tier is used for data that is rarely accessed or modified but needs to be accessible without delay. Data in this tier should be stored for at least 90 days. The Cold tier pricing model has lower storage capacity costs but higher access costs compared to cool and hot tiers.
- **Archive**   This access tier is used to archive data for long-term storage that is accessed rarely, can tolerate several hours of retrieval latency, and will remain in the Archive

tier for at least 180 days. This tier is the most cost-effective option for storing data, but accessing that data is more expensive than accessing data in other tiers. Blob rehydration might take up to 15 hours before the blob is accessible.

New blobs will default to the access tier that is set at the storage account level, though you can override that at the blob level by setting a different access tier, including the archive tier.

> ***NOTE*** **ARCHIVE TIER SUPPORTABILITY**
>
> Currently, the Archive tier is not supported for ZRS, GZRS, or RA-GZRS accounts.

## Create an Azure storage account

To create a storage account using the Azure portal, type **storage accounts** in the search box. On the Storage Accounts blade, click Create to open the Create A Storage Account blade (see Figure 2-1). You must choose a unique name for the storage account. Storage account names must be globally unique and may only contain lowercase characters and digits. Select the Azure region (Location), the performance tier, and replication mode for the account. The blade adjusts based on the settings you choose so that you cannot select an unsupported feature combination.

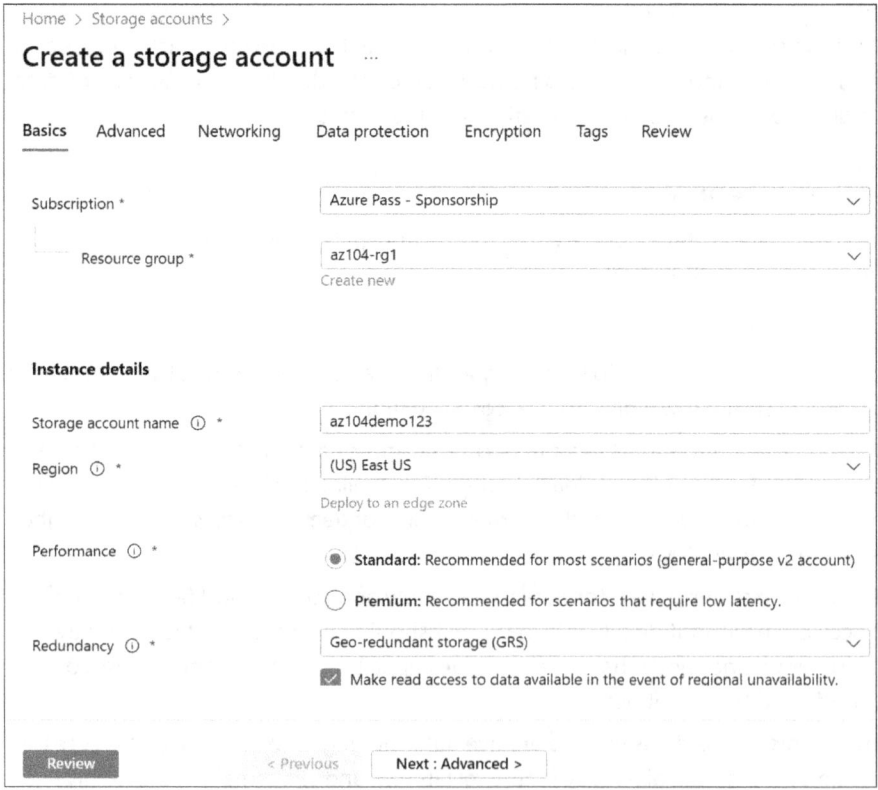

**FIGURE 2-1** Creating an Azure storage account using the Azure portal

The Advanced tab of the Create A Storage Account blade is shown in Figure 2-2. This tab defines additional security settings, hierarchical namespace support, and access protocols.

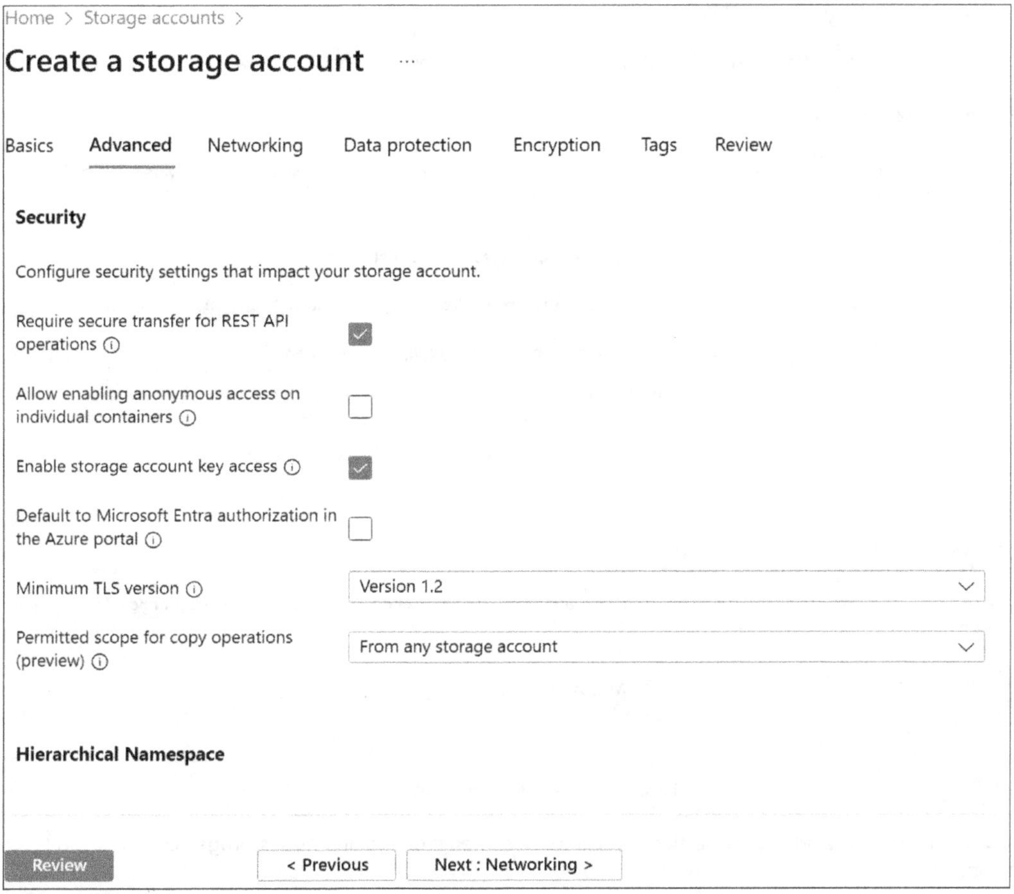

Home > Storage accounts >

# Create a storage account ...

Basics    **Advanced**    Networking    Data protection    Encryption    Tags    Review

**Security**

Configure security settings that impact your storage account.

Require secure transfer for REST API
operations ⓘ                                        ☑

Allow enabling anonymous access on
individual containers ⓘ                             ☐

Enable storage account key access ⓘ               ☑

Default to Microsoft Entra authorization in        ☐
the Azure portal ⓘ

Minimum TLS version ⓘ            | Version 1.2                                    ⌄ |

Permitted scope for copy operations  | From any storage account                   ⌄ |
(preview) ⓘ

**Hierarchical Namespace**

| Review |        | < Previous |    | Next : Networking > |

**FIGURE 2-2**   The advanced settings that can be set when creating an Azure storage account using the portal

The Networking tab of the Create A Storage Account blade is shown in Figure 2-3. On this tab, choose to maintain storage account access either publicly by choosing Enable Public Access From All Networks or privately by choosing Disable Public Access And Use Private Access.

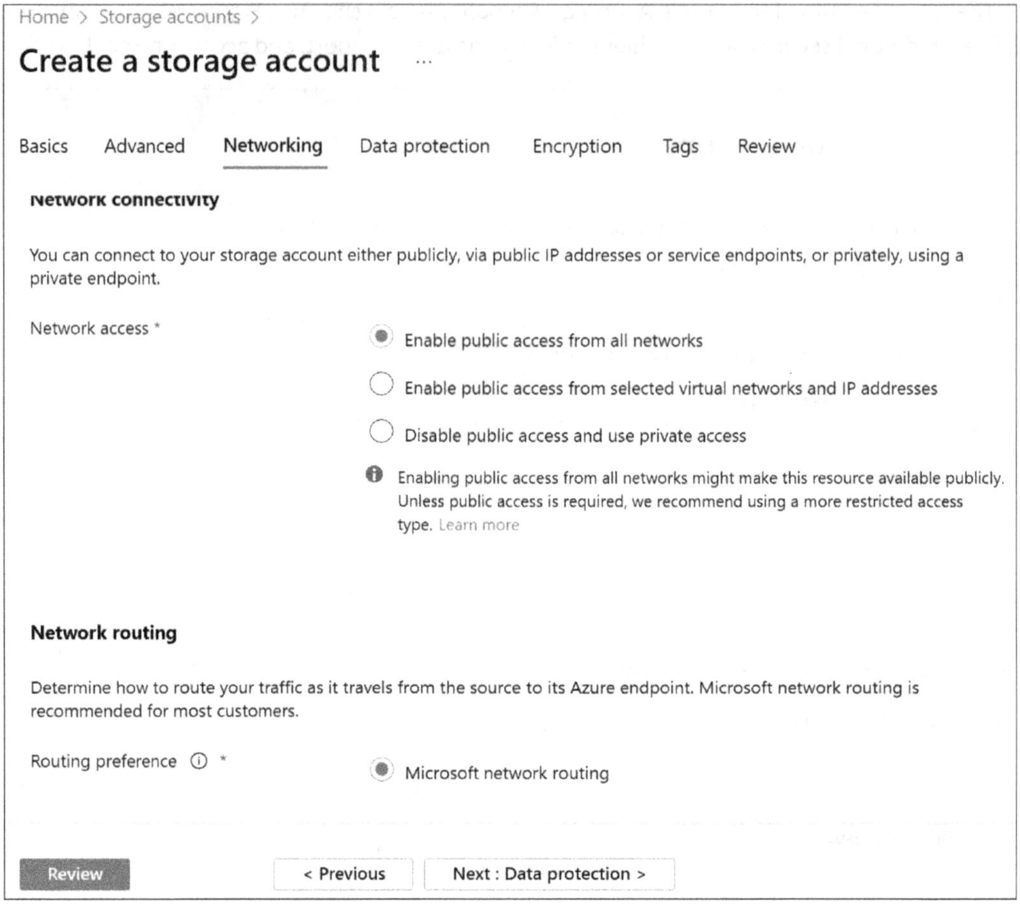

**FIGURE 2-3** The networking properties that can be set when creating an Azure storage account using the portal

The Data Protection tab provides options for configuring the recovery, tracking, and access control of the storage account. This includes soft delete options, retention periods, blob versioning, and version-level immutability support. Figure 2-4 shows the Data Protection tab.

The Encryption tab provides options for configuring the encryption type, support for customer-managed keys, and infrastructure encryption. By default, storage accounts are encrypted using Microsoft-managed keys. However, you can configure customer-managed keys to encrypt data using your own keys. Figure 2-5 shows the Encryption tab.

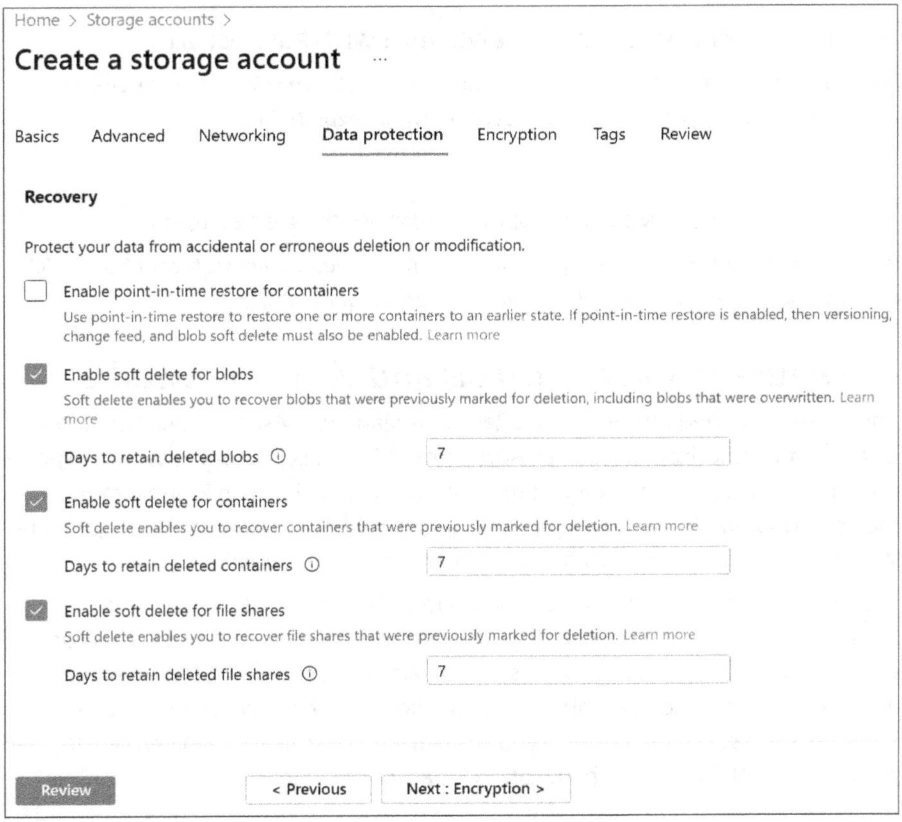

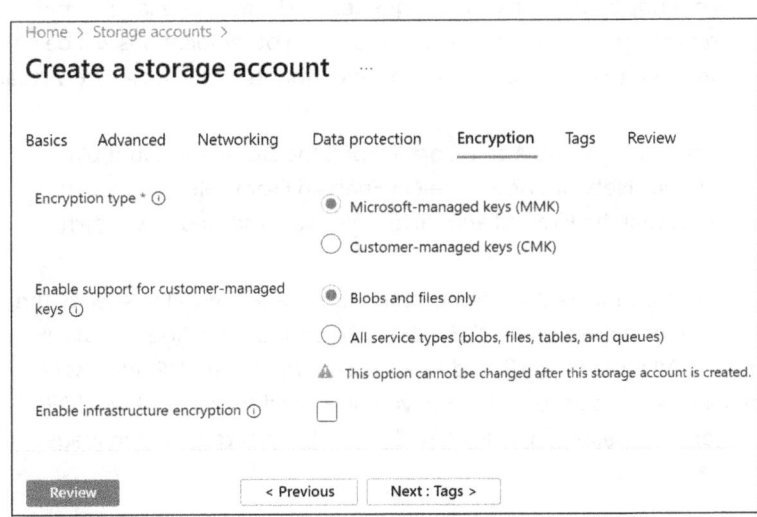

**FIGURE 2-4** The data protection properties that can be set when creating an Azure storage account using the portal

Home > Storage accounts >

# Create a storage account    ...

Basics    Advanced    Networking    Data protection    **Encryption**    Tags    Review

Encryption type * ⓘ        ● Microsoft-managed keys (MMK)

                          ○ Customer-managed keys (CMK)

Enable support for customer-managed    ● Blobs and files only
keys ⓘ
                          ○ All service types (blobs, files, tables, and queues)

                          ⚠ This option cannot be changed after this storage account is created.

Enable infrastructure encryption ⓘ    ☐

[ Review ]        [ < Previous ]    [ Next : Tags > ]

**FIGURE 2-5** The encryption properties that can be set when creating an Azure storage account using the portal

## Configure Azure Storage firewalls and virtual networks

Storage accounts are managed through Azure Resource Manager. Management operations are authenticated and authorized using Microsoft Entra ID RBAC. Each storage service exposes its own endpoint used to manage the data in that storage service (blobs in Blob Storage, entities in tables, and so on). These service-specific endpoints are not exposed through Azure Resource Manager; instead, they are (by default) internet-facing endpoints.

Access to these internet-facing storage endpoints must be secured, and Azure Storage provides several ways to do so. In this section, you will review the network-level access controls: the storage firewall and service endpoints. This section also discusses Blob Storage access levels. The following sections then describe the application-level controls: shared access signatures and access keys. In later sections, you will learn about Azure Storage replication and how to leverage Microsoft Entra ID authentication for a storage account.

### Storage firewall

Using the storage firewall, you can limit access to specific IP addresses or an IP address range. It applies to all storage services endpoints (blobs, tables, queues, and files). For example, by limiting access to the IP address range of your company, access from other locations will be blocked. Service endpoints are used to restrict access to specific subnets within an Azure virtual network.

To configure the storage firewall using the Azure portal, open the storage account blade and click Networking. Under Public Network Access, select Enabled From Selected Virtual Networks And IP Addresses to reveal the Firewall and Virtual Networks settings, as shown in Figure 2-6.

When accessing the storage account via the internet, use the storage firewall to specify the internet-facing source IP addresses (for example, 32.54.231.0/24, as shown in Figure 2-6) which will make the storage requests. All internet traffic is denied, except the defined IP addresses in the storage firewall. You can specify a list of either individual IPv4 addresses or IPv4 CIDR address ranges. (CIDR notation is explained in Skill 4.1 in Chapter 4, "Configure and manage virtual networking.")

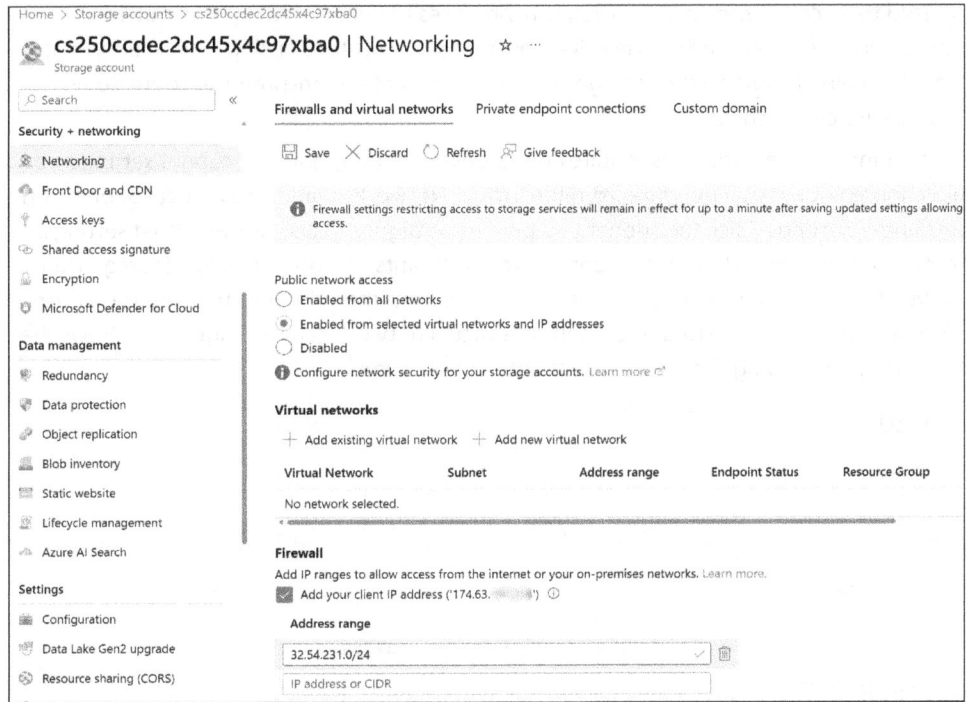

**FIGURE 2-6** Configuring a storage account firewall and virtual network service endpoint access

The storage firewall includes an option to allow access from trusted Microsoft services. As an example, these services include Azure Backup, Azure Site Recovery, Azure Networking, and more. For example, it will allow access to storage for NSG flow logs if Allow Trusted Microsoft Services To Access This Account is selected. Separately, you can enable Allow Read Access To Storage Logging From Any Network or Allow Read Access To Storage Metrics From Any Network to allow read-only access to storage metrics and logs.

> **NOTE  ADDRESS SPACE FOR A STORAGE FIREWALL**
>
> When creating a storage firewall, you must use public internet IP address space. You cannot use IPs in the private IP address space. Additionally, you cannot use /32 or /31 as a CIDR range, you must specify the individual IP addresses for individual or small ranges.

## Virtual network service endpoints

In some scenarios, a storage account is only accessed from within an Azure virtual network. In this case, it is desirable from a security standpoint to block all internet access. Configuring virtual network service endpoints for your Azure storage account, you can remove access from the public internet and only allow traffic from a virtual network for improved security.

Another benefit of using service endpoints is optimized routing. Service endpoints create a direct network route from the virtual network to the storage service. If forced tunneling is being used to force internet traffic to your on-premises network or to another network appliance, requests to Azure Storage will follow that same route. By using service endpoints, you can use a direct route to the storage account instead of the on-premises route, so no additional latency is incurred.

Configuring service endpoints requires two steps. First, to update the subnet settings, you should choose your virtual network from the Virtual Networks blade. Then select Subnets on the left under Settings. Click the subnet you plan to configure to access the subnet settings. After selecting the desired subnet, under Service Endpoints, choose Microsoft.Storage from the Services drop-down menu. This creates the route from the subnet to the storage service but does not restrict which storage account the virtual network can use. Figure 2-7 shows the subnet settings, including the service endpoint configuration.

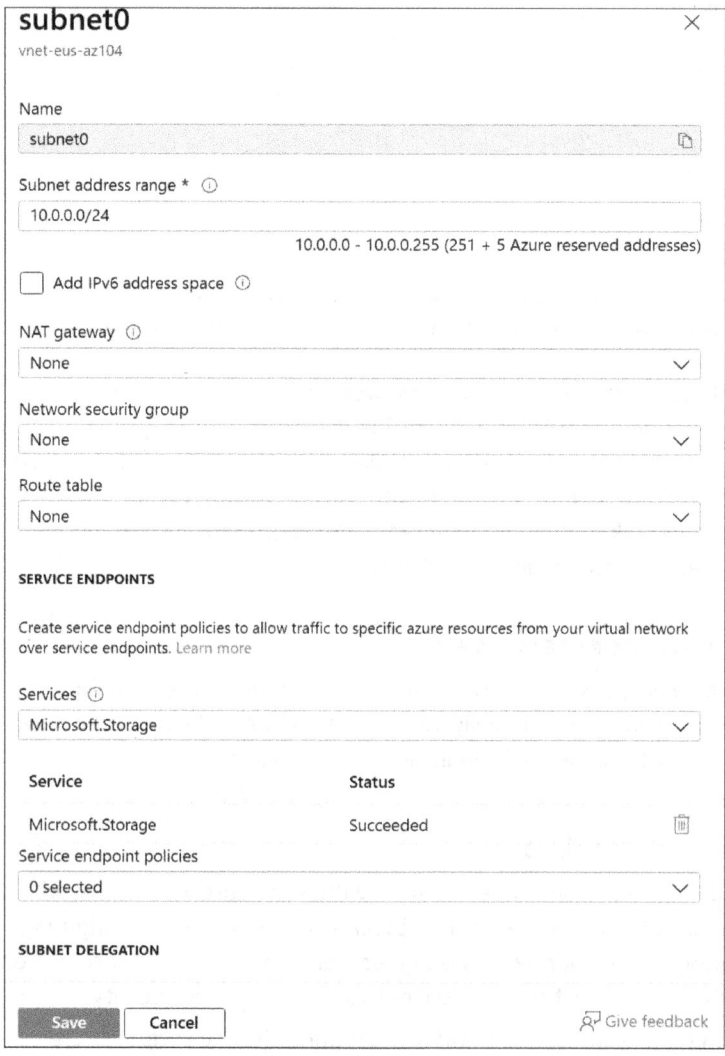

FIGURE 2-7   Configuring a subnet with a service endpoint for Azure Storage

The second step is to configure which virtual networks can access a particular storage account. From the storage account blade, click Networking. Under Public Network Access, click Enabled From Selected Virtual Networks And IP Addresses to reveal the Firewall and Virtual Network settings, as shown previously in Figure 2-1. Under Virtual Networks, select Add Existing Virtual Network to add the virtual networks and subnets that should have access to this storage account.

## Blob Storage access levels

Storage accounts support an additional access control mechanism that is limited only to Blob Storage. By default, no public read access is enabled for anonymous users, and only users with rights granted through RBAC or with the storage account name and key will have access to the stored blobs. To enable anonymous user access, you must enable Allow Blob Anonymous Access (shown in Figure 2-8) and configure the container access level (shown in Figure 2-9).

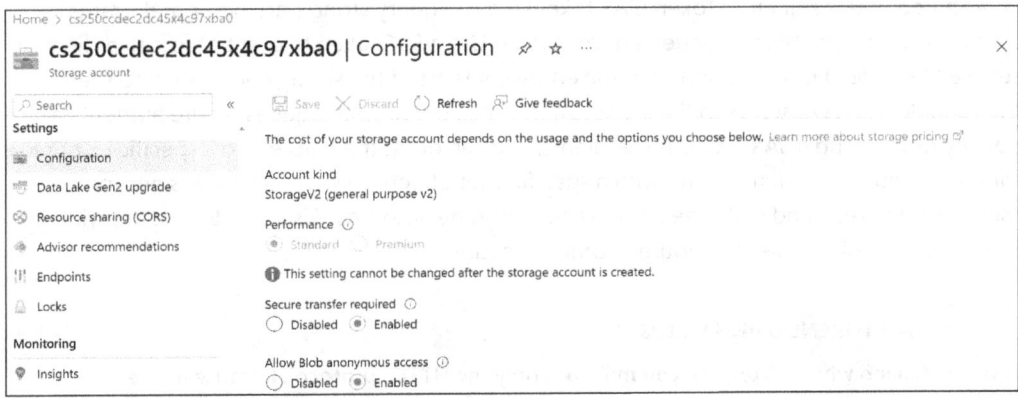

**FIGURE 2-8** Storage account configuration

The anonymous access level for a container can be specified during creation, or modified after it has been created. The supported levels of blob containers are as follows:

- **Private** Only principals with permissions can access the container and its blobs. Anonymous access is denied.
- **Blob** Only blobs within the container can be accessed anonymously.
- **Container** Blobs and their containers can be accessed anonymously.

You can change the access level through the Azure portal, Azure PowerShell, Azure CLI, programmatically using the REST API, or by using Azure Storage Explorer. The access level is configured separately on each blob container.

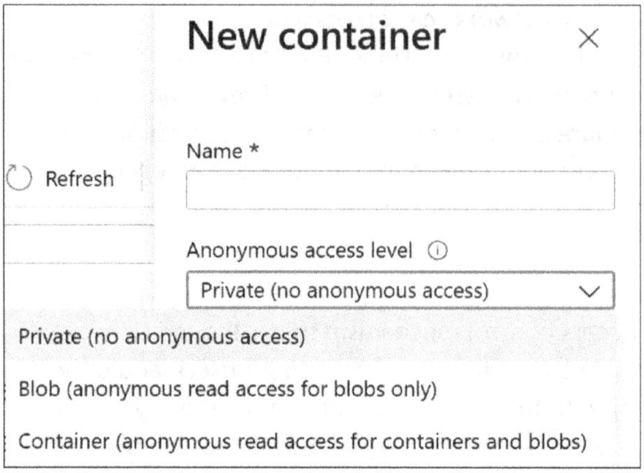

**FIGURE 2-9** Blob Storage access levels

A shared access signature token (SAS token) is a URI query string parameter that grants access to containers, blobs, queues, and/or tables. Use a SAS token to grant access to a client or service that should not have access to the entire contents of the storage account (and therefore, should not have access to the storage account keys) but still requires secure authentication. By distributing a SAS URI to these clients, you can grant them access to a specific resource, for a specified period of time, and with a specified set of permissions. SAS tokens are commonly used to read and write the data to users' storage accounts. Also, SAS tokens are widely used to copy blobs or files to another storage account.

> **NOTE** **SAS TOKENS USING HTTPS**
>
> When dealing with SAS tokens, you must use only the HTTPS protocol. Because active SAS tokens provide direct authentication to your storage account, you must use a secure connection, such as HTTPS, to distribute SAS token URIs.

## Create and use shared access signature (SAS) tokens

There are a few different ways you can create a SAS token. A SAS token is a way to granularly control how a client can access data in an Azure storage account. You can also use an account-level SAS to access the account itself. You can control many things, such as what services and resources the client can access, what permission the client has, how long the token is valid for, and more.

This section examines how to create SAS tokens using various methods. The simplest way to create one is by using the Azure portal. Browse to the Azure storage account and open the Shared Access Signature blade (see Figure 2-10). You can check the services, resource types, and permissions based on specific requirements, along with the duration for the SAS token validity and the IP addresses that are providing access. Lastly, you have an option to choose which key you want to use as the signing key for this token.

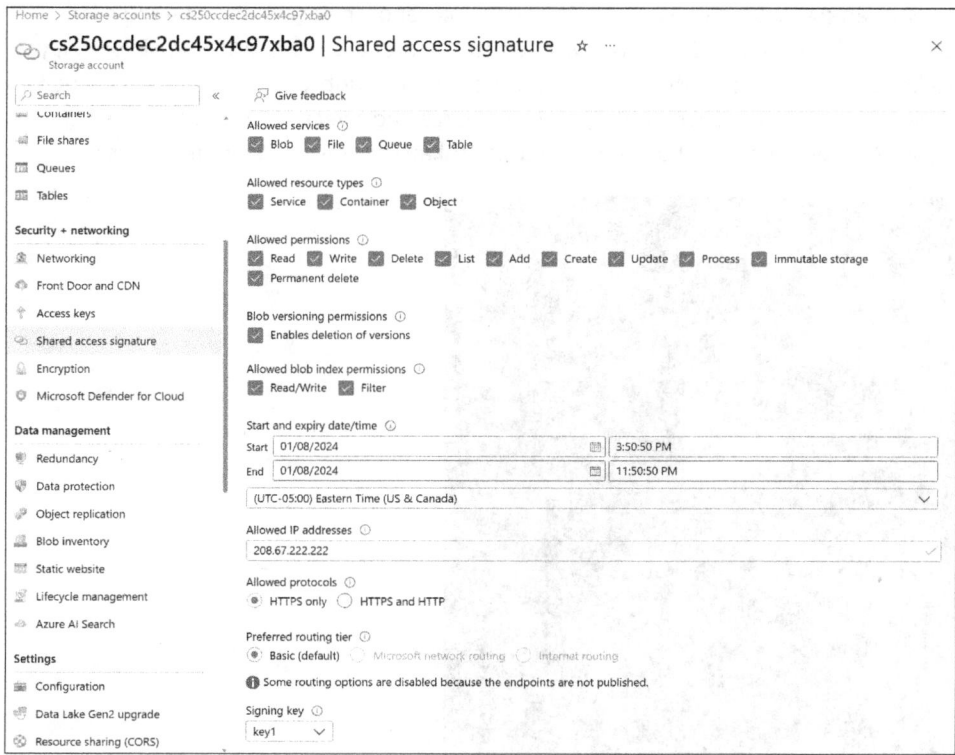

**FIGURE 2-10** Creating a shared access signature using the Azure portal

Once the token is generated, it will be listed along with connection string and SAS URLs, as shown in Figure 2-11.

**Generate SAS and connection string**

Connection string

BlobEndpoint=https://cs250ccdec2dc45x4c97xba0.blob.core.windows.net/;QueueEndpoint=https://cs250ccdec2dc45x4c97xba0.qu...

SAS token ⓘ

?sv=2022-11-02&ss=bfqt&srt=sco&sp=rwdlacupiytfx&se=2024-01-09T04:50:50Z&st=2024-01-08T20:50:50Z&sip=208.67.222.222...

Blob service SAS URL

https://cs250ccdec2dc45x4c97xba0.blob.core.windows.net/?sv=2022-11-02&ss=bfqt&srt=sco&sp=rwdlacupiytfx&se=2024-01-09...

File service SAS URL

https://cs250ccdec2dc45x4c97xba0.file.core.windows.net/?sv=2022-11-02&ss=bfqt&srt=sco&sp=rwdlacupiytfx&se=2024-01-09T0...

Queue service SAS URL

https://cs250ccdec2dc45x4c97xba0.queue.core.windows.net/?sv=2022-11-02&ss=bfqt&srt=sco&sp=rwdlacupiytfx&se=2024-01-0...

Table service SAS URL

https://cs250ccdec2dc45x4c97xba0.table.core.windows.net/?sv=2022-11-02&ss=bfqt&srt=sco&sp=rwdlacupiytfx&se=2024-01-09...

**FIGURE 2-11** Generated SAS token with connection string and SAS URLs

Also, you can create SAS tokens using Storage Explorer or the command-line tools (or programmatically using the REST APIs/SDK). To create a SAS token using Storage Explorer, you need to first select the resource (storage account, container, blob, and so on) for which the SAS token needs to be created. Then right-click the resource and select Get Shared Access Signature. Figure 2-12 demonstrates how to create a SAS token using Azure Storage Explorer.

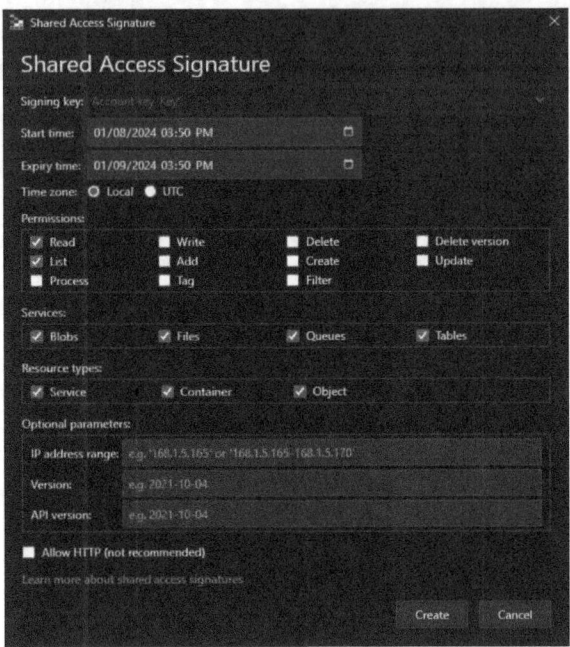

**FIGURE 2-12**   Creating a shared access signature using Azure Storage Explorer

> **NEED MORE REVIEW?   AZURE STORAGE EXPLORER**
>
> Azure Storage Explorer is a free download from Microsoft that enables convenient cloud storage management from your device. Learn more about Azure Storage Explorer at *https://azure.microsoft.com/en-us/products/storage/storage-explorer/.*

## Use shared access signatures

Each SAS token is a query string parameter that can be appended to the full URI of the blob or other storage resource for which the SAS token was created. Create the SAS URI by appending the SAS token to the full URI of the blob or other storage resource.

The following example shows the combination in more detail. Suppose the storage account name is examref, the blob container name is examrefcontainer, and the blob path is sample-file.png. The full URI to the blob in storage is

```
https://examrefstorage.blob.core.windows.net/examrefcontainer/sample-file.png
```

The combined URI with the generated SAS token is

```
https://examrefstorage.blob.core.windows.net/examrefcontainer/sample-file.png?sv=2024-
01-02&ss=bfqt&srt=sco&sp=rwdlacupx&se=2024-02-02T08:50:14Z&st=2024-01-01T00:50:14Z&spr=h
ttps&sig=65tNhZtj2luOtih8HQtK7aEL9YCIpGGprZocXjiQ%2Fko%3D
```

Currently, stored access policy is not supported for account-level SAS.

> **NEED MORE REVIEW?  ACCOUNT LEVEL SAS**
>
> You can learn more about the account level SAS at *https://learn.microsoft.com/en-us/rest/api/*
> *storageservices/create-account-sas*.

## Use user delegation SAS

You can also create user delegation SAS using Microsoft Entra ID credentials. The user delegation SAS is only supported by Blob Storage, and it can grant access to containers and blobs. Currently, SAS is not supported for user delegation SAS.

> **NEED MORE REVIEW?  USER DELEGATION SAS**
>
> You can learn more about the user delegation SAS at *https://learn.microsoft.com/en-us/rest/*
> *api/storageservices/create-user-delegation-sas*.

# Configure stored access policies

A SAS token incorporates the access parameters (start and end time, permissions, and so on) as part of the token. The parameters cannot be changed without generating a new token, and the only way to revoke an existing token before its expiry time is to regenerate the storage account key used to generate the token or to delete the blob. In practice, these limitations can make standard SAS tokens difficult to manage.

Stored access policies allow the parameters for a SAS token to be decoupled from the token itself. The access policy specifies the start time, end time, and access permissions, and the access policy is created independently of the SAS tokens. SAS tokens are generated that reference the stored access policy instead of embedding the access parameters explicitly.

With this arrangement, the parameters of existing tokens can be modified by simply editing the stored access policy. Existing SAS tokens remain valid and use the updated parameters. You can revoke the SAS token by deleting the access policy, renaming it (changing the identifier), or changing the expiry time.

> **NOTE  STORED ACCESS POLICY EFFECT**
>
> It can take up to 30 seconds for a stored access policy to take effect, and users might see an
> HTTP 403 when attempting access during that time.

Figure 2-13 shows the creation of stored access policies in the Azure portal.

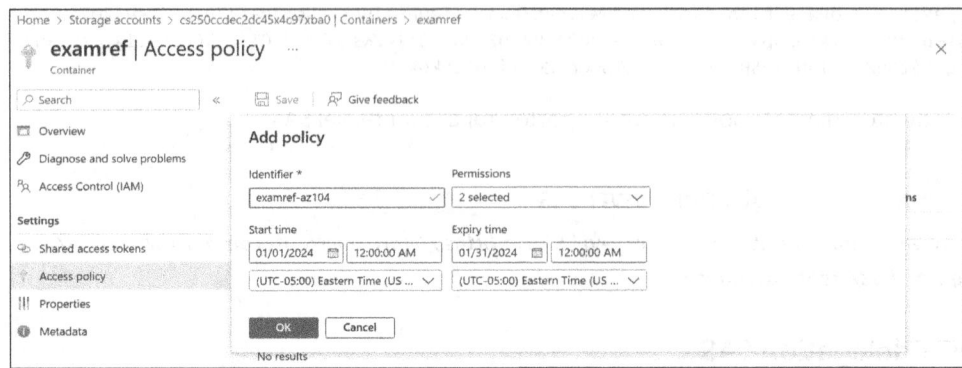

**FIGURE 2-13**   Creating stored access policies using the Azure portal

Figure 2-14 shows stored access policies being created in Azure Storage Explorer.

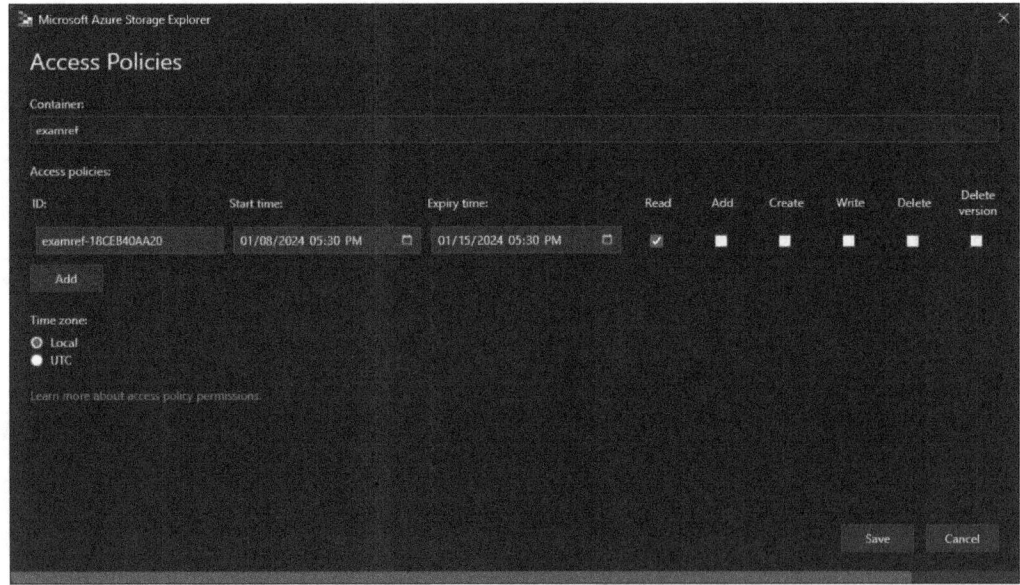

**FIGURE 2-14**   Creating stored access policies using Azure Storage Explorer

To use the created policies, reference them by name when creating a SAS token using Storage Explorer or when creating a SAS token using PowerShell or the CLI tools.

> **NOTE**   **MAXIMUM ACCESS POLICIES**
>
> You can have a maximum of only five access policies on a container, table, queue, or file share.

# Manage access keys

The simplest way to manage access to a storage account is to use access keys. With the storage account name and an access key to the Azure storage account, you have full access to all data in all services within the storage account. You can create, read, update, and delete containers, blobs, tables, queues, and file shares. In addition, you have full administrative access to everything other than the storage account itself. (You cannot delete the storage account or change settings on the storage account, such as its type.)

Applications will use the storage account name and key for access to Azure Storage. Sometimes, this is to grant access by generating a SAS token, and sometimes, it is for direct access with the name and key.

To access the storage account name and key, open the storage account from within the Azure portal and click Access Keys. Figure 2-15 shows the primary and secondary access keys for a storage account.

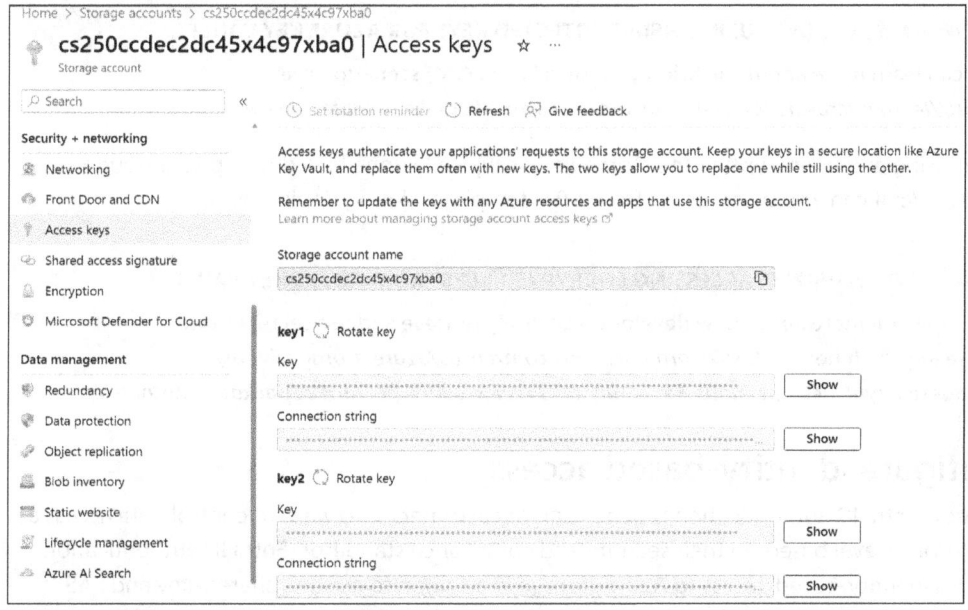

**FIGURE 2-15**   Access keys for an Azure storage account

Each storage account has two access keys. This means you can modify applications to use the second key instead of the first and then regenerate the first key. This technique is known as "key rolling" or "key rotation." You can reset the primary key with no downtime for applications that directly access storage using an access key.

Storage account access keys can be regenerated using the Azure portal or the command-line tools. In PowerShell, this is accomplished with the New-AzStorageAccountKey cmdlet; with Azure CLI, you will use the az storage account keys renew command.

## Managing access keys in Azure Key Vault

It is important to protect the storage account access keys because they provide full access to the storage account. Azure Key Vault helps safeguard cryptographic keys and secrets used by cloud applications and services, such as authentication keys, storage account keys, data encryption keys, and certificate private keys.

Keys in Azure Key Vault can be protected in software or by using hardware security modules (HSMs). HSM keys can be generated in place or imported. Importing keys is often referred to as bring your own key, or BYOK.

Accessing and unencrypting the stored keys is typically done by a developer, although keys from Key Vault can also be accessed from ARM templates during deployment.

# Configure identity-based access

Microsoft Entra ID authentication is beneficial for customers who want to control data access at an enterprise level based on their security and compliance standards. Entra ID authentication provides identity-based access to Azure storage in addition to existing shared-key and SAS token authorization mechanisms for Azure Storage (Blob and Queue). Azure blobs, files, and queues are supported by Entra ID authentication.

Entra ID authentication enables customers to leverage RBAC in Azure for granting the required permissions to a security principal (users, groups, and applications) down to the scope of an individual blob container or queue. While authenticating a request, Entra ID returns an OAuth 2.0 token to security principal, which can be used for authorization against Azure Storage.

Entra ID authorization can be implemented in many ways, such as assigning RBAC roles to a security principal (users, groups, and applications), using a managed identity, or creating shared access signatures signed by Entra ID credentials.

If an application is running from within an Azure entity such as an Azure VM, a virtual machine scale set, or an Azure Functions app, it can use a managed identity to access a storage account.

**NEED MORE REVIEW?** **AUTHORIZING ACCESS**

More information about authorizing access to data with managed identities for Azure resources can be found at *https://learn.microsoft.com/en-us/azure/storage/blobs/authorize-access-azure-active-directory*.

## RBAC roles for blobs and queues

There are several built-in RBAC roles available in Azure for authorizing access to Blob and Queue Storage:

- **Storage Blob Data Owner**   Sets ownership and manages POSIX access control for Azure Data Lake Storage Gen2
- **Storage Blob Data Contributor**   Grants read/write/delete permissions for Blob Storage
- **Storage Blob Data Reader**   Grants read-only permissions for Blob Storage
- **Storage Queue Data Contributor**   Grants read/write/delete permissions for Queue Storage
- **Storage Queue Data Reader**   Grants read-only permissions for Queue Storage
- **Storage Queue Data Message Processor**   Grants peek, retrieve, and delete permissions to messages in queues
- **Storage Queue Data Message Sender**   Grants add permissions to messages in queues
- **Storage Table Data Contributor**   Allows read, write, and delete access to tables and entities
- **Storage Table Data Reader**   Provides read-only access to tables and entities

**NEED MORE REVIEW?** **BUILT-IN ROLE DETAILS**

For more information about built-in roles, see *https://learn.microsoft.com/en-us/azure/role-based-access-control/built-in-roles#storage*.

## Resource scope for blobs and queues

It is also important to determine the scope of the access for the security principal before you assign an RBAC role. You can narrow the scope to the container, queue, or table level. Here are the valid scopes:

- **Container**  The role assignment will be applicable at the container level. All the blobs inside the container, the container properties, and the metadata will inherit the role assignment when this scope is selected.
- **Queue**  The role assignment will be applicable at the queue level. All the messages inside the queue, as well as queue properties and metadata, will inherit the role assignment when this scope is selected.
- **Table**  The role assignment will be applicable at the table level. All tables and entities within the storage account will be accessible based on the role assignment with this scope.
- **Storage account**  The role assignment will be applicable at the storage account level. All the containers, blobs, queues, and messages within the storage account will inherit the role assignment when this scope is selected.
- **Resource group**  The role assignment will be applicable at the resource group level. All the containers or queues in all the storage accounts in the resource group will inherit the role assignment when this scope is selected.
- **Subscription**  The role assignment will be applicable at the subscription level. All the containers or queues in all the storage accounts in all the resource groups in the subscription will inherit the role assignment when this scope is selected.

## Entra ID authentication and authorization in the Azure portal

In the following example, you will learn how to configure the Entra ID authentication method to allow users to access the blob data.

In Figure 2-16, you can see the examref container has one blob named SampleFile.txt. Also, notice that the authentication method is currently set as Access Key.

**FIGURE 2-16**  The Overview blade of examrefcontainer

Click Switch To Microsoft Entra User Account to change the authentication method. You will see a warning message indicating that you do not have permission to list the data (see Figure 2-17).

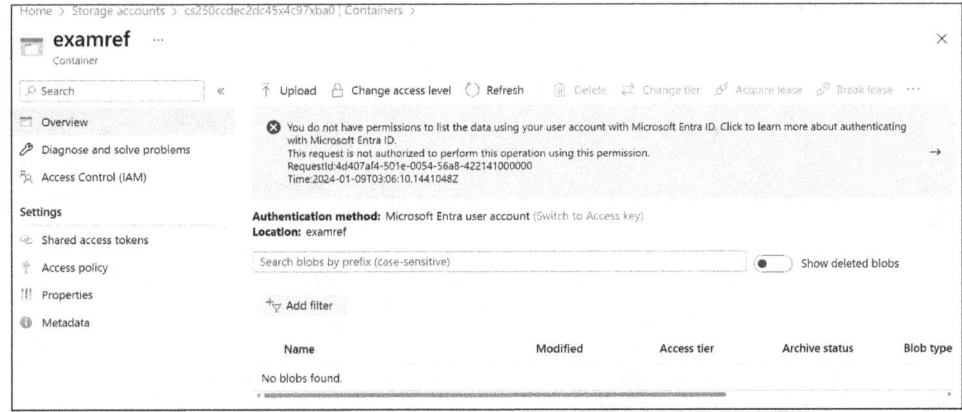

**FIGURE 2-17** Warning message that you don't have permission

Now you'll assign the Storage Blob Data Reader role to the logged-in user at the container level.

1. Open the Access Control (IAM) blade for the container and select Add, Add Role Assignment.
2. On the Role tab, select the Storage Blob Data Reader role, and then click Next.
3. On the Members tab, select your user account.
4. Click Review + Assign twice to apply the role assignment (see Figure 2-18).

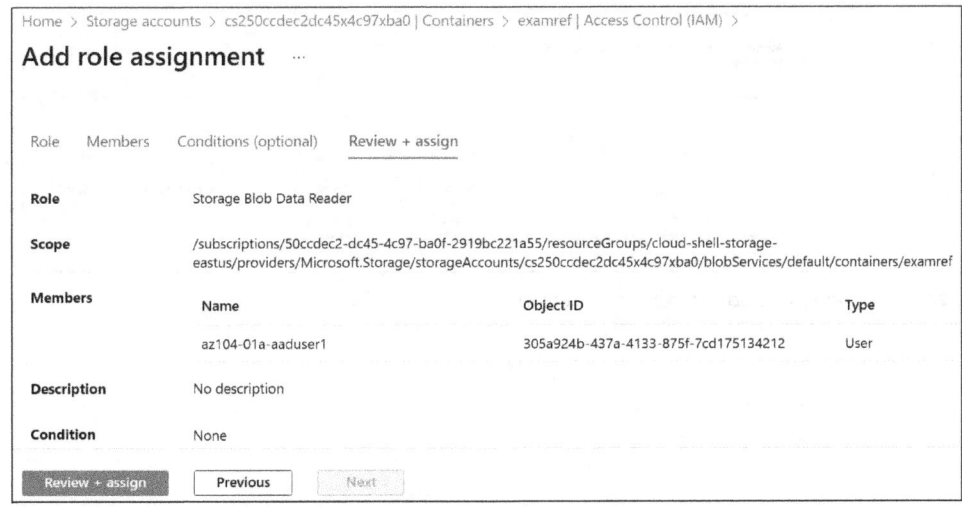

**FIGURE 2-18** Storage Blob Data Reader role assignment

You should now see the user with the role Storage Blob Data Reader, which appears under the Role heading (see Figure 2-19).

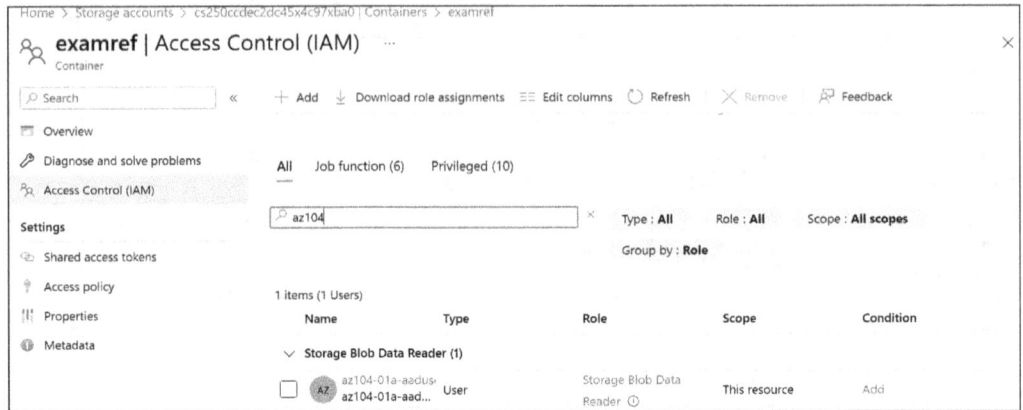

**FIGURE 2-19** Role assignments for examrefcontainer

If you navigate to the Overview blade of examref now, you will see the SampleFile.txt blob with the authentication method shown as Microsoft Entra User Account (see Figure 2-20).

> **NOTE    RBAC ROLES TAKING EFFECT**
>
> Sometimes, RBAC roles take up to five minutes to propagate the role assignments.

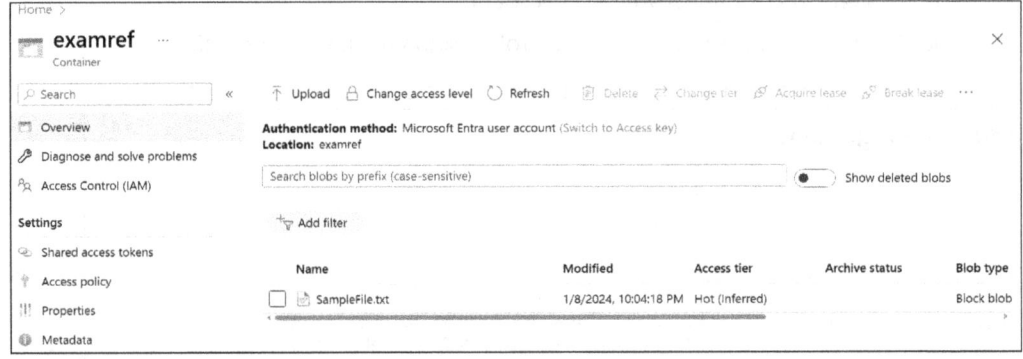

**FIGURE 2-20** The Overview blade of examrefcontainer

# Skill 2.2: Configure and manage storage accounts

Azure storage accounts include multiple replication options to determine how the data is stored in the primary Azure region and secondary replicated region. The storage account can be configured to duplicate data in only the primary region or replicated to a secondary Azure region. Storage accounts also provide encryption options to use the default Microsoft-managed key, or to bring your own key using Azure Key Vault. In a later part of this skill, you will also learn how to use tools like Azure Storage Explorer and AzCopy.

> **This skill covers how to:**
> - Configure Azure storage redundancy
> - Configure object replication
> - Configure storage account encryption
> - Manage data using Azure Storage Explorer
> - Manage data using AzCopy

## Configure Azure storage redundancy

The data in your Azure storage accounts is always replicated for durability and high availability. The built-in storage replication options were discussed at a high level in Table 2-2. It's important to understand when each replication option should be used and the level of availability you require for your scenario. Table 2-3 describes the scenarios and expected availability for each of the replication options.

Storage accounts can be modified between the LRS, GRS, and RA-GRS replication modes. Azure will replicate the data asynchronously in the background as required.

You can set the replication mode for a storage account after it is created through the Azure portal by clicking the Redundancy link on the storage account and selecting an option from the Redundancy drop-down menu (see Figure 2-21). Be aware that changing the replication option can also result in a significant price difference depending on the options that you select.

**TABLE 2-3** Durability and availability for various replication options

| Scenario | LRS | ZRS | GRS | RA-GRS | GZRS | RA-GZRS |
|---|---|---|---|---|---|---|
| Supported storage account types | GPv21, GPv12, blob | GPv2 | GPv1, GPv2, blob | GPv1, GPv2, blob | GPv2 | GPv2 |
| Server or other failure within a data center | Available | Available | Available | Available | Available | Available |
| Failure affecting an entire data center (such as a fire) | Not available | Available | Available | Available | Available | Available |
| Failure affecting all data centers in a region (such as a major hurricane) | Not available | Not available | Failover to secondary region | Read access only until failed over | Failover to secondary region | Read access only until failed over |
| Designed durability | At least 99.999999999 percent | At least 99.9999999999 percent | At least 99.99999999999999 percent | At least 99.99999999999999 percent | At least 99.99999999999999 percent | At least 99.99999999999999 percent |
| Availability SLA for read requests | At least 99.9 percent (99 percent for cool access tier) | At least 99.9 percent (99 percent for cool access tier) | At least 99.9 percent (99 percent for cool access tier) | At least 99.99 percent (99.9 percent for cool access tier) | At least 99.99 percent (99.9 percent for cool access tier) | At least 99.99 percent (99.9 percent for cool access tier) |
| Availability SLA for write requests | At least 99.9 percent (99 percent for cool access tier) | At least 99.9 percent (99 percent for cool access tier) | At least 99.9 percent (99 percent for cool access tier) | At least 99.9 percent (99 percent for cool access tier) | At least 99.9 percent (99 percent for cool access tier) | At least 99.9 percent (99 percent for cool access tier) |

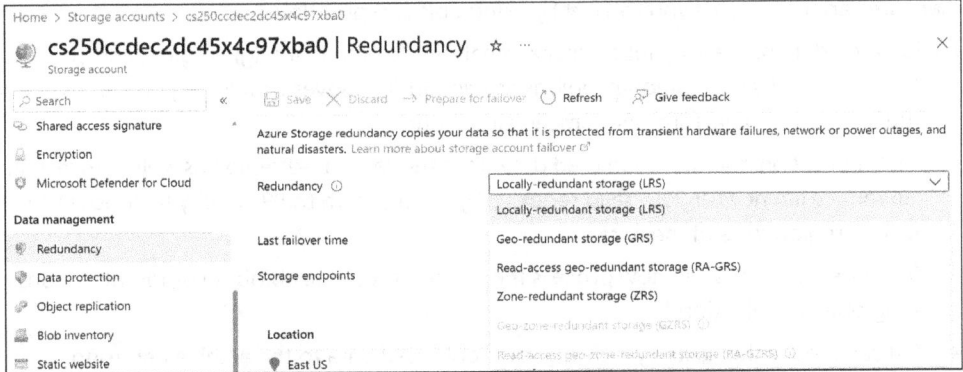

**FIGURE 2-21** The Redundancy blade of an Azure storage account

**NEED MORE REVIEW?** **MORE EXAMPLES WITH POWERSHELL**

There are many variations for using the async copy service with PowerShell. For more information, see the following: *https://learn.microsoft.com/en-us/powershell/module/az.storage/start-azstorageblobcopy?view=azps-11.2.0.*

**NEED MORE REVIEW?** **MORE EXAMPLES WITH CLI**

There are many variations for using the async copy service with the Azure CLI. For more information, see *https://learn.microsoft.com/en-us/cli/azure/storage/blob/copy?view=azure-cli-latest.*

## Configure object replication

Azure Storage blob object replication provides asynchronous replication of block blobs from one storage account to another. The blobs are replicated based on the defined replication rules.

Using object replication requires that the blob versioning options are enabled for both the source and destination storage accounts. Additionally, the source storage account must have the blob change feed enabled. If these settings are not enabled, they will be enabled automatically when you create the first replication rule. These settings might increase the costs of the storage account.

**NOTE** **BLOB VERSIONING AND BLOB CHANGE FEED**

Blob versioning captures the state of a blob when it is modified or deleted; Azure Storage creates a new version ID for a blob with each change. The blob change feed provides all the changes with the blobs and its metadata in the form of transactional logs.

There are various benefits you can get by using object replication:

- For large data processing jobs, you can analyze the data in a single region, and you can distribute results to additional regions as needed. This saves processing time and compute resources to perform the same in all regions.

- With replication, the users can read data from the replicated region as well. Hence, you can reduce latency for your read requests by giving them the flexibility to choose the nearest region to read the data.

- Compute workloads can now process the same sets of block blobs in different regions using object replication.

- You can reduce the costs by moving your replicated data to the archive tier using lifecycle management policies.

Keep in mind that object replication performs multiple read and write transactions against the source and destination accounts. This can incur additional costs.

To set up the object replication rules, open the storage account, browse to Object Replication under Data Management, and click Create Replication Rules (see Figure 2-22). You can define up to 10 replication rules per policy using the Azure portal, or up to 1,000 if you use JSON to upload the rules.

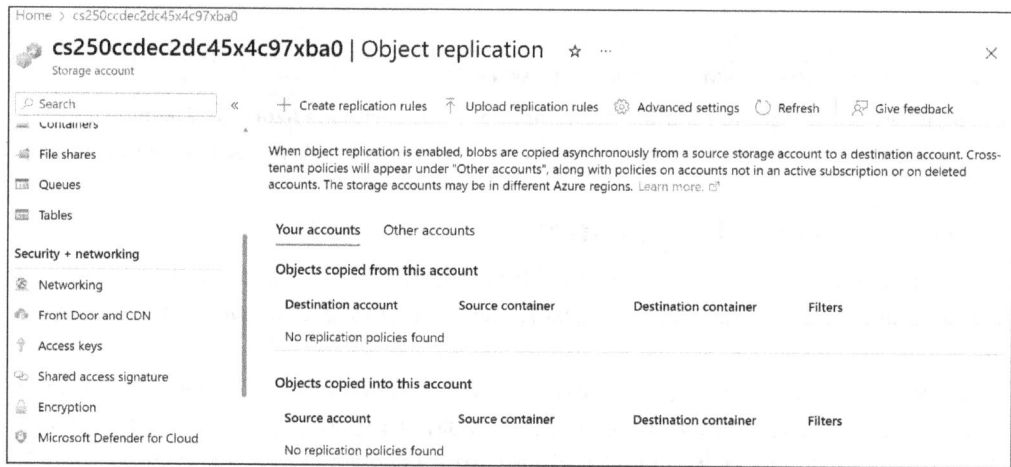

**FIGURE 2-22**   Create Replication Rules option on the Object Replication blade

You need to select the destination subscription and destination storage account that will be used for replication. You also need to select the source containers and destination containers in a pair. You can limit the replication scope with filters by specifying the prefix match for blobs. See Figure 2-23.

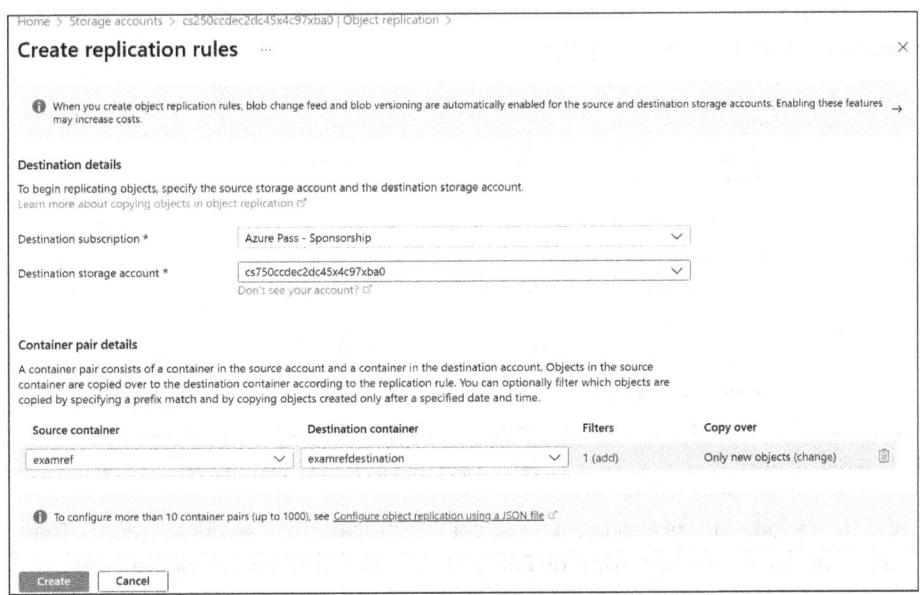

**Create replication rules** ...                                                                                    ✕

ℹ️ When you create object replication rules, blob change feed and blob versioning are automatically enabled for the source and destination storage accounts. Enabling these features →
may increase costs.

**Destination details**

To begin replicating objects, specify the source storage account and the destination storage account.
Learn more about copying objects in object replication ↗

Destination subscription *            Azure Pass - Sponsorship                                        ∨

Destination storage account *         cs750ccdec2dc45x4c97xba0                                         ∨
                                      Don't see your account? ↗

**Container pair details**

A container pair consists of a container in the source account and a container in the destination account. Objects in the source
container are copied over to the destination container according to the replication rule. You can optionally filter which objects are
copied by specifying a prefix match and by copying objects created only after a specified date and time.

| Source container | Destination container | Filters | Copy over |
|---|---|---|---|
| examref ∨ | examrefdestination ∨ | 1 (add) | Only new objects (change)  🗑 |

ℹ️ To configure more than 10 container pairs (up to 1000), see Configure object replication using a JSON file ↗

[ Create ]    [ Cancel ]

**FIGURE 2-23**   The Create Replication Rules blade with destination and container pair details

You can also control how objects are copied to the destination container using three options: Everything, Only New Objects, and Custom. If you select Everything, then all the blobs matching the filters will be copied to the destination container, but if you select Only New Objects, only the newly added blobs matching the filters will be copied to the destination container. If you select Custom, you will have a chance to manually specify a date and time to copy the blobs created later, as shown in Figure 2-24.

**Copy over**                                        ✕

To manage how many objects are copied to the destination
container, specify that objects are copied based on when they were
added to the source container.

Copy over
◯ Everything
◯ Only new objects
◉ Custom

Copy objects that were created starting from *
[ MM/DD/YYYY        📅 ] [ h:mm:ss AM/PM ]
[ (UTC-05:00) Eastern Time (US & Canada)          ∨ ]

**FIGURE 2-24**   Copy Over option for a replication rule

Once created, the consolidated view of replication rules can be viewed by visiting the Object Replication blade. You can also right-click a rule and select Edit Rules or Download Rules, as shown in Figure 2-25. The downloaded rules can be edited and reused using the Upload Replication Rules options instead of re-creating them.

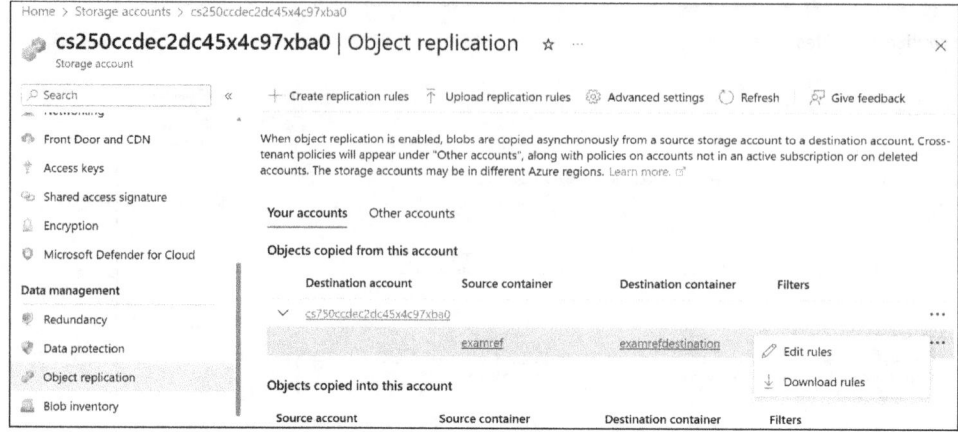

**FIGURE 2-25** Replication rules

To check the replication status of a source blob, select the source blob to see its properties. The Object Replication section shows the replication Policy ID, Rule ID, and Status (see Figure 2-26).

Home > Storage accounts > cs250ccdec2dc45x4c97xba0 | Containers > examref >

## file.pdf ...
Blob

🖫 Save   ✕ Discard   ⬇ Download   ↻ Refresh   🗑 Delete   ⮂ Change tier   ⬔ Acquire lease   ⬔ Break lease   🗛 Give feedback

| | |
|---|---|
| VERSION ID | 2024-01-10T19:05:44.2602638Z |
| TYPE | Block blob |
| SIZE | 2.55 MiB |
| ACCESS TIER | Hot (Inferred) |
| ACCESS TIER LAST MODIFIED | N/A |
| ARCHIVE STATUS | - |
| REHYDRATE PRIORITY | - |
| SERVER ENCRYPTED | true |
| ETAG | 0x8BDC120F257DDC70 |
| VERSION-LEVEL IMMUTABILITY POLICY | Disabled |
| CACHE-CONTROL | |
| CONTENT-TYPE | application/pdf |
| CONTENT-MD5 | oWXEg4s9e3TtXpK64mKueA... |
| CONTENT-ENCODING | |
| CONTENT-LANGUAGE | |
| CONTENT-DISPOSITION | |
| LEASE STATUS | Unlocked |
| LEASE STATE | Available |
| LEASE DURATION | - |
| COPY STATUS | - |
| COPY COMPLETION TIME | - |

Undelete

Object replication

| Policy ID | Rule ID | Status |
|---|---|---|
| b4d2dc32-d839-4b22-9b0f-4caf9... | 58260d32-1f42-42e5-a8bb-89a5b... | Complete |

**FIGURE 2-26** Replication status

There are certain limitations with blob object replication that are crucial to review before implementation:

- Object replication doesn't work with the Archive tier.

- Blob snapshots and immutable snapshots are not supported with object replication.

- Object replication doesn't work with accounts with a hierarchical namespace (Azure Data Lake Storage Gen2).

- Because block blob data is replicated asynchronously, there is no SLA on when accounts are in sync. However, you can check the replication status of a blob.

- The source account can only have a maximum of two destination accounts.

- Once you create a replication policy, the destination container is read-only, and you can no longer perform write operations against it.

## Configure storage account encryption

Data in an Azure storage account is encrypted using AES 256-bit encryption and is FIPS 140-2 compliant. Encryption in an Azure storage account is enabled automatically and cannot be disabled.

By default, Microsoft manages the keys used to encrypt and decrypt the data. In this scenario, Microsoft is responsible for key storage, rotation, control, and scope. If your organization has business or compliance requirements to manage these components, the Azure storage account can be configured to use customer-managed keys. This key would be used to encrypt the data for blob and file storage only.

Using customer-managed keys requires two additional Azure components: Azure Key Vault and a managed identity. The Azure Key Vault acts as the secure repository to store the key that you select for encryption operations in the storage account. The managed identity is used by the storage account to access and retrieve the key from the key vault.

To configure encryption for an Azure storage account, select Encryption from the Storage Account menu. On the Encryption tab, select Customer-managed Keys. This will show the additional configuration items, where you can select an existing key vault, identity type, and key. Figure 2-27 displays the Encryption page.

## Manage data using Azure Storage Explorer

Azure Storage Explorer is a cross-platform application designed to help you quickly manage one or more Azure storage accounts. It can be used with all storage services: Blob Storage, Azure Tables, Queue Storage, and Azure Files. In addition, Azure Storage Explorer also supports the CosmosDB and Azure Data Lake Storage services.

You can install Azure Storage Explorer by navigating to its landing page at *https://azure. microsoft.com/features/storage-explorer/* and selecting your operating system (Windows, macOS, or Linux).

In addition, a browser-based storage explorer with similar functionality is integrated into the Azure portal. To access it, click Storage Browser from the Storage Account blade.

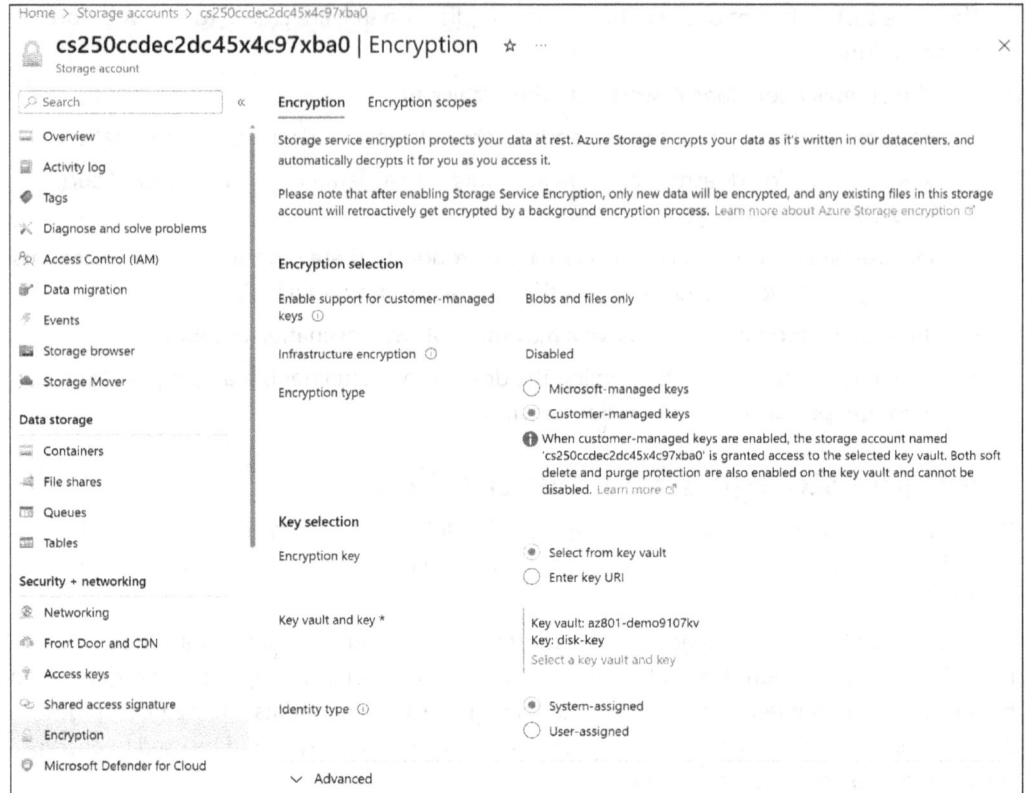

**FIGURE 2-27** Storage account encryption

## Connect Storage Explorer to Storage Accounts

After Storage Explorer is installed, you can connect to Azure Storage in a number of different ways (shown in Figure 2-28):

- **Subscription** This option allows you to sign in using a work or Microsoft account and access all your storage accounts via role-based access control.

- **Storage Account or service** This option requires you to have access to the storage account name and key. These values can also be accessed from the Azure portal under Access Keys.

- **Blob container or directory** Attach directly to only the blob service endpoint to interact with a container or directory.

- **ADLS Gen2 container or directory** Attach directly to only an Azure Data Lake Storage Gen2 container in a storage account.

- **File share** Attach directly to an Azure Files share that has already been created in a storage account.
- **Queue** Attach directly to the queue service endpoint of an existing storage account.
- **Table** Attach directly to the table service endpoint of an existing storage account.
- **Local Storage Emulator** Allows you to connect to the local Azure Storage emulator as part of the Microsoft Azure SDK.

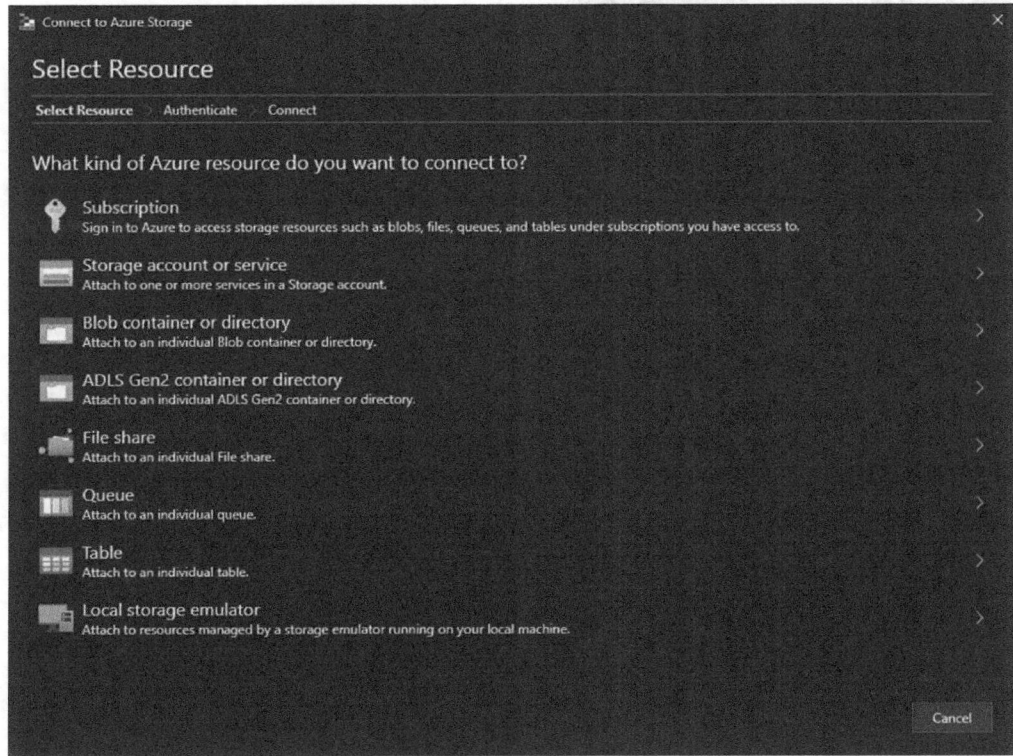

**FIGURE 2-28** Connecting to an Azure storage account using Azure Storage Explorer

After connecting, filter which subscriptions to use. Once you select a subscription, all the supported services within the subscription will be made available. Figure 2-29 shows an expanded Azure Storage container named examref.

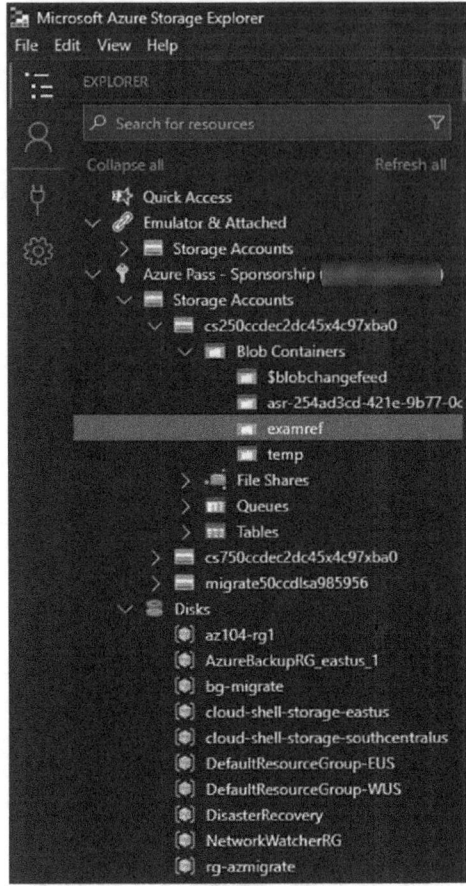

**FIGURE 2-29** Azure Storage Explorer showing an Azure storage account beneath the subscription

## Use Storage Explorer

Using Storage Explorer, you can manage each of the storage services: Blob Storage, Azure Tables, Queue Storage, and Azure Files. Table 2-4 summarizes the supported operations for each service.

**TABLE 2-4** Storage Explorer operations

| Storage service | Supported operations |
|---|---|
| **Blob** | **Blob containers** Create, rename, copy, delete, control public access level, manage leases, and create and manage shared access signatures and access policies<br>**Blobs** Upload, download, manage folders, rename and delete blobs, copy blobs, create and manage blob snapshots, change blob access tier, and create and manage shared access signatures and access policies |
| **Table** | **Tables** Create, rename, copy, delete, and create and manage shared access signatures and access policies<br>**Table entities** Import, export, view, add, edit, delete, and query |

| Storage service | Supported operations |
|---|---|
| Queue | **Queues**   Create, delete, and manage shared access signatures and access policies<br>**Messages**   Add, view, dequeue, and clear all messages |
| Files | **File Shares**   Create, rename; copy, delete, create and manage snapshots, connect VM to file share, and create and manage shared access signatures and access policies<br>**Files**   Upload folders or files, download folders or files, manage folders, copy, rename, and delete |

In each case, Azure Storage Explorer provides an intuitive GUI interface for each operation.

## Copy storage blobs

The Azure Storage Explorer can be used to copy storage blobs. To copy between storage accounts, navigate to the source storage account, select one or more files, and click Copy on the toolbar. Next, navigate to the destination storage account, expand the container that you want to copy to, and click Paste on the toolbar. In Figure 2-30, the SampleFile.txt blob was copied from the examref container to the temp container using this technique.

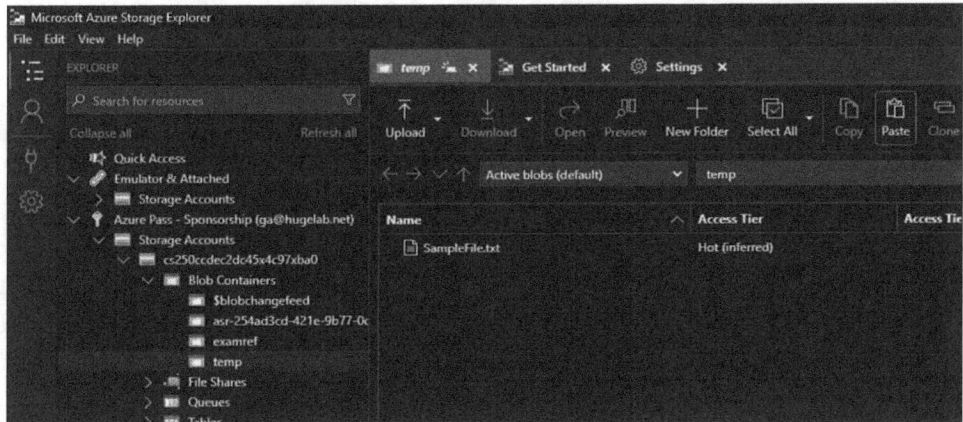

**FIGURE 2-30**   Using the async blob copy service with Storage Explorer

# Manage data by using AzCopy

AzCopy is a command-line utility that you can use to perform large-scale bulk transfer of data to and from Azure Storage. AzCopy performs all the operations asynchronously and can run simultaneously. Moreover, it is also fault-tolerant, so if the operation is interrupted for some reason, it can resume from where it left off once the issue is resolved.

With the latest version of AzCopy, you can take incremental backups of blobs and keep them synchronized in order to contain the same version of data. AzCopy can be added to the system path, so that you can run AzCopy from any folder from your system while using it in Windows PowerShell. Otherwise, you have to change the directory to where the AzCopy executable is stored every time. You can see a list of commands using azcopy -h.

AzCopy needs an authentication to Azure Storage before it runs any operations within the session. It can be achieved by running the azcopy login command and signing in. AzCopy also supports other authorizations, such as service principal, SAS token, access key, managed identity, and so on. For example, run this command to authenticate using service principal:

```
azcopy login --service-principal --application-id <application-id>
--tenant-id=<tenant-id>
```

## Upload and download data using AzCopy

You can upload data to Azure Blob Storage using AzCopy. The only condition is that the storage account and destination container should already exist. In the following example, the CreateUserTemplate.csv file will be copied to the destcontainer. This example assumes that the CSV file is in the location that you are running the *azcopy* command.

```
azcopy copy "CreateUserTemplate.csv" "https://examref.blob.core.windows.net/
destcontainer/CreateUserTemplate.csv"
```

If you are using a SAS token, the syntax would be

```
azcopy copy "CreateUserTemplate.csv" "https://examref.blob.core.windows.net/
destcontainer/CreateUserTemplate.csv?<sas token>"
```

You can upload multiple files with folder structures using the --recursive=true option with AzCopy.

```
azcopy copy "CreateUserTemplate.csv" "https://examref.blob.core.windows.net/
destcontainer/CreateUserTemplate.csv?<sas token>"
```

You can also download the data from Azure Blob Storage using AzCopy. In the following example, the CreateUserTemplate.csv file will be downloaded from the srccontainer:

```
azcopy copy "https://examref.blob.core.windows.net/srccontainer/CreateUserTemplate.csv "
"CreateUserTemplate.csv"
```

## Async blob copy

The AzCopy application can also be used to copy between storage accounts. The following example shows how to copy the blob from the source storage account's container to the destination storage account's container using a SAS token:

```
AzCopy copy "https://examref.blob.core.windows.net/srccontainer/[blob-path]?<sas token>"
"https://examrefdest.blob.core.windows.net/destcontainer/[blob-path]?<sas token>"
```

> **NEED MORE REVIEW?  AZCOPY**
>
> AzCopy version 10 is multi-platform, and works with Windows, Linux, and macOS. For more information on AzCopy, see *https://learn.microsoft.com/en-us/azure/storage/common/ storage-use-azcopy-v10*.

## Sync blob copy

You can use the azcopy sync command to synchronize copies between two blob containers. This command synchronizes the contents of a destination container with a source container by copying blobs if the last modified time of a blob in the destination is earlier than that of the corresponding blob in the source. By default, the recursive flag is true for the sync command and copies all subdirectories:

```
azcopy sync "https://examref.blob.core.windows.net/srccontainer/?<sas token>"
"https://examref.blob.core.windows.net/destcontainer/"
```

> **NOTE  DELETE DESTINATION FLAG**
>
> You can use --delete-destination flag with the azcopy sync command if you want to delete blobs in the destination that don't exist in the source. It can be set to true, false, or prompt. Using prompt will prompt you for deletions to make it safer.

# Skill 2.3: Configure Azure Files and Azure Blob Storage

Azure Files is a fully managed file share service that offers endpoints for the Server Message Block (SMB) protocol, Network File System (NFS) protocol, and the Azure Files REST API. This allows you to create one or more file shares in the cloud (with a default maximum size of 5 TiB per share). You can enable large file share for a storage account and create file shares up to 100 TiB. Also, if you are using the Premium SKU, you get 100 TiB by default. Azure Files can be used as you would a regular Windows file server, such as for shared storage or for new uses such as part of a lift-and-shift migration strategy.

## Create and configure a file share in Azure Storage

There are several common use cases for Azure Files. A few examples include the following:

- Migration of existing applications that require a file share for storage
- Shared storage of files, such as web content, log files, application configuration files, or even installation media
- Replacement of an existing file server

Figure 2-31 shows the hierarchy of files stored in Azure Files.

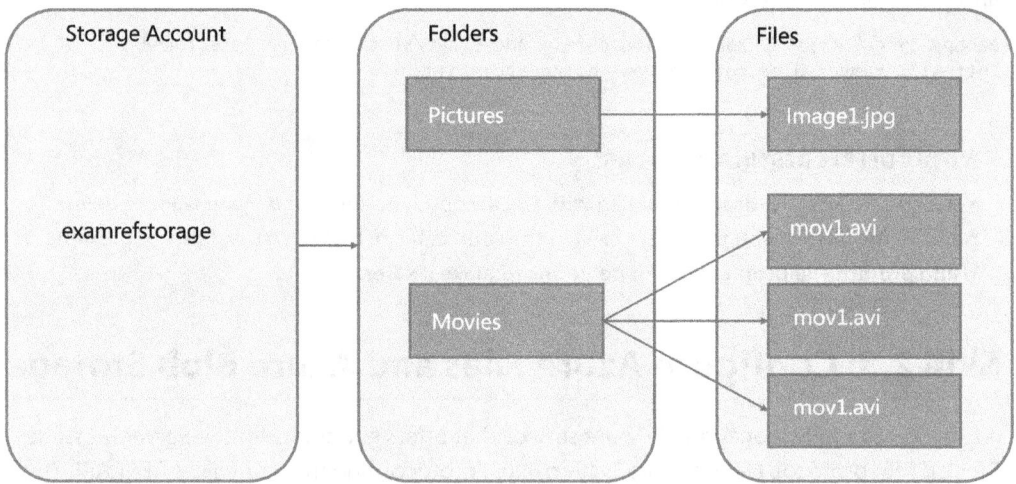

**FIGURE 2-31** Azure Files entities and relationship hierarchy

### Create a file share

To create a new file share using the Azure portal, open an Azure storage account, click File Shares, and then click File Share. In the New File Share blade shown in Figure 2-32, you must provide the file share name and the quota size, which can be a maximum size of 5 TiB. Large file shares and premium storage accounts support up to 100 TiB.

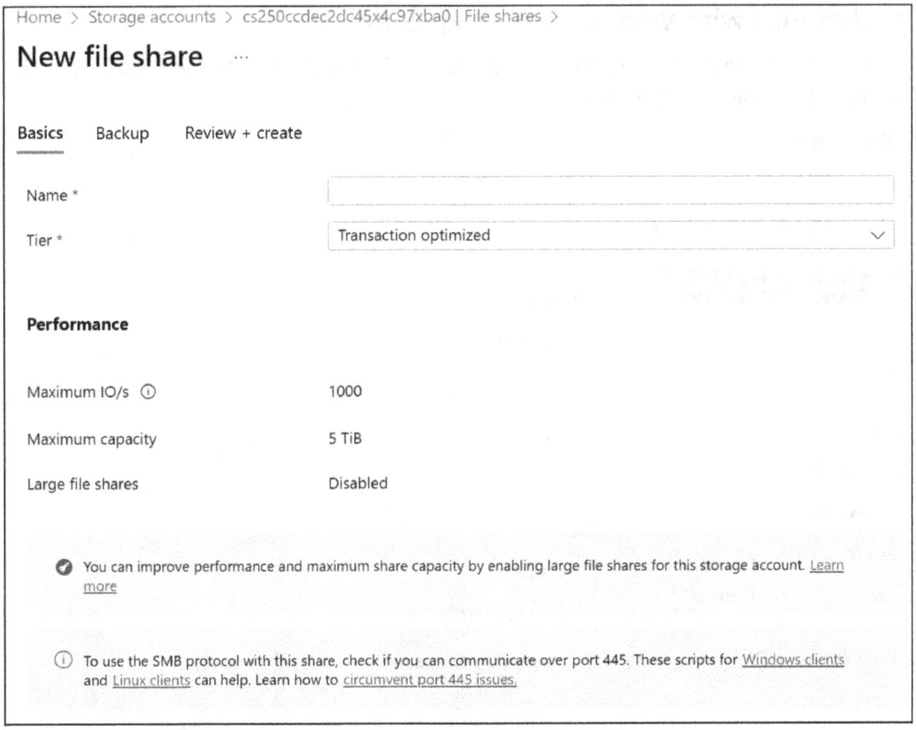

## New file share  ...

Basics    Backup    Review + create

Name *    [                                      ]

Tier *    [ Transaction optimized                  ⌄ ]

**Performance**

Maximum IO/s  ⓘ          1000

Maximum capacity          5 TiB

Large file shares         Disabled

> ✓ You can improve performance and maximum share capacity by enabling large file shares for this storage account. Learn more

> ⓘ To use the SMB protocol with this share, check if you can communicate over port 445. These scripts for Windows clients and Linux clients can help. Learn how to circumvent port 445 issues.

**FIGURE 2-32**   Adding a new share with Azure Files

## Connect to Azure Files outside of Azure

Because Azure Files provides support for SMB 3.0, it is possible to connect directly to a file share from a computer running outside of Azure. In this case, remember to open the outbound TCP port 445 in your local network. Many companies block 445 because of the insecure nature of SMB 1.0. Please check your network connections if you have problems connecting. Alternatively, you can leverage a virtual private network or ExpressRoute where port 445 can't be unblocked. Note that Windows 7 and Windows Server 2008 R2 do not support SMB 3.0.

> *NEED MORE REVIEW?*  **HOW TO REMOVE SMB V1**
>
> In order to disable SMB v1 from your environment, you can disable the smb1protocol feature. See the following link for more details: *https://learn.microsoft.com/en-us/windows-server/storage/file-server/troubleshoot/detect-enable-and-disable-smbv1-v2-v3?tabs=server#how-to-gracefully-remove-smb-v1-in-windows-81-windows-10-windows-2012-r2-windows-server-2016-and-windows-server-2019.*

## Connect and mount with Windows File Explorer

There are several ways to mount a file share from Windows. The first is to use the Map Network Drive feature within Windows File Explorer.

1. Open File Explorer.

2. Right-click This PC.

3. Select Map Network Drive, as shown in Figure 2-33.

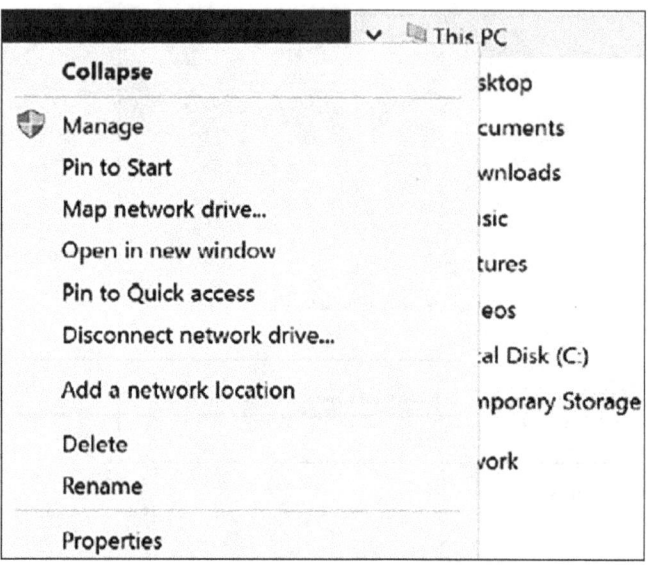

**FIGURE 2-33**   The Map Network Drive option from the This PC context menu

4. In the Map Network Drive dialog box, specify the following configuration options, as shown in Figure 2-34:

   - For Folder, enter **\\[name of storage account].files.core.windows.net\[name of share]**.

   - Select the Connect Using Different Credentials option.

5. When you click Finish, you will be prompted to enter your username and password to access the file share, as shown in Figure 2-35. The username should be in the format Azure\[name of storage account], and the password should be the access key for the Azure storage account.

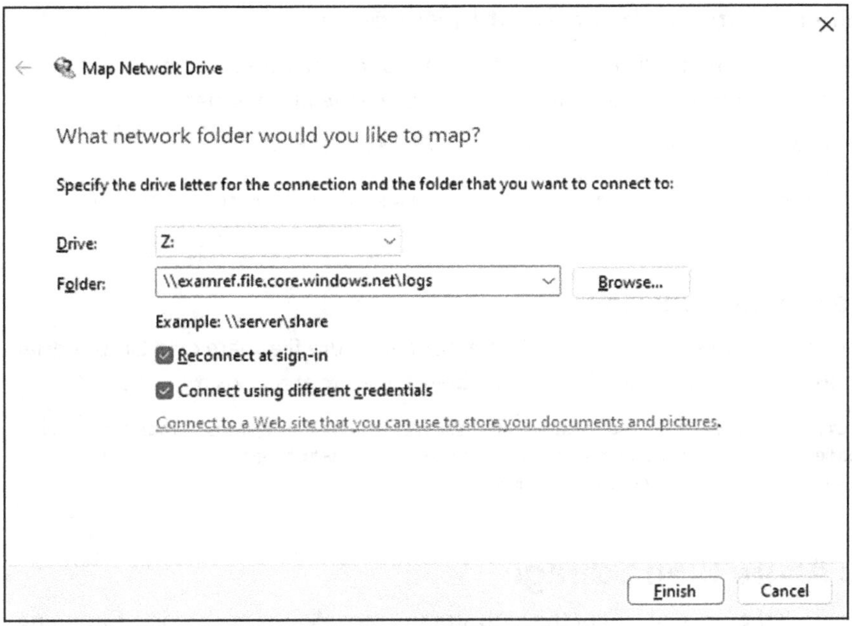

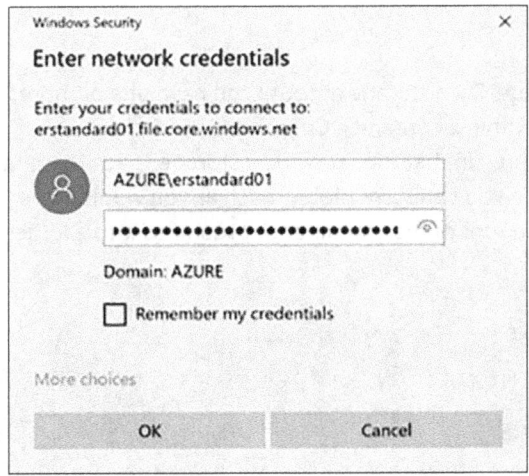

**FIGURE 2-34** Mapping a network drive to an Azure file share

**FIGURE 2-35** Specifying credentials for the Azure file share

## Connect and mount with the net use command

You can also mount the Azure file share using the Windows net use command as the following example demonstrates:

```
net use x \\examref.file.core.windows.net\logs  /u:AZURE\examuser01
r21Dk4qgY1HpcbriySWrBxnXnbedZLmnRK3N49PfaiL1t3ragpQaIB7FqK5zbez/sMnDEzEu/dgA9Nq/W7IF4A==
```

### Automatically reconnect after reboot in Windows

To make the file share automatically reconnect and map to the drive after Windows is rebooted, use the following command (ensuring you replace the placeholder values):

```
cmdkey /add:<storage-account-name>.file.core.windows.net /user:AZURE\<storage-
account-name> /pass:<storage-account-key>
net use Z: \\\\<storage-account-name>.file.core.windows.net\\<file-share-name> /
persistent:yes
```

### Connect and mount from Linux

Use the mount command (elevated with sudo) to mount an Azure file share on a Linux virtual machine. In this example, the logs file share would be mapped to the /logs mount point.

```
sudo mount -t cifs //<storage-account-name>.file.core.windows.net/logs /logs -o
vers=3.0,username=<storage-account-name>.,password=<storage-account
-key>,dir_mode=0777,file_mode=0777,sec=ntlmssp
```

## Configure Azure Blob Storage

This section describes the key features of the blob storage provided by each storage account. Azure Blob Storage is used for large-scale storage of arbitrary data objects, such as media files, log files, and so on.

### Blob containers

Figure 2-36 shows the layout of the blob storage. Each storage account can have one or more blob containers and all blobs must be stored within a container. Containers are similar in concept to a folder on the drive of your computer, in that they provide a storage space for data in your storage account. Within each container, you can store blobs, much as you would store files on a hard drive. Blobs can be placed at the root of the container or organized into a folder hierarchy.

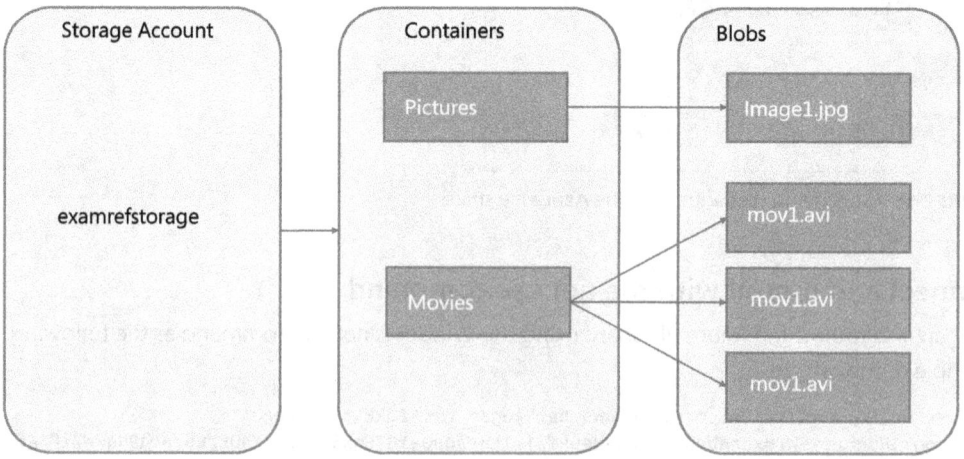

**FIGURE 2-36**  Azure storage account entities and hierarchy relationships

Each blob has a unique URL. The format of this URL is https://[account name].blob.core.windows.net/[container name]/[blob path and name].

Optionally, you can create a container at the root of the storage account, by specifying the special name $root for the container name. You can store blobs in the root of the storage account and reference them with URLs such as https://[account name].blob.core.windows.net/fileinroot.txt.

## Blob types

Blobs come in three types, and it is important to understand when each type of blob should be used and what the limitations are for each.

- **Page blobs**  Optimized for random-access read and write operations. Page blobs are used to store virtual disk (VHD) files, which use unmanaged disks with Azure virtual machines. The maximize page blob size is 8 TiB.

- **Block blobs**  Optimized for efficient uploads and downloads, for video, images, and other general-purpose file storage. The maximum block blob size depends on the service version API that is used, with a maximum capacity of approximately 190 TiB.

- **Append blobs**  Optimized for append operations. Updating or deleting existing blocks in the blob is not supported. Up to 50,000 blocks can be added to each append blob, and each block can be up to 4MiB in size, giving a maximum append blob size of slightly more than 195 GiB. Append blobs are most commonly used for log files.

Blobs of all three types can share a single blob container.

**EXAM TIP**

The type of the blob is set at creation and cannot be changed after the fact. A common problem that might show up on the exam is if a .vhd file was accidently uploaded as a block blob instead of a page blob. The blob must be deleted first and reuploaded as a page blob before it can be mounted as an operating system or data disk to an Azure VM.

*NEED MORE REVIEW?*  **BLOB TYPES**

You can learn more about the intricacies of each blob type here: *https://learn.microsoft.com/en-us/rest/api/storageservices/understanding-block-blobs--append-blobs--and-page-blobs.*

## Manage blobs and containers (Azure portal)

You can create and manage containers through the Azure portal, Azure Storage Explorer, third-party storage tools, or through the command-line tools. To create a container in the Azure management portal, open a storage account by selecting All Services, Storage Accounts, and then choosing your storage account. Within the storage account blade, open the Containers blade, and then click + Container. The New Container options are shown in Figure 2-37. See Skill 2.1 for more information on setting the public access level.

**FIGURE 2-37** Creating a container using the Azure management portal

After a container is created, you can also use the portal to upload blobs to the container, as demonstrated in Figure 2-38. Click Upload in the container and then browse to the blob to upload. If you click Advanced, you can select the blob type (Blob, Page, or Append), the block size, and optionally, a folder to which the blob is to be uploaded.

**FIGURE 2-38** Uploading a blob to a storage account container

---

**NEED MORE REVIEW?**  **MANAGING BLOB STORAGE WITH POWERSHELL**

The Azure PowerShell cmdlets offer a rich set of capabilities for managing blobs in storage. You can learn more about their capabilities here: *https://learn.microsoft.com/en-us/azure/ storage/blobs/storage-quickstart-blobs-powershell*.

**NEED MORE REVIEW?** **MANAGING BLOB STORAGE WITH THE AZURE CLI**

The Azure CLI also offers a rich set of capabilities for managing blobs in storage. You can learn more about their capabilities here: *https://learn.microsoft.com/en-us/azure/storage/blobs/storage-quickstart-blobs-cli*.

## Manage blobs and containers (Storage Explorer)

Azure Storage Explorer provides rich functionality for managing storage data, including blobs and containers. To create a container, expand the Storage Accounts node, expand the storage account you want to use, and right-click the Blob Containers node. Choose Create Blob Container from the drop-down menu, as shown in Figure 2-39.

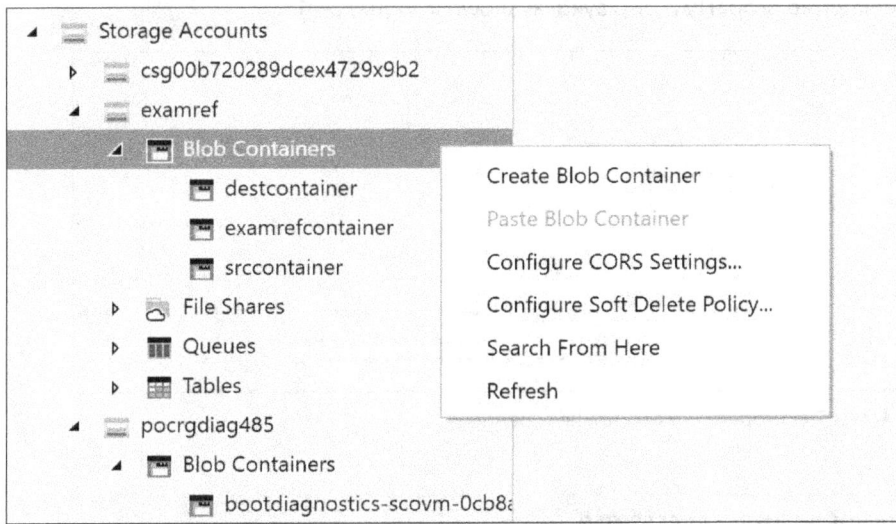

**FIGURE 2-39** Creating a container using the Azure Storage Explorer

Azure Storage Explorer provides the ability to upload a single file or multiple files at once. Use the Upload Folder feature to upload the entire contents of a local folder, re-creating the hierarchy in the Azure storage account. Figure 2-40 shows the two upload options.

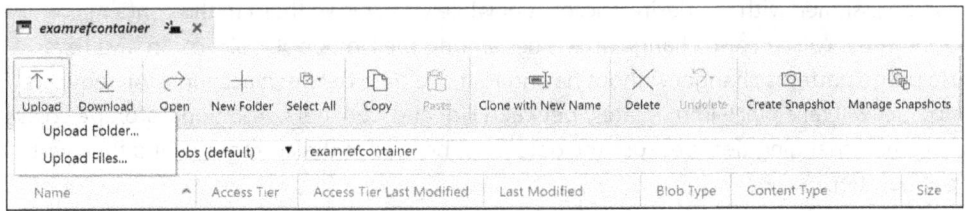

**FIGURE 2-40** Uploading files and folders using Azure Storage Explorer

# Configure storage tiers

As discussed in skill 2.1, Azure Blob Storage supports four access tiers: Hot, Cool, Cold, and Archive. Each represents a trade-off of availability and cost. There is no trade-off on the durability (probability of data loss), which is defined by the SKU and replication option of the storage account that is selected.

## Account-level tiering

The storage account blobs can coexist between three tiers within the same account. If any blob does not have an assigned tier, it infers the access tier from the account access tier setting by default. In such a scenario, you will see that the access tier inferred blob property is set to Inferred, and the access tier blob property matches the account level tier. In the Azure portal, the access tier inferred property is displayed, as shown in Figure 2-41.

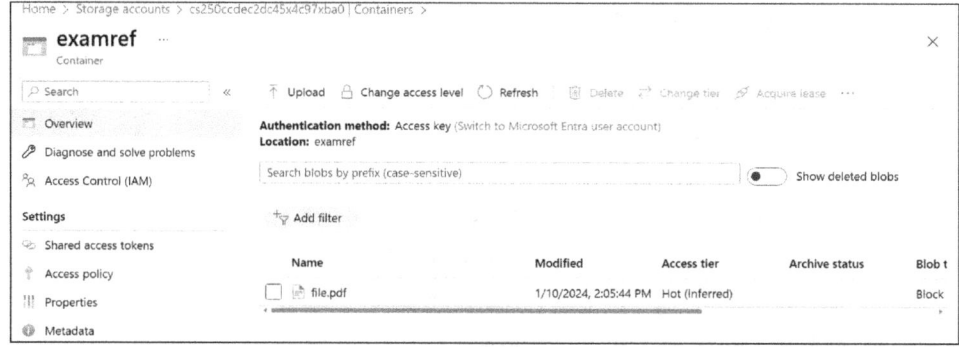

**FIGURE 2-41**  Access Tier property for account level tiering

**NOTE   CHANGE ACCOUNT ACCESS TIER**

Changes to the account access tier apply to all access tier inferred objects stored in the account that don't have an explicit tier set.

## Blob-level tiering

Blobs can be assigned with the desired access tier while you upload them to the container (see Figure 2-42). You can also change access tier among the Hot, Cool, Cold, or Archive tiers (because usage patterns change) without having to move data between accounts. All requests to change tier will take place immediately between Hot and Cool tiers. Additional storage costs will occur when changing tiers because the data must be read at the current tier and then written at the new tier.

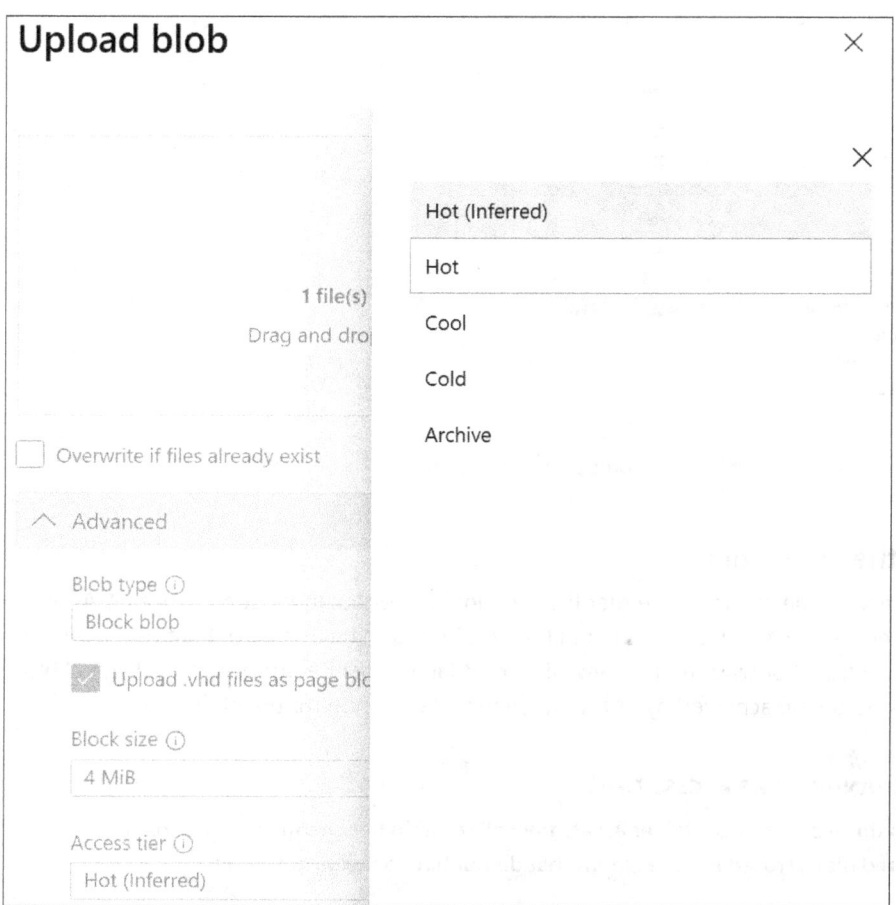

**FIGURE 2-42** Changing the access tier while uploading the blobs to a container

When the access tier is changed, the access tier's Last Modified property will be updated with the time when the most recent change was made to the tier (see Figure 2-43).

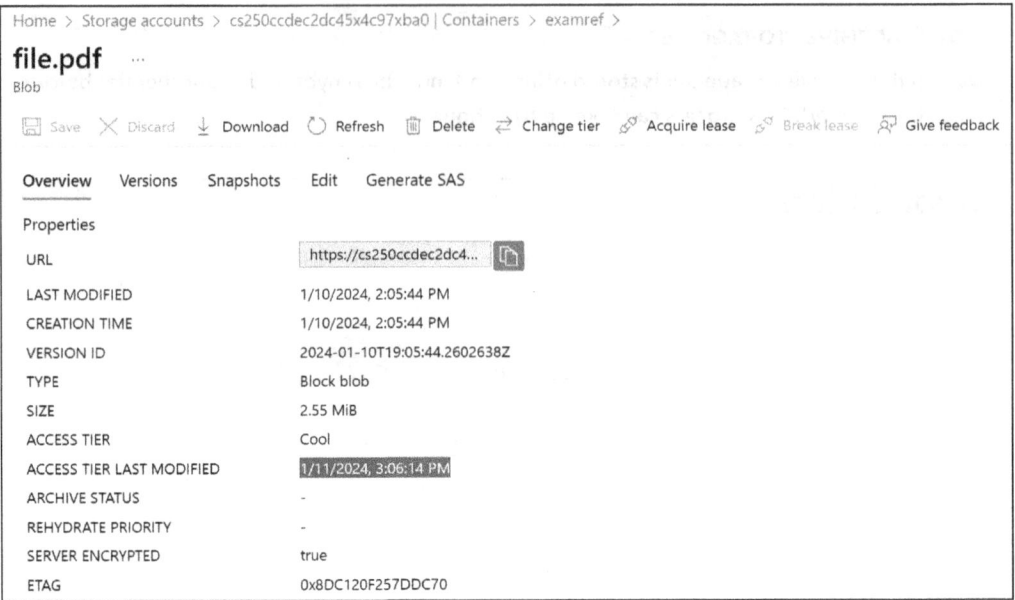

Home > Storage accounts > cs250ccdec2dc45x4c97xba0 | Containers > examref >

**file.pdf** ...
Blob

🖫 Save   ✕ Discard   ↓ Download   ○ Refresh   🗑 Delete   ⇄ Change tier   ⌕ Acquire lease   ⌕ Break lease   🗨 Give feedback

Overview   Versions   Snapshots   Edit   Generate SAS

Properties

| | |
|---|---|
| URL | https://cs250ccdec2dc4... 📋 |
| LAST MODIFIED | 1/10/2024, 2:05:44 PM |
| CREATION TIME | 1/10/2024, 2:05:44 PM |
| VERSION ID | 2024-01-10T19:05:44.2602638Z |
| TYPE | Block blob |
| SIZE | 2.55 MiB |
| ACCESS TIER | Cool |
| ACCESS TIER LAST MODIFIED | 1/11/2024, 3:06:14 PM |
| ARCHIVE STATUS | - |
| REHYDRATE PRIORITY | - |
| SERVER ENCRYPTED | true |
| ETAG | 0x8DC120F257DDC70 |

**FIGURE 2-43**   Access Tier Last Modified property for the blob

## Change the access tier

The access tier can be changed at either the account level or the individual blob level. At the account level, you can set the access tier in the Configuration blade (the default option unless assigned it explicitly) or by using the new Lifecycle Management feature. At an individual blob level, the same can be achieved by using the Change Tier option for the blob.

> **NOTE   CHANGING THE ACCESS TIER**
>
> Changing the account or container access tier will result in tier change charges for access tier-inferred blobs stored in the account that do not have an explicit tier set.

To make a change at the account level, browse to the storage account, open the Configuration blade, and change Access Tier (Default) to Cool or Hot (see Figure 2-44).

Similarly, to make a change at blob level, browse to a blob, click Change Tier, and select the access tier from the menu. Your options are Hot, Cool, Cold, or Archive (refer to Figure 2-42).

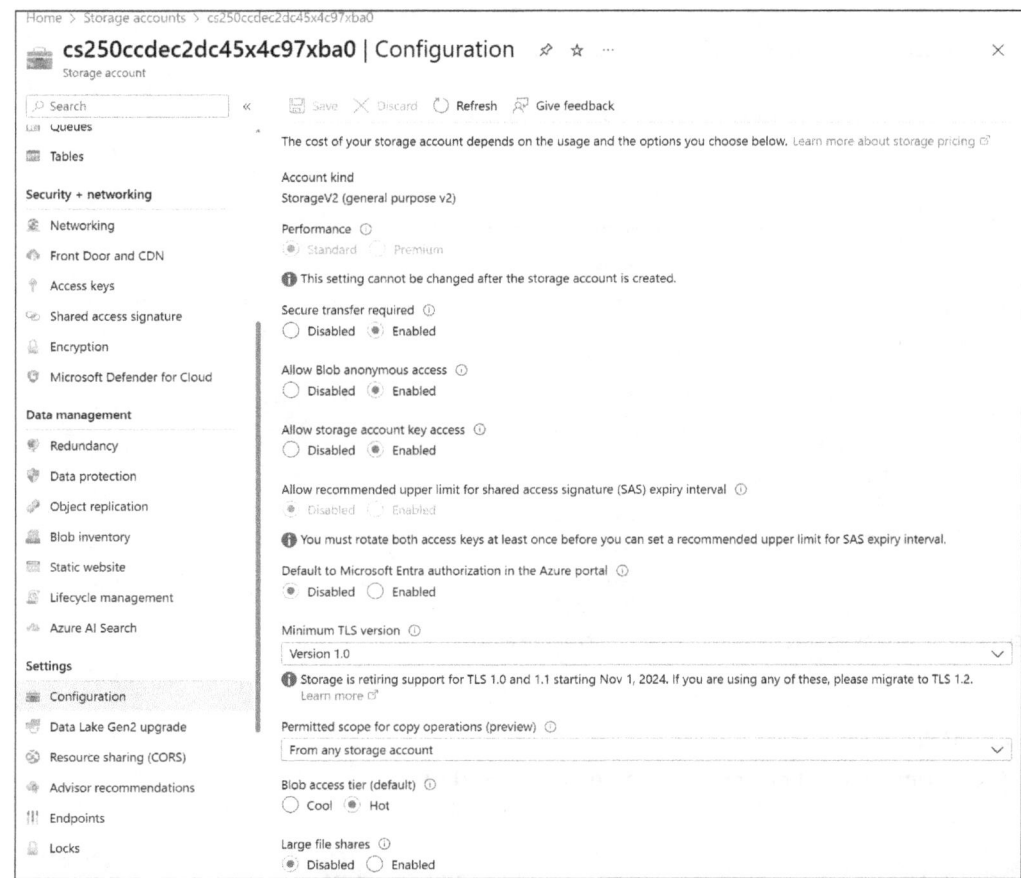

**cs250ccdec2dc45x4c97xba0** | Configuration

Storage account

Search «    💾 Save  ✕ Discard  🔄 Refresh  📝 Give feedback

Queues

Tables

The cost of your storage account depends on the usage and the options you choose below. Learn more about storage pricing 🔗

**Security + networking**

🔷 Networking

Account kind
StorageV2 (general purpose v2)

🔷 Front Door and CDN

Performance ⓘ

🔑 Access keys

● Standard  ○ Premium

🔷 Shared access signature

ⓘ This setting cannot be changed after the storage account is created.

🔒 Encryption

Secure transfer required ⓘ
○ Disabled  ● Enabled

🛡 Microsoft Defender for Cloud

Allow Blob anonymous access ⓘ
○ Disabled  ● Enabled

**Data management**

🔷 Redundancy

Allow storage account key access ⓘ
○ Disabled  ● Enabled

🔷 Data protection

Allow recommended upper limit for shared access signature (SAS) expiry interval ⓘ
● Disabled  ○ Enabled

🔷 Object replication

🔷 Blob inventory

ⓘ You must rotate both access keys at least once before you can set a recommended upper limit for SAS expiry interval.

📄 Static website

Default to Microsoft Entra authorization in the Azure portal ⓘ
● Disabled  ○ Enabled

🔷 Lifecycle management

🔷 Azure AI Search

Minimum TLS version ⓘ

**Settings**

| Version 1.0 | ⌄ |

🔷 Configuration

ⓘ Storage is retiring support for TLS 1.0 and 1.1 starting Nov 1, 2024. If you are using any of these, please migrate to TLS 1.2. Learn more 🔗

🔷 Data Lake Gen2 upgrade

🔷 Resource sharing (CORS)

Permitted scope for copy operations (preview) ⓘ

🔷 Advisor recommendations

| From any storage account | ⌄ |

Endpoints

Blob access tier (default) ⓘ
○ Cool  ● Hot

🔒 Locks

Large file shares ⓘ
● Disabled  ○ Enabled

**FIGURE 2-44**   Change the access tier

# Configure soft delete, versioning, and snapshots

Azure Storage provides several options for recovering data that might be accidentally deleted or overwritten. These options include snapshots, soft delete, and version tracking for blobs and files.

## Soft delete

The default behavior when you delete a blob is that the blob is deleted and lost forever. With soft delete, you can recover your data when blobs or blob snapshots are deleted even in the event of an overwrite. This feature must be enabled on the Azure storage account, and a retention period must be set for how long the deleted data is available (see Figure 2-45).

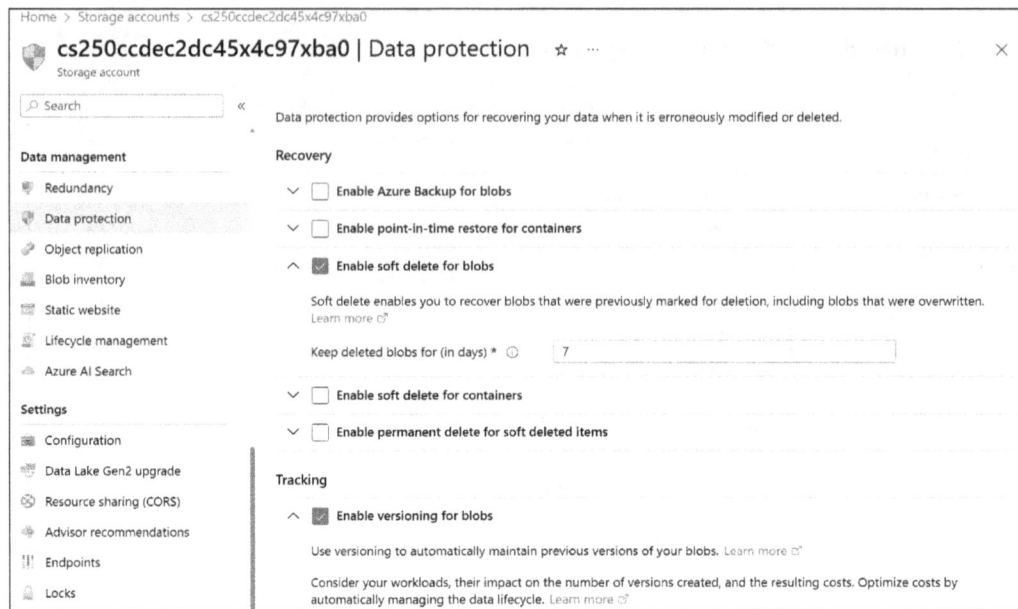

**cs250ccdec2dc45x4c97xba0 | Data protection** ☆ ⋯

Storage account

🔍 Search ≪

Data protection provides options for recovering your data when it is erroneously modified or deleted.

**Data management**

  Redundancy

  Data protection

  Object replication

  Blob inventory

  Static website

  Lifecycle management

  Azure AI Search

**Settings**

  Configuration

  Data Lake Gen2 upgrade

  Resource sharing (CORS)

  Advisor recommendations

  Endpoints

  Locks

**Recovery**

∨ ☐ Enable Azure Backup for blobs

∨ ☐ Enable point-in-time restore for containers

∧ ☑ Enable soft delete for blobs

Soft delete enables you to recover blobs that were previously marked for deletion, including blobs that were overwritten. Learn more

Keep deleted blobs for (in days) * ⓘ     [ 7 ]

∨ ☐ Enable soft delete for containers

∨ ☐ Enable permanent delete for soft deleted items

**Tracking**

∧ ☑ Enable versioning for blobs

Use versioning to automatically maintain previous versions of your blobs. Learn more

Consider your workloads, their impact on the number of versions created, and the resulting costs. Optimize costs by automatically managing the data lifecycle. Learn more

**FIGURE 2-45** Enabling soft delete on an Azure storage account

**EXAM TIP**

The maximum retention period for soft delete is 365 days.

**NEED MORE REVIEW?** **SOFT DELETE FOR AZURE STORAGE BLOBS**

You can learn more about using soft delete with Azure Blob Storage here: *https://learn.microsoft.com/en-us/azure/storage/blobs/soft-delete-blob-overview*.

Soft delete can also be configured for SMB-based Azure Files shares. It operates the same as blobs where data can be recovered if accidentally deleted. Soft delete is configured at the storage account level for all shares in the account. To configure soft delete for Azure Files, navigate to the storage account and select File Shares. On the File Shares blade, select Disabled next to Soft Delete. In the Soft Delete dialog box, set the toggle to Enabled and then configure the retention policy. The default retention policy is 7 days, but can be configured from 1 to 365 days, as shown in Figure 2-46.

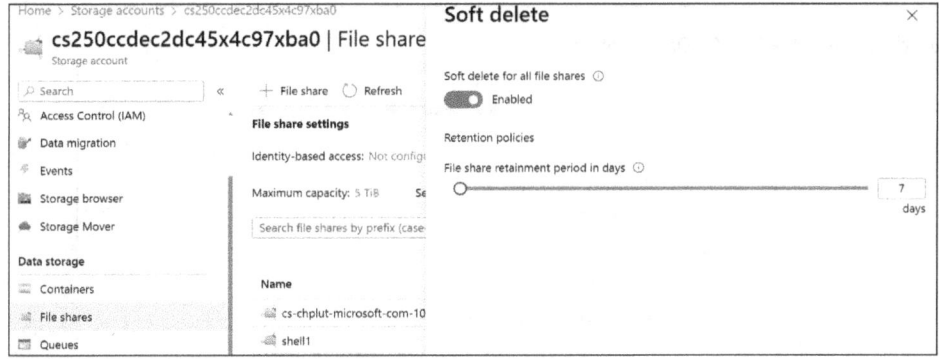

**FIGURE 2-46**  Soft delete for Azure files

## Versioning

An Azure storage account provides two primary options for tracking blob objects: versioning and change feed. Versioning can be used to automatically retain previous versions of a blob object. You can choose whether to keep all versions of a blob or, to help with managing storage costs, delete versions after a specific number of days. The retention period for versions can be defined between 1 and 365 days, with the default set to 7 days.

Blob change feeds keep a log of all create, modify, and delete operations to the blob objects in the storage account. You can configure the account to keep all logs, or to delete logs after a certain number of days. This can be configured from 1 day to 146,000 days, with a default of 7 days.

To configure both versioning and blob change feed, navigate to the Data Protection blade of a storage account. Figure 2-47 displays this blade with the default retention periods configured.

## Snapshots

Snapshots are point-in-time restore options for containers and file shares. To use snapshots with blob containers, the blob versioning and blob change feed options must be enabled. Using a snapshot with a blob container will revert all objects in the container to the version, or point in time, that you select. Snapshots for blob storage are configured on the Data Protection blade of the storage account using the Enable Point-In-Time Restore option, also shown in Figure 2-47.

Snapshots can also be configured for Azure Files at the file share level. Unlike blobs, which are kept with the storage account, using snapshots for file shares requires an Azure Recovery Services vault and backup policy to retain the previous file version data. As with blobs, using snapshots with file shares also requires that soft delete be enabled for the file shares in the account.

To configure snapshots on a file share, navigate to File Shares from the storage account, and then select a configured file share. From the file share, select Snapshots. You will be prompted to either configure or select an Azure Recovery Services vault and associated backup policy and frequency for the files, as shown in Figure 2-48.

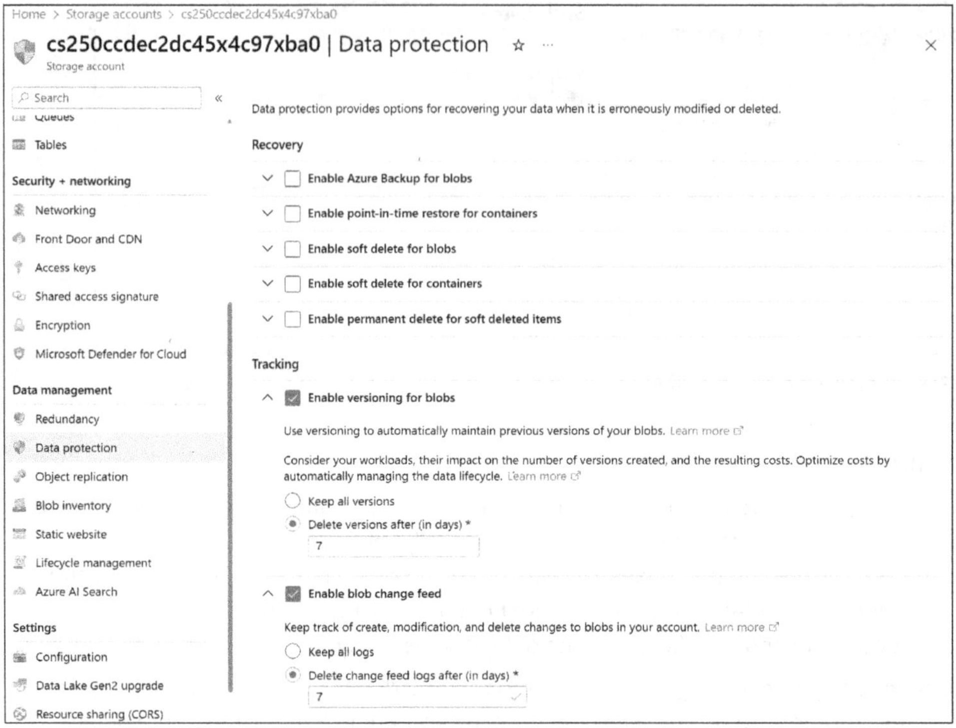

**FIGURE 2-47** Data protection versioning options

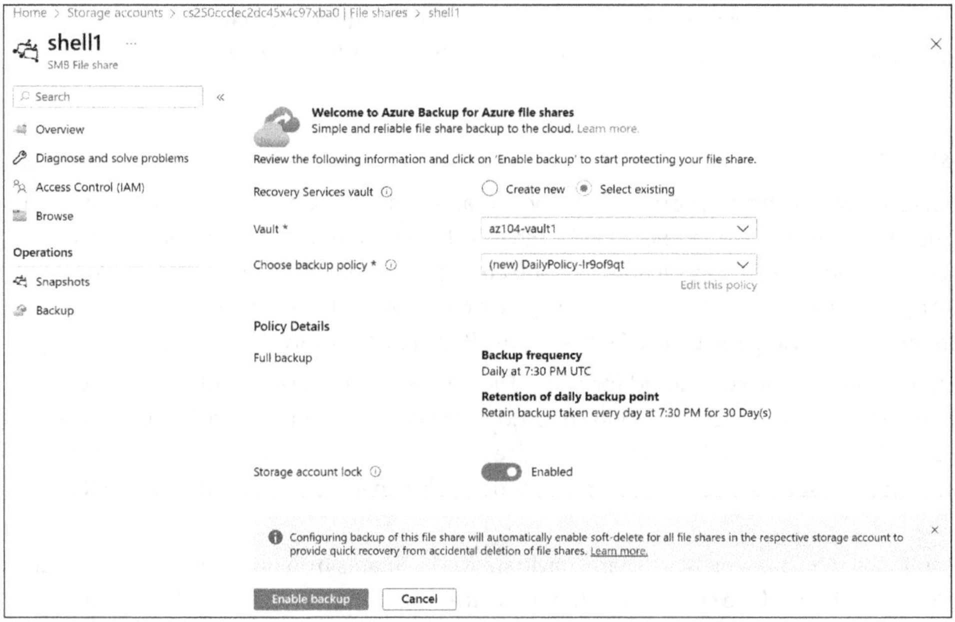

**FIGURE 2-48** Snapshots – Azure file shares

# Configure blob lifecycle management

Azure Storage has a lifecycle management capability, which can be used to transition data to lower-access tiers automatically based on pre configured rules. You can also delete the data at the end of its lifecycle. These rules can be executed against the storage account once per day. Specific blobs and containers can be targeted using filter sets.

To configure the lifecycle management rules, open the storage account, browse to Lifecycle Management under Data Management, and click Add A Rule (see Figure 2-49). You can define up to 100 rules.

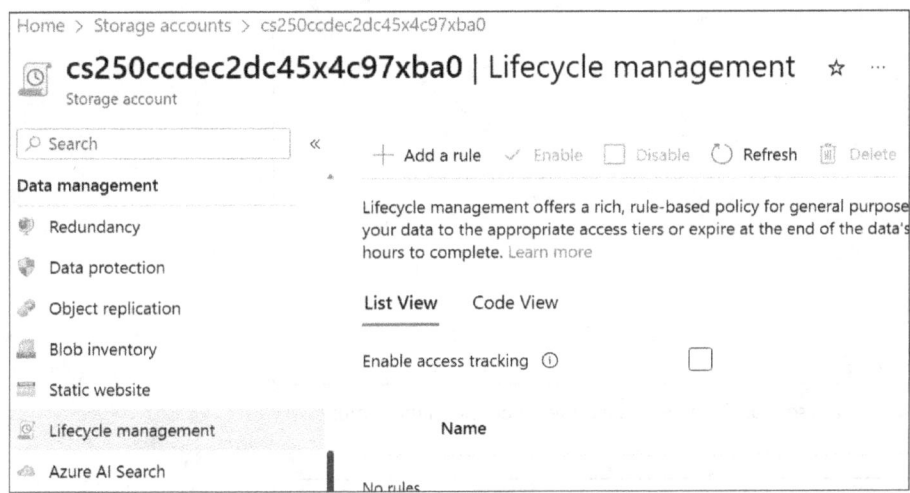

**FIGURE 2-49**   Add A Rule option on the Lifecycle Management blade

You can limit the rule scope with a filter set by selecting Limit Blobs With Filters, as shown in Figure 2-50. You can also select the Blob Type and Blob Subtype that should be applicable to this rule. You can also select Append Blobs, Snapshots, and Versions. Then click Next.

You can configure rules on the Base Blobs page to define the blob lifecycle policy. You can create multiple if-then blocks to define the several conditions. For example, you can move blob data to cool storage if it is not modified for the specified number of days. Similarly, you can also create rules to move blobs to cold or archive storage, or delete them if not modified for the defined number of days. In Figure 2-51, the condition has been created for 30 days, and all four actions are shown in the drop-down menu. Click Next to configure the filter set.

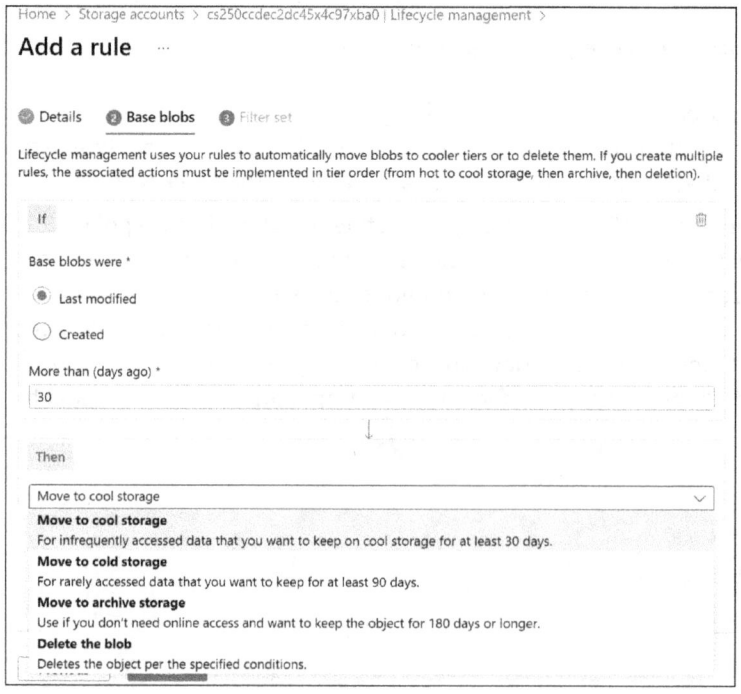

**Add a rule**  ...

**①  Details**     ②  Base blobs     ③  Filter set

A rule is made up of one or more conditions and actions that apply to the entire storage account. Optionally, specify that rules will apply to particular blobs by limiting with filters.

Rule name *

ExamRefRule

Rule scope *

○ Apply rule to all blobs in your storage account

◉ Limit blobs with filters

Blob type *

☑ Block blobs

☐ Append blobs

Blob subtype *

☑ Base blobs

☐ Snapshots

☐ Versions

Previous     **Next**

**FIGURE 2-50**    Details section for a new rule under Lifecycle Management

**Add a rule**  ...

Details     **②  Base blobs**     ③  Filter set

Lifecycle management uses your rules to automatically move blobs to cooler tiers or to delete them. If you create multiple rules, the associated actions must be implemented in tier order (from hot to cool storage, then archive, then deletion).

If                                                                        🗑

Base blobs were *

◉ Last modified

○ Created

More than (days ago) *

30

↓

Then

Move to cool storage                                                     ⌄

**Move to cool storage**
For infrequently accessed data that you want to keep on cool storage for at least 30 days.
**Move to cold storage**
For rarely accessed data that you want to keep for at least 90 days.
**Move to archive storage**
Use if you don't need online access and want to keep the object for 180 days or longer.
**Delete the blob**
Deletes the object per the specified conditions.

**FIGURE 2-51**    Base Blobs tab for a rule on the Lifecycle Management blade

On the Filter Set tab, you can specify the prefix to find items within the container. You need to specify the container name/prefix. For example, you could choose data/cost where data is the name of the container, and cost is the prefix, as shown in Figure 2-52. You can also use the Blob Index Match if you have indexed the items with keys and values in your containers. You can specify up to 10 prefixes per rule.

Now click Add to create the rule.

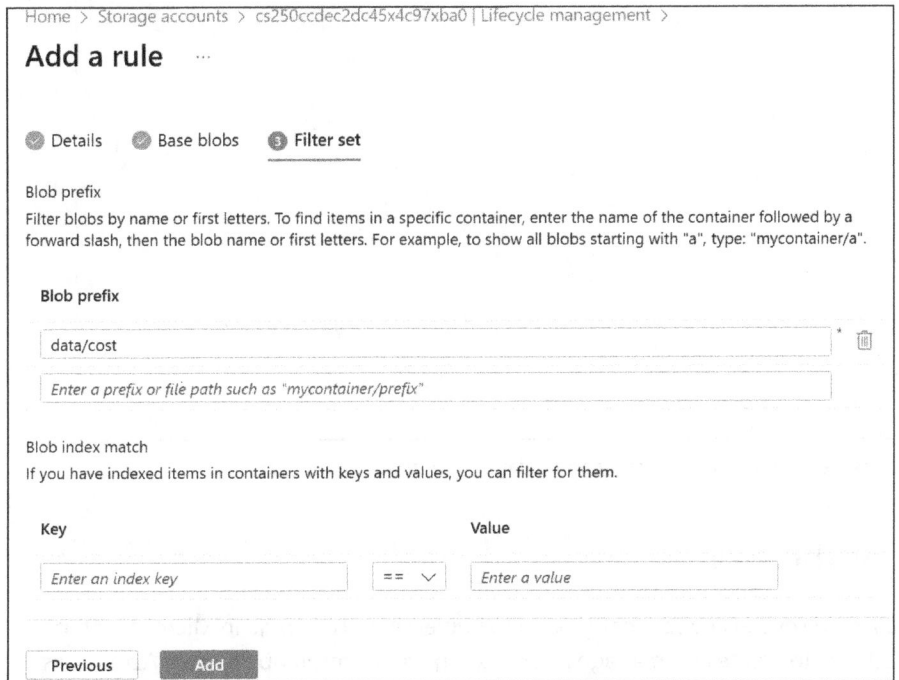

FIGURE 2-52   Filter set for a rule on the Lifecycle Management blade

Once created, the consolidated view of code can be viewed on the Code View tab, as shown in Figure 2-53.

> **NOTE   LIFECYCLE MANAGEMENT EFFECT**
> The policy can take up to 24 hours to go into effect, and then the action can take an additional 24 hours to run. Overall, it takes up to 48 hours for policy actions to complete once you set up lifecycle management.

You can delete a rule anytime if it isn't required anymore using the Lifecycle Management blade.

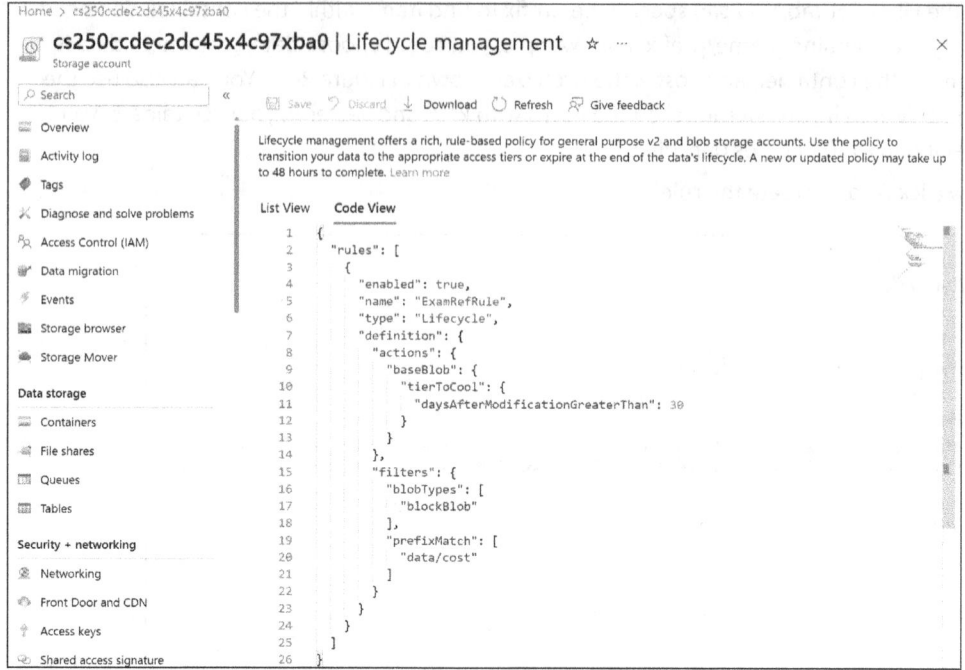

**FIGURE 2-53** Code view for the lifecycle management configuration

# Chapter summary

This chapter covered several key services related to implementing storage in Microsoft Azure. Topics included how to create and manage Azure storage accounts, Blob Storage, Azure Files, importing and exporting data, Storage Explorer, AzCopy, lifecycle management, and object replication.

Here are some of the key takeaways from this chapter:

- Azure storage accounts provide four separate services: Blob Storage, Table Storage, Queue Storage, and Azure Files. It is important to understand the usage scenarios of each service.

- Standard storage accounts use magnetic drives and provide the lowest cost per GB. This type of account is best suited for applications that require bulk storage or where data is accessed infrequently.

- Premium storage accounts use solid state drives and offer consistent, low-latency performance. This type of account can only be used with Azure virtual machine disks and are best for I/O-intensive applications, like databases.

- Storage accounts must specify a replication mode. The options are locally redundant, zone-redundant, geo-redundant, read-access geo-redundant storage, geo zone-redundant, and read-access geo zone-redundant.

- Blob Storage supports three types of blobs (block, page, and append blobs), and four access tiers (Hot, Cool, Cold, and Archive).

- There are three kinds of storage accounts: General-Purpose V1, General-Purpose V2, and Blob Storage. The availability of features varies between storage account types. However, General-purpose v1 is considered legacy and is not recommended.

- Azure Storage can be managed through several tools directly from Microsoft: the Azure portal, PowerShell, CLI, Storage Explorer, and AzCopy. It is important to know when to use each tool.

- Access to storage accounts can be controlled using several techniques. Among them are Entra ID authentication; storage account name and key; SAS; SAS with access policy; and using the storage firewall and virtual network service endpoints. Access to Blob Storage can also be controlled using the public access level of the storage container.

- You can also use AzCopy to copy files between storage accounts or from outside publicly accessible locations to your Azure storage account.

- Azure Storage has a lifecycle-management capability, and it can be used to transition data to lower-access tiers automatically based on preconfigured rules. You can also delete the data at the end of its lifecycle. These rules can be executed against the storage account once per day. Specific blobs and containers can be targeted using filter sets.

- Azure Storage also provides blob object replication capabilities that provide asynchronous replication of block blobs from one storage account to another. The blobs are replicated based on the defined replication rules.

- You can leverage object replication only when blob versioning is enabled for both the source and destination storage accounts, and the blob change feed is enabled for the source storage account.

# Thought experiment

In this thought experiment, apply what you have learned about this objective. You can find answers to these questions in the next section.

You are asked to design an Azure Storage solution for a large insurance company. The company wants the data to be accessible based on the role of individual users within the organization. Various departments have their separate datasets that they access on a daily basis. The company wants to restrict users from modifying the data from other departments, but all the users must be able to access the data across departments.

Also, there is a requirement to store that data forever with minimal cost possible. The data is rarely used after two years from the date it was last modified.

1. What steps should you take to assign the storage access based on their departments?
2. What changes need to be made in order to keep storing data forever with minimal cost?

# Thought experiment answers

This section contains the solution to the thought experiment for the chapter.

To solve this problem, you can leverage various capabilities of Azure Storage, such as Entra authentication with role-based access control and lifecycle management for the blob storage.

1. Create an Azure storage account and create a container for each department to store its data. Next, assign the Storage Blob Data Reader role for all the department groups but assign the Storage Blob Data Contributor role for each department group. This allows users to access all department data, but they can modify only their department data.

2. Create a rule on the Lifecycle Management blade for the storage account and select the Apply Rule To All Blobs In Your Storage Account option. Then, add an if-then block to move data to the archive tier after 730 days (two years). This will let you store the data forever with minimal cost in the archive tier.

# CHAPTER 3

# Deploy and manage Azure compute resources

Microsoft Azure offers many features and services that can be used to create inventive solutions for almost any IT problem. Some of the most common services for designing these solutions are Microsoft Azure Virtual Machines (VM) and Virtual Machine Scale Sets (VMSS). Virtual machines are among the key compute options for deploying workloads in Microsoft Azure.

The flexibility of virtual machines makes them a common option for many workloads. For example, you have a choice of server operating systems with various supported versions of Windows and Linux distributions. Azure virtual machines also provide you full control over the operating system along with advanced configuration options for networking and storage. In addition to VM capabilities, scale sets provide the unique ability to scale out certain types of workloads to handle large processing problems, and they optimize cost by running instances only when needed.

In addition to this, you have other compute services for container environments, such as Azure Kubernetes Service (AKS), Azure Container Apps (ACA), and Azure Container Instances (ACI). With the wide adoption of containerized workloads across many IT companies, Microsoft is heavily investing in enhancing their current product set to support container-based workloads. Azure also has other PaaS options, such as Azure App Service and its App Service plans to manage and host web applications.

In this chapter, you will learn the ins and outs of deploying and managing these compute resources in Azure. The chapter covers creation through the Azure portal and the command-line tools, automation with templates, and core management tasks.

## Skills covered in this chapter:

- Skill 3.1: Automate deployment of resources
- Skill 3.2: Create and configure virtual machines
- Skill 3.3: Provision and manage containers
- Skill 3.4: Create and configure Azure App Service

# Skill 3.1: Automate deployment of resources

The ability to provision virtual machines on demand using the Azure portal is incredibly powerful. The true power of the cloud, however, is the ability to automatically deploy one or more resources defined in code, such as a script or a template. Use cases such as defining an application configuration and automatically deploying it on demand help teams to be more agile by providing developer, test, or production environments in a fast and repeatable fashion. Because the configuration is stored as code, changes to infrastructure can also be tracked in a version control system. In this skill, you will learn some of the core capabilities for automating workload deployments in Azure.

**This skill covers how to:**

- Interpret an Azure Resource Manager (ARM) template
- Modify an existing ARM template
- Deploy resources from a template
- Export a deployment template
- Interpret and modify a Bicep file

## Interpret an Azure Resource Manager template

Azure Resource Manager (ARM) templates are authored using JavaScript Object Notation (JSON) and provide the ability to define the configuration of resources, such as virtual machines, storage accounts, and so on, in a *declarative* manner. Templates go beyond just providing the ability to create the resources; you can also customize and create dependencies between some resources, such as virtual machines. This makes it possible to create templates that have capabilities for orchestrated deployments of completely functional solutions.

The Azure team maintains a list of ARM templates with examples for most resources. This list is located at *https://azure.microsoft.com/resources/templates/* and is backed by a source code repository in GitHub. If you want to go directly to the source to file a bug, you can access it at *https://github.com/Azure/azure-quickstart-templates*.

The basic structure of a resource manager template has most of the following elements:

```
{
  "$schema": "https://schema.management.azure.com/schemas/2019-04-01/deploymentTemplate.json#",
  "contentVersion": "1.0.0.0",
  "parameters": { },
  "variables": { },
  "functions": [ ],
  "resources": [ ],
  "outputs": { }
}
```

- **$schema**  The JSON schema file is the reference to the standard structure defined for an ARM template, which can help you determine when something is wrong with your template in comparison to the schema file syntax. The JSON schema is used by features, such as code completion or IntelliSense, so you can make changes in the templates easily.

  For resource group targeted deployments, use

  `https://schema.management.azure.com/schemas/2019-04-01/deploymentTemplate.json#`

- **contentVersion**  This provides source control to track the changes made in your template. You can provide any value for this element. When deploying resources using the template, this value can be used to make sure that the right template is being used.

- **parameters**  Using parameters, you can define the values that are passed at runtime without changing the exact template file. The parameters can be changed by the azuredeploy.parameters.json file or in the PowerShell script that is used to deploy your template. The parameters are key elements when dealing with nested templates to pass the values from parent template to the child templates.

- **variables**  This defines values that are used in your template to simplify template language. Mostly, variables are hard-coded values, but they also can be created dynamically using parameters or standard template functions.

- **functions**  Users can create functions that can be used within the template. The complex expressions that are being used multiple times in the template can be defined as a function once. You need to create your own namespace and create member functions as needed. You cannot access variables or any other user-defined functions within your function.

- **resources**  This contains resources that are deployed or updated in a resource group. You can define the *condition* to control the provisioning of each resource. Also, the dependsOn value determines which resources must be deployed first before a specific resource.

- **outputs**  Here, you can define the type of values that are returned after deployment. This section is used to keep track of resources that are being deployed or updated.

## Define a virtual network

This skill is focused on learning how to deploy Windows and Linux virtual machines. A prerequisite of deploying a virtual machine is having a virtual network. Listing 3-1 shows how to define the structure of the virtual network using several variables that describe the address space and subnet allocation.

**LISTING 3-1**  Variables for virtual network creation

```
"ExamRefRGPrefix": "10.0.0.0/16",
"ExamRefRGSubnet1Name": "FrontEndSubnet",
"ExamRefRGSubnet1Prefix": "10.0.0.0/24",
"ExamRefRGSubnet2Name": "BackEndSubnet",
"ExamRefRGSubnet2Prefix": "10.0.1.0/24",
"ExamRefRGSubnet1Ref": "[concat(variables('vnetId'), '/subnets/',
 variables('ExamRefRGSubnet1Name'))]",
"VNetId": "[resourceId('Microsoft.Network/virtualNetworks', variables('VirtualNetworkN
ame'))]",
"VirtualNetworkName": "ExamRefVNET",
```

After the variables are defined, you can add the virtual network resource to the resource's element in your template. Listing 3-2 shows a portion of an ARM template for a virtual network named ExamRefVNET, with an address space of 10.0.0.0/16 and two subnets: FrontEndSubnet 10.0.0.0/24 and BackEndSubnet 10.0.1.0/24. Note that this is only an example and would not validate as a complete template, and that the syntax to read the value of variables—[variables('variablename')]—is used heavily when authoring templates. The virtual network's location is set based on the return value of the built-in resourceGroup() function, which returns information about the resource group the resource is being created or updated in.

**LISTING 3-2**  Template structure for creating a virtual network

```
{
    "name": "[variables('VirtualNetworkName')]",
    "type": "Microsoft.Network/virtualNetworks",
    "location": "[resourceGroup().location]",
    "apiVersion": "2023-06-01",
    "dependsOn": [],
    "properties": {
      "addressSpace": {
        "addressPrefixes": [
          "[variables('ExamRefRGPrefix')]"
        ]
      },
      "subnets": [
        {
          "name": "[variables('ExamRefRGSubnet1Name')]",
          "properties": {
            "addressPrefix": "[variables('ExamRefRGSubnet1Prefix')]"
          }
        },
        {
          "name": "[variables('ExamRefRGSubnet2Name')]",
          "properties": {
            "addressPrefix": "[variables('ExamRefRGSubnet2Prefix')]"
```

```
            }
          }
        ]
      }
    }
```

## Define a network interface

Every virtual machine has one or more network interfaces. To create one with a template, add a variable to the variables section to store the network interface resource name as the following snippet demonstrates:

```
"VMNicName": "VMNic"
```

Listing 3-3 defines a network interface named `WindowsVMNic`. This resource has a dependency on the `ExamRefVNET` virtual network. This dependency will ensure that the virtual network is created prior to the network interface creation when the template is deployed and is a critical feature of orchestration of resources in the correct order. The network interface is associated to the subnet by referencing the `ExamRefRGSubnet1Ref` variable.

**LISTING 3-3** Creating a network interface

```
{
    "name": "[variables('VMNicName')]",
    "type": "Microsoft.Network/networkInterfaces",
    "location": "[resourceGroup().location]",
    "apiVersion": "2023-06-01",
    "dependsOn": [
      "[resourceId('Microsoft.Network/virtualNetworks', 'ExamRefVNET')]"
    ],
    "properties": {
      "ipConfigurations": [
        {
          "name": "ipconfig1",
          "properties": {
            "privateIPAllocationMethod": "Dynamic",
            "subnet": {
              "id": "[variables('ExamRefRGSubnet1Ref')]"
            }
          }
        }
      ]
    }
}
```

---

**EXAM TIP**

To specify a static private IP address in template syntax, specify an address from the assigned subnet using the `privateIpAddress` property and set the `privateIpAllocation` method to `Static`.

```
"privateIpAddress": "10.0.0.10",
"privateIpAllocationMethod": "Static,
```

---

## Add a Public IP address

To add a public IP address to the virtual machine, you must make several modifications. The first is to define a parameter that the user will use to specify a unique DNS name for the public IP. The following code goes in the parameters block of a template:

```
"VMPublicIPDnsName": {
  "type": "string",
  "minLength": 1
}
```

The second modification is to add the public IP resource itself. Before adding the resource, add a new variable in the variables section to store the name of the public IP resource.

```
"VMPublicIPName": "VMPublicIP"
```

Listing 3-4 shows a public IP address resource with the public IP allocation method set to `Dynamic` (it can also be set to `Static`). The `domainNameLabel` property of the IP address `dnsSettings` element is populated by the parameter. This makes it easy to specify a unique value for the address at deployment time.

**LISTING 3-4**   Creating a network interface

```
{
      "name": "[variables('VMPublicIPName')]",
      "type": "Microsoft.Network/publicIPAddresses",
      "location": "[resourceGroup().location]",
      "apiVersion": "2023-06-01",
      "dependsOn": [ ],
      "properties": {
        "publicIPAllocationMethod": "Dynamic",
        "dnsSettings": {
          "domainNameLabel": "[parameters('VMPublicIPDnsName')]"
        }
      }
    }
```

The next modification is to update the network interface resource that the public IP address is associated with. The network interface must now have a dependency on the public IP address to ensure it is created before the network interface. The following example shows the addition to the `dependsOn` array:

```
"dependsOn": [
  "[resourceId('Microsoft.Network/virtualNetworks', 'ExamRefVNET')]",
  "[resourceId('Microsoft.Network/publicIPAddresses',
variables('VMPublicIPName'))]"
  ],
```

The `ipConfigurations -> properties` element must also be modified to reference the `publicIPAddress` resource. See Listing 3-5.

**LISTING 3-5**   IP configurations

```
"ipConfigurations": [
      {
        "name": "ipconfig1",
        "properties": {
```

```
        "privateIPAllocationMethod": "Dynamic",
        "subnet": {
          "id": "[variables('ExamRefRGSubnet1Name')]"
        },
        "publicIPAddress": {
          "id": "[resourceId('Microsoft.Network/publicIPAddresses',
variables('VMPublicIPName'))]"
        },
      }
    }
  ]
```

## Define a virtual machine resource

Before creating the virtual machine resource, you will add several parameters and variables to define. Each virtual machine requires administrative credentials. To enable a user to specify the credentials at deployment time, add two additional parameters for the administrator account and the password.

```
"VMAdminUserName": {
  "type": "string",
  "minLength": 1
},
"VMAdminPassword": {
  "type": "string",
  "minLength": 1
}
```

Several variables are needed to define the configuration of the virtual machine resource. The following variables define the VM name, operating system image, and the VM size. These should be inserted into the variables section of the template.

```
"VMName": "MyVM",
"VMImagePublisher": "MicrosoftWindowsServer",
"VMImageOffer": "WindowsServer",
"VMOSVersion": "WS2019-Datacenter",
"VMOSDiskName": "VM2OSDisk",
"VMSize": "Standard_D2_v2",
"VM2ImagePublisher": "MicrosoftWindowsServer",
"VM2ImageOffer": "WindowsServer",
"VM2OSDiskName": "VM2OSDisk",
"VMSize": "Standard_D2_v2"
```

The VM has a dependency on the network interface. It doesn't have to have a dependency on the virtual network because the network interface itself does. This VM is using managed disks, so there are no references to storage accounts for the VHD file. Listing 3-6 shows a sample virtual machine resource.

**LISTING 3-6**  Virtual machine resource

```
{
    "name": "[parameters('VMName')]",
    "type": "Microsoft.Compute/virtualMachines",
    "location": "[resourceGroup().location]",
```

```
  "apiVersion": "2023-06-01",
  "dependsOn": [
    "[resourceId('Microsoft.Network/networkInterfaces', variables('VMNicName'))]"
  ],
  "properties": {
    "hardwareProfile": {
      "vmSize": "[variables('vmSize')]"
    },
    "osProfile": {
      "computerName": "[variables('VMName')]",
      "adminUsername": "[parameters('VMAdminUsername')]",
      "adminPassword": "[parameters('VMAdminPassword')]"
    },
    "storageProfile": {
      "imageReference": {
        "publisher": "[variables('VMImagePublisher')]",
        "offer": "[variables('VMImageOffer')]",
        "sku": "[variables('VMOSVersion')]",
        "version": "latest"
      },
      "osDisk": {
        "createOption": "FromImage"
      }
    },
    "networkProfile": {
      "networkInterfaces": [
        {
          "id": "[resourceId('Microsoft.Network/networkInterfaces',
                  variables('VMNicName'))]"
        }
      ]
    }
  }
}
```

There are several properties of a virtual machine resource that are critical to its configuration:

- **hardwareProfile**   This element is where you set the size of the virtual machine. Set the vmSize property to the desired size, such as Standard_D2_v2.

- **osProfile**   This element at a basic level is where you set the computerName and adminUsername properties. The adminPassword property is required if you do not specify an SSH key. This element also supports other properties, including windowsConfiguration, linuxConfiguration, and secrets.

- **osProfile, windowsConfiguration**   While the example doesn't use this configuration, this element provides the ability to set advanced properties on Windows VMs:

  - **provisionVMAgent**   This is enabled by default, but you can disable it. Specify whether extensions can be added.

  - **enableAutomaticUpdates**   Specify whether Windows updates are enabled.

  - **timeZone**   Specify the time zone for the virtual machine.

- **additionalUnattendContent**  Pass unattended install configuration for additional configuration options.
- **winRM**  Configure Windows PowerShell remoting.
- **provisionVMAgent**  Enabled by default, but you can disable. Specify whether extensions can be added.
- **disablePasswordAuthentication**  If set to true, you must specify an SSH key.
- **Ssh, publicKeys**  Specify the public key to use for authentication with the VM.
- **osProfile, secrets**  This element secrets is used for deploying certificates that are in Azure Key Vault.
- **storageProfile**  This element is where the operating system image is specified, and the operating system and data disk configuration are set.
- **networkProfile**  This element is where the network interfaces for the virtual machine are specified.

---

*NEED MORE REVIEW?*  **RESOURCE MANAGER TEMPLATE SCHEMA**

Reading through the Azure Resource Manager template schema is a great way to learn the capabilities of templates. The latest virtual machine schema is published at *https://learn. microsoft.com/en-in/azure/templates/microsoft.compute/2022-03-01/virtualmachines.*

---

# Modify an existing ARM template

Often you will need to modify a template that you have previously used to change the configuration. As previously mentioned, one of the key benefits of using templates to describe your infrastructure (commonly referred to as *Infrastructure as Code*) is so you can modify it and deploy it in a versioned manner. To accommodate this behavior, ARM supports two different deployment modes: Complete and Incremental.

---

*NEED MORE REVIEW?*  **INFRASTRUCTURE AS CODE**

Infrastructure as Code (known as IaC) is a descriptive model for managing the infrastructure. More information can be found at *https://learn.microsoft.com/en-us/devops/deliver/ what-is-infrastructure-as-code.*

---

## ARM template Complete mode

In Complete mode, Azure Resource Manager deletes resources that exist in the resource group that are not in the template. This is helpful if you need to remove a resource from Azure and you want to make sure your template matches the deployment. You can remove the resource from the template, deploy using Complete mode, and it will be removed. You should consider using the *what-if* parameter when using Complete mode as a test before actually submitting it for deployment.

## ARM template Incremental mode

In Incremental mode, Azure Resource Manager leaves unchanged any resources that exist in the resource group but aren't in the template. It will update the resources in the resource group if the settings in the template differ from what is deployed. Incremental mode can have unintended impacts on resource properties. If your template doesn't cover all the properties of a resource, then at the time of deployment, unspecified properties will be reset to default values that can potentially affect the environment.

Incremental is the default mode for the Azure portal and when you are deploying through the command-line tools or Visual Studio. To use Complete mode, you must use the REST API or the command-line tools with the `-Mode`/`--mode` parameter set to `Complete`.

The following example deploys a template in Complete mode using PowerShell:

```
New-AzResourceGroupDeployment `
  -Mode Complete `
  -Name simpleVMDeployment `
  -ResourceGroupName ExamRefRG `
  -TemplateFile C:\ARMTemplates\deploy.json
```

The next example deploys a template in Complete mode using the Azure CLI:

```
az group deployment create \
  --name simpleVMDeployment \
  --mode Complete \
  --resource-group ExamRefRG \
  --template-file deploy.json
```

## Configure a VHD template

It is assumed that you already know the structure of the ARM template. For detailed structure and syntax, please refer to *https://learn.microsoft.com/en-us/azure/azure-resource-manager/templates/syntax*.

In the `storageProfile` section of a virtual machine resource, you can specify the `imageReference` element that references an image from the Azure Marketplace:

```
"imageReference": {
    "publisher": "[variables('VMImagePublisher')]",
    "offer": "[variables('VMImageOffer')]",
    "sku": "[parameters('VMOSVersion')]",
    "version": "latest"
}
```

You also can specify a generalized VHD that you have previously created. To specify a user image, you must specify the osType property (Windows or Linux), the URL to the VHD itself, and the URL to where the disk will be created in Azure Storage (osDiskVhdName). The following alternative code snippet demonstrates this. (This sample does not build on the previous example.)

```
"storageProfile": {
    "osDisk": {
        "name": "[concat(variables('vmName'),'-osDisk')]",
        "osType": "[parameters('osType')]",
        "caching": "ReadWrite",
        "image": {
            "uri": "[parameters('vhdUrl')]"
        },
        "vhd": {
            "uri": "[variables('osDiskVhdName')]"
        },
        "createOption": "FromImage"
    }
}
```

For context, the following vhdUrl parameter and osDiskVhdName variable is shown:

```
"vhdUrl": {
    "type": "string",
      "metadata": {
          "description": "VHD Url..."
      }
  }
"osDiskVhdName": "[concat('http://',parameters('userStorageAccountName'),
'.blob.core.windows.net/',parameters('userStorageContainerName'),'/',
parameters('vmName'),'osDisk.vhd')]"
```

See the following for a complete template example: *https://learn.microsoft.com/en-us/partner-center/marketplace/azure-vm-image-test*.

## Deploy resources from a template

You can deploy templates using the Azure portal, the command-line tools, or directly using the REST API. You'll start with deploying a template that creates a virtual machine using the Azure portal. To deploy a template from the Azure portal, search for **Deploy a custom template**. On the Custom Deployment blade, choose to either build your own template in the portal, or select a common template from the gallery, as shown in Figure 3-1.

From there, you have the option to build your own template using the editor in the Azure portal (you can paste your own template in or upload from a file using this option, too) or choose from one of the most common templates. Last of all, you can search the existing samples in the Quickstart samples repository in GitHub and choose one of them as a starting point.

Click Build Your Own Template In The Editor to paste in template code directly. You can author and then deploy templates using the Azure portal for simple testing. Figure 3-2 shows the Edit Template blade.

**FIGURE 3-1** Template deployment options

**FIGURE 3-2** Editing a template using the Azure portal editor

Clicking Save on the Select A Template tab takes you to the Basics tab shown in Figure 3-3, where you can specify the resource group and any parameters required to deploy the template.

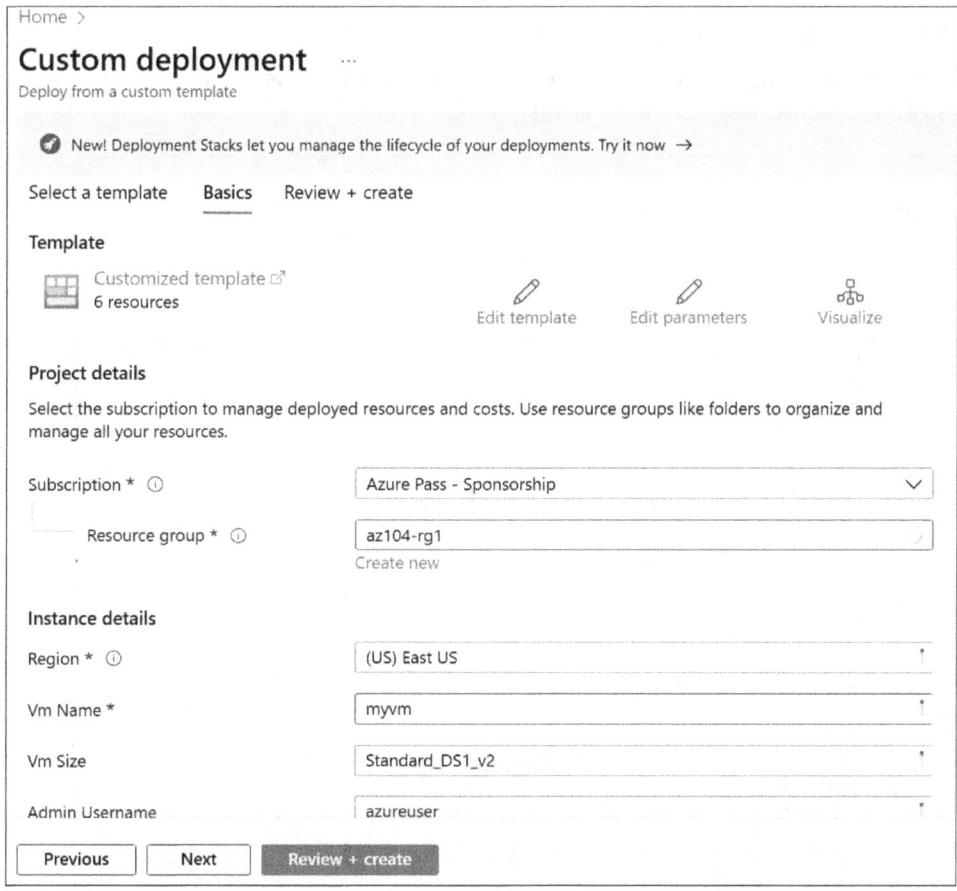

**FIGURE 3-3**  The Custom Deployment blade

Click Edit Parameters to edit a JSON view of the parameters for the template, as shown in Figure 3-4. This file can also be downloaded and is used to provide different behaviors for the template at deployment time without modifying the entire template.

Common examples of using a parameters file:

- Defining different instance sizes or SKUs for resources based on the intended usage (small instances for test environments for example)
- Defining different number of instances
- Defining different regions
- Defiing different credentials

It is recommended that you use the securestring type for the parameters when passing confidential data, such as passwords and secrets.

```
Home >

Edit parameters  ...

↑ Load file   ↓ Download

 1  {
 2    "$schema": "https://schema.management.azure.com/schemas/2019-04-01/deploymentParameters.json#",
 3    "contentVersion": "1.0.0.0",
 4    "parameters": {
 5      "vmName": {
 6        "value": "myvm"
 7      },
 8      "vmSize": {
 9        "value": "Standard_DS1_v2"
10      },
11      "adminUsername": {
12        "value": "azureuser"
13      },
14      "adminPassword": {
15        "value": null
16      },
17      "imageSku": {
18        "value": "2019-Datacenter"
19      },
20      "dscConfigurationUrl": {
21        "value": null
22      },
23      "dscConfigurationScript": {
24        "value": null
25      },
26      "dscConfigurationFunction": {

Save    Discard
```

FIGURE 3-4 Editing template parameters using the Azure portal

**NEED MORE REVIEW?  ARM TEMPLATE BEST PRACTICES**

Recommended practices for working with ARM templates can be found at
*https://learn.microsoft.com/en-us/azure/azure-resource-manager/templates/best-practices.*

The last step to creating a template using the Azure portal is to click Review + Create. This will validate the parameters that you have customized and provide a summary of what will be deployed. Click Create to trigger the deployment.

The Azure command-line tools can also deploy resources using templates. The template files can be located locally on your file system, accessed via HTTP/HTTPS, or uploaded to your cloud shell environment. Common deployment models include storing the templates in a source code repository or an Azure storage account to make it easy for others to deploy the template.

**EXAM TIP**

The parameters of a template can be passed to the `New-AzResourceGroupDeployment` cmdlet using the `TemplateParameterObject` parameter for values that are defined directly in the script as .json. The `TemplateParameterFile` parameter can be used for values stored in a local .json file. The `TemplateParameterUri` parameter for values that are stored in a .json file at an HTTP endpoint.

**EXAM TIP**

The parameters of a template can be passed to the `az group deployment create` command using the parameters section for values that are defined directly in the script as .json. The template-file parameter can be used for values stored in a local .json file. The `template-uri` parameter can be used for values that are stored in a .json file at an HTTP endpoint.

## Export a deployment template

An existing deployment can be exported as a template that you can use to regenerate the environment or to just gain a better understanding of how the deployment is configured. There are two ways of exporting a template from a deployment.

The first way is to export the actual template used by the Azure portal for the deployment. When you use the Azure portal to deploy a resource, it still creates and submits the ARM template to the service for deployment. On the Review tab of most deployments, there is a Download A Template For Automation option, as shown in Figure 3-5 for a storage account.

This method exports the template exactly as you customized it in the Azure portal, including the values for parameters and variables during the original execution. This approach does not capture any changes made to the deployment after it was deployed.

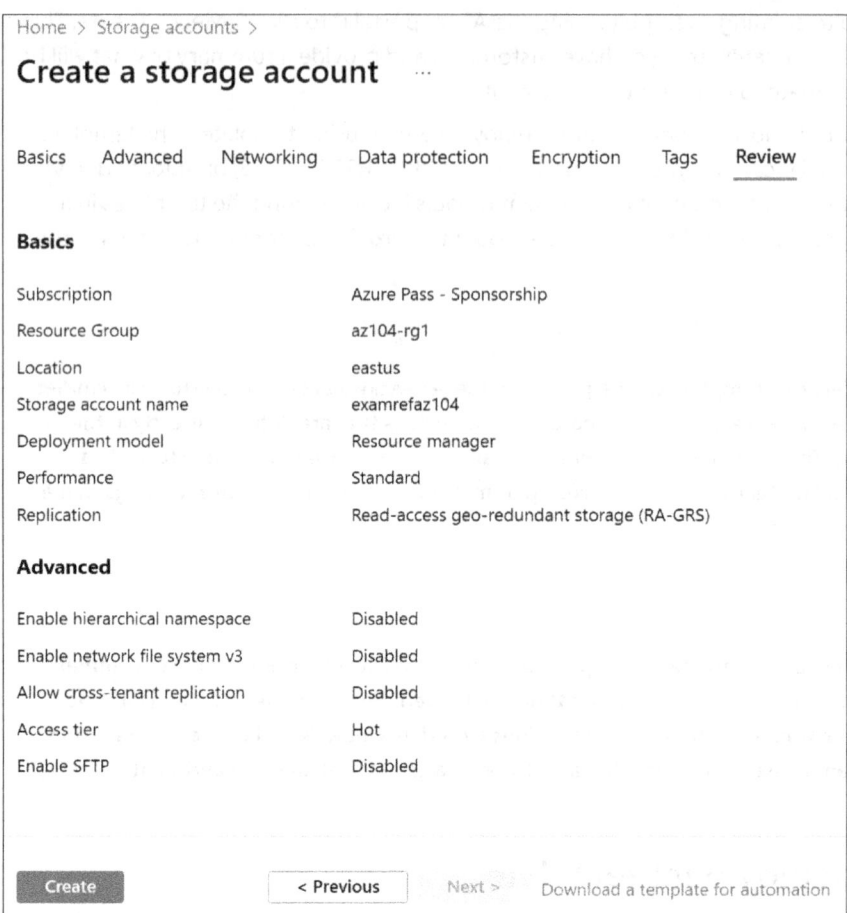

**FIGURE 3-5**  The Review tab of a deployment with the option to download the template

Clicking Download A Template For Automation opens the Template deployment view, as shown in Figure 3-6. From here, click Download to download the template locally, click Deploy to deploy the template, or click Add To Library to save to your template gallery for later deployment.

The second approach to generating an ARM template is to use the Export Template menu option for a resource or the resource group. It generates a template that represents the current state of the context you selected. The state might have been updated by multiple templates, or it might have been updated by changes from the Azure portal or changes via the REST API or command line. It might include many hard-coded values and not as many parameters as you would expect in a template that was designed for reusability. This template is useful for redeploying to the same resource group because of the hard-coded values. Using it for other new deployments typically requires a significant amount of editing to customize parameters and values. You can access this template by navigating to the resource or resource group and clicking Export Template on the left, as shown in Figure 3-7.

**FIGURE 3-6** The Template view for a deployment

**FIGURE 3-7** The Export Template blade of a resource group

# Interpret and modify a Bicep file

Bicep is a domain-specific language that uses an easy-to-read, declarative syntax to deploy Azure resources. Bicep is not a replacement for ARM templates but builds on top of ARM with its features and benefits. Bicep has integration as an extension with Visual Studio Code, which provides IntelliSense to help you author new Bicep files.

## Compare ARM templates and Bicep

ARM templates are built in JSON which make them verbose and tough to read. The following 29-line code block with quotes, brackets, and commas presents an ARM template that deploys a storage account.

```
{
  "$schema": "https://schema.management.azure.com/schemas/2019-04-01/deploymentTemplate.
json#",
  "contentVersion": "1.0.0.0",
  "parameters": {
    "location": {
      "type": "string",
      "defaultValue": "[resourceGroup().location]"
    },
    "storageAccountName": {
      "type": "string",
      "defaultValue": "[format('toylaunch{0}', uniqueString(resourceGroup().id))]"
    }
  },
  "resources": [
    {
      "type": "Microsoft.Storage/storageAccounts",
      "apiVersion": "2021-06-01",
      "name": "[parameters('storageAccountName')]",
      "location": "[parameters('location')]",
      "sku": {
        "name": "Standard_LRS"
      },
      "kind": "StorageV2",
      "properties": {
        "accessTier": "Hot"
      }
    }
  ]
}
```

Compare that to the next 14-line block of code that represents the same storage account, but as a Bicep file. The format is easier to read and write when creating files for Infrastructure as Code (IaC) or other automation.

```
param location string = resourceGroup().location
param storageAccountName string = 'toylaunch${uniqueString(resourceGroup().id)}'
resource storageAccount 'Microsoft.Storage/storageAccounts@2021-06-01' = {
  name: storageAccountName
  location: location
  sku: {
    name: 'Standard_LRS'
```

```
  }
  kind: 'StorageV2'
  properties: {
    accessTier: 'Hot'
  }
}
```

Bicep files still have parameters and variables, but also allow you to include loops, conditional values, deployment scopes, and more in the deployment process.

## Install the Bicep tools

The Bicep tools are available as extensions in both Visual Studio Code and Visual Studio. With this extension, you can use the features of Visual Studio Code, such as IntelliSense, to help author your Bicep files. Figure 3-8 shows the Bicep extension for Visual Studio Code.

**FIGURE 3-8**   The Bicep extension for Visual Studio Code

## Author a Bicep file

After you install the Bicep extension for Visual Studio or Visual Studio Code, any time that you edit or create a new file with the .bicep extension, the Bicep extension will be available to assist in authoring your file.

For example, to use Bicep IntelliSense to begin creating a Bicep file for a storage account, simply type **storage**. The IntelliSense menu will provide an option of "res-storage." Select the "res-storage" item and press Tab to automatically complete the required fields for the storage account. You could then change these values to parameters or variables for other use in the file. Figure 3-9 displays the IntelliSense option in Visual Studio Code.

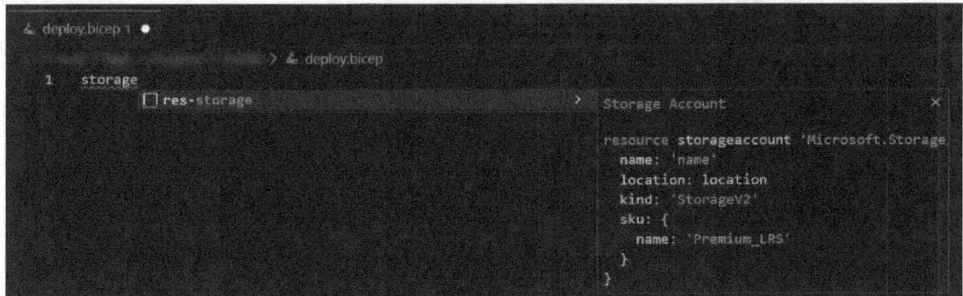

**FIGURE 3-9**   The Bicep IntelliSense option for storage accounts

## Deploy a Bicep file

After you have created the Bicep file with the parameters, variables, resources, and other components that you might need to deploy in your Azure subscription, you can submit the Bicep file for deployment. Bicep will then translate your file to an ARM template and submit the template for deployment.

You can deploy the Bicep file directly from either Visual Studio or Visual Studio Code by right-clicking in the .bicep file, as shown in Figure 3-10. This assumes that you are authenticated to your Azure subscription from Visual Studio or Visual Studio Code.

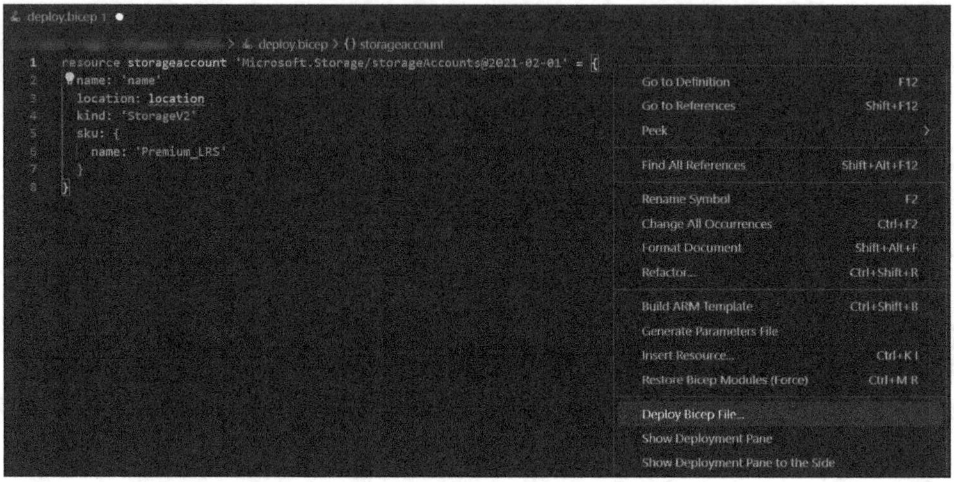

**FIGURE 3-10**  The Bicep context menu with the Deploy Bicep File option

You can also deploy a Bicep file from the CLI or PowerShell by submitting the file for deployment. If you plan to use the cloud shell, you must first upload the bicep file to the cloud shell environment. The following Azure CLI command deploys the deploy.bicep file to a resource group named az104-rg1.

```
az deployment group create --resource-group 'az104-rg1' --template-file deploy.bicep
```

From PowerShell, use the `New-AzResourceGroupDeployment` cmdlet to deploy a Bicep file. The following command also deploys the file to the same resource group.

```
New-AzResourceGroupDeployment -ResourceGroupName 'az104-rg1' -TemplateFile ./deploy.bicep
```

# Skill 3.2: Create and configure virtual machines

There are multiple ways to create and configure virtual machines, depending on your intended use. The easiest way to create an individual virtual machine is to use the Azure portal. If you have a need for automated provisioning (or you just enjoy the command line), the Azure

PowerShell cmdlets and the Azure command-line interface (CLI) are a good fit and supported across multiple platforms. For more advanced automation—even including orchestration of multiple virtual machines—Azure Resource Manager templates and Bicep files can also be used. Each method brings its own capabilities and tradeoffs, and it is important to understand which tool should be used in the right scenario. This section covers various aspects and features to efficiently manage VMs and supporting resources in an Azure environment.

**This skill covers how to:**
- Create a virtual machine
- Configure Azure Disk Encryption
- Move VMs from one resource group or subscription to another
- Manage VM sizes
- Manage VM disks
- Deploy VMs to availability sets and zones
- Deploy and configure Virtual Machine Scale Sets

# Create a virtual machine

Virtual machines in Azure are one of the core compute options available, and offer the most flexibility and control in your environment. As with on-premises VMs, you are responsible for managing the VM, accounts, data security, patches and updates, and more.

Before you create a VM, there are several parameters that you should consider. Your organization might have other business or technical requirements to meet that aren't specifically mentioned here. The minimum considerations include

- Naming the VM resources
- The Azure region to deploy to
- The size and SKU of the VM
- Subscription quotas or limits
- VM operating system requirements
- VM configuration
- Related resources

When you use the Azure portal to deploy a VM, it separates the required information into a series of tabs in the Create A Virtual Machine blade. The tabs for creating a VM include

- Basics
- Disks
- Networking

- Management
- Monitoring
- Advanced
- Tags

## Basics

To use the Azure portal to create a virtual machine, search for **Virtual Machines**. On the Virtual Machines blade, select Create and then select Azure Virtual Machine. The Basics tab will appear, where you can configure the following information:

- **Subscription** The Azure subscription that the VM will deploy to.
- **Resource Group** The resource group that the VM will logically be associated with.
- **VM Name** The name of the virtual machine resource.
- **Region** The Azure region where the VM will be deployed.
- **Availability Options** The target redundancy for the VM. These are discussed more later in this skill.
- **Security Type** Choose between standard, trusted launch, or a confidential VM.
- **Image** The operating system image to use for the VM.
- **VM Architecture** Choose between Arm64 or x64 processor architecture.
- **Run With Azure Spot Discount** Whether the VM is an Azure Spot instance. Azure Spot instances are less expensive but could be deallocated by Azure. These are typically used in dev/test or stateless compute environments.
- **Size** The SKU of the VM, which determines the CPUs, Memory, Disk, and Networking capabilities. The SKU is the primary cost for running the VM.
- **Authentication Details** Either username and password or SSH key information to authenticate to the VM.
- **Inbound Ports** Whether the default network security group (NSG) that is created has any configured allow rules.

Figure 3-11 shows the completed fields for a new VM named az104-vm1.

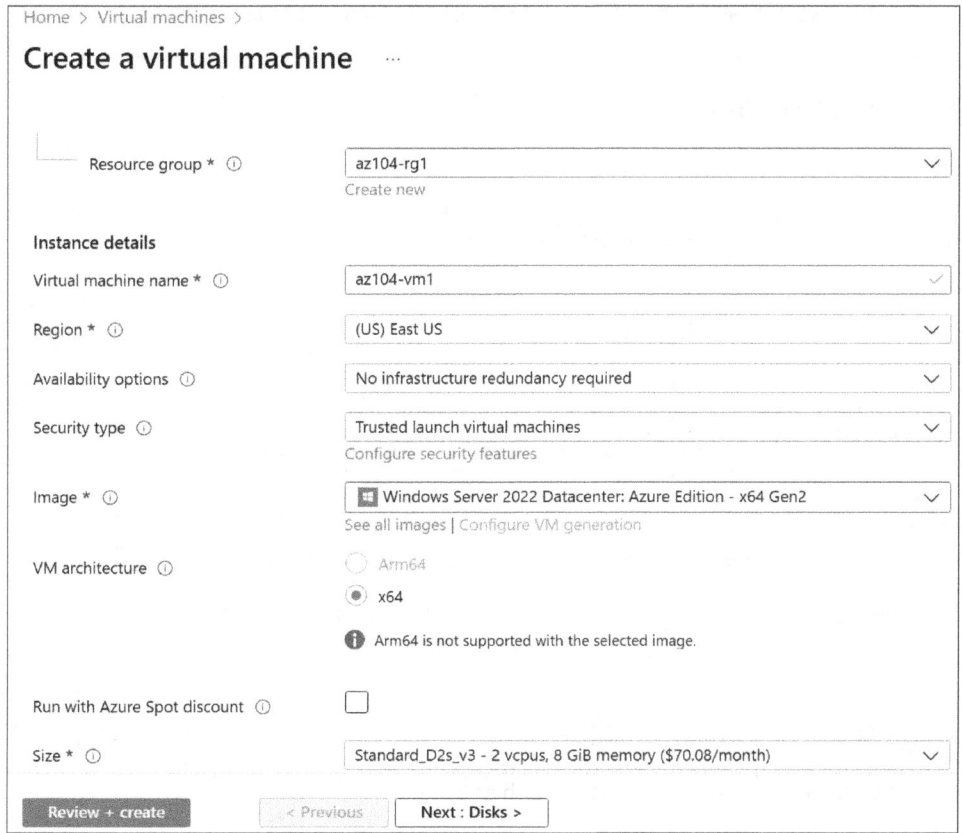

**FIGURE 3-11** The Basics tab of the Create A Virtual Machine blade

## Disks

On the Disks tab of the Create A Virtual Machine blade, configure the size and performance of the operating system disk, as well as any data disks you might need to add to the VM. The options for VM disks include

- **OS Disk Size** The capacity of the operating system disk.
- **OS Disk Type** The performance tier of the operating system disk.
- **Delete With VM** Whether the disk is deleted with the VM and is recommended for resource lifecycle management.
- **Key Management** How the encryption key for the disk is stored. Disk encryption is discussed later in this skill.
- **Ultra Disk Compatibility** Whether the operating system disk uses an ultra disk for performance.
- **Data Disks** Additional data disks available to the VM with their own size and performance configurations.

Figure 3-12 displays the Disks tab of the Create A Virtual Machine blade.

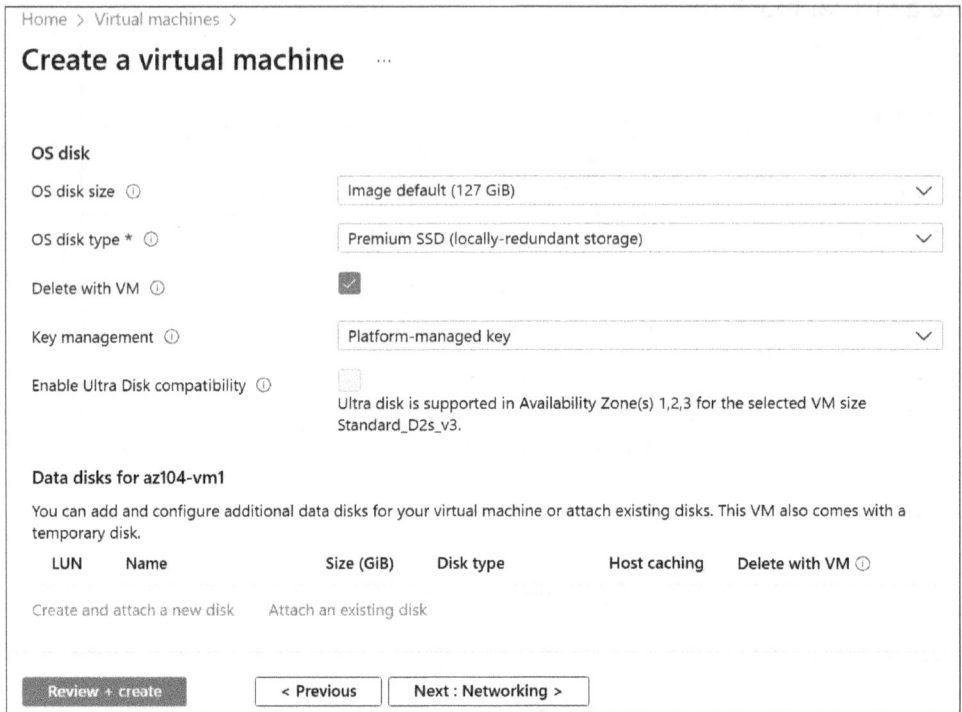

**FIGURE 3-12** The Disks tab of the Create A Virtual Machine blade

Depending on the VM SKU and Azure region that you select, disks have three primary performance options:

- Premium SSD
- Standard SSD
- Standard HDD

The disk that you select will have performance, cost, and availability impacts on the virtual machine. For example, a Standard HDD will cost less but will have less performance and target SLA than a VM using a Premium SSD for the operating system disk.

## Networking

The Networking tab of the Create A Virtual Machine blade defines the connectivity options for the virtual machine. If a virtual network already exists in the resource group that you select, it will automatically select that network and the first subnet from that network to deploy the VM to. If this is your first VM or network, it will be populated with a new virtual network to create.

The options for VM networking include

- **Virtual Network** The network that the default NIC of the VM will be associated with.
- **Subnet** The subnet of the network that the NIC of the VM will be associated with.

- **Public IP**   Whether the VM requires a public IP address directly on the NIC.
- **NIC Network Security Group**   The type of NSG that is associated with the NIC, if any.
- **Public Inbound Ports**   Whether any allow rules are added to the NSG during deployment.
- **Delete Public IP And NIC When VM Is Deleted**   Sets up networking resource lifecycle management with the VM.
- **Enable Accelerated Networking**   Available depending on the SKU of the VM.
- **Load Balancing**   Whether to integrate this VM with a load balancing option during deployment.

Figure 3-13 displays the Networking tab of the Create A Virtual Machine blade.

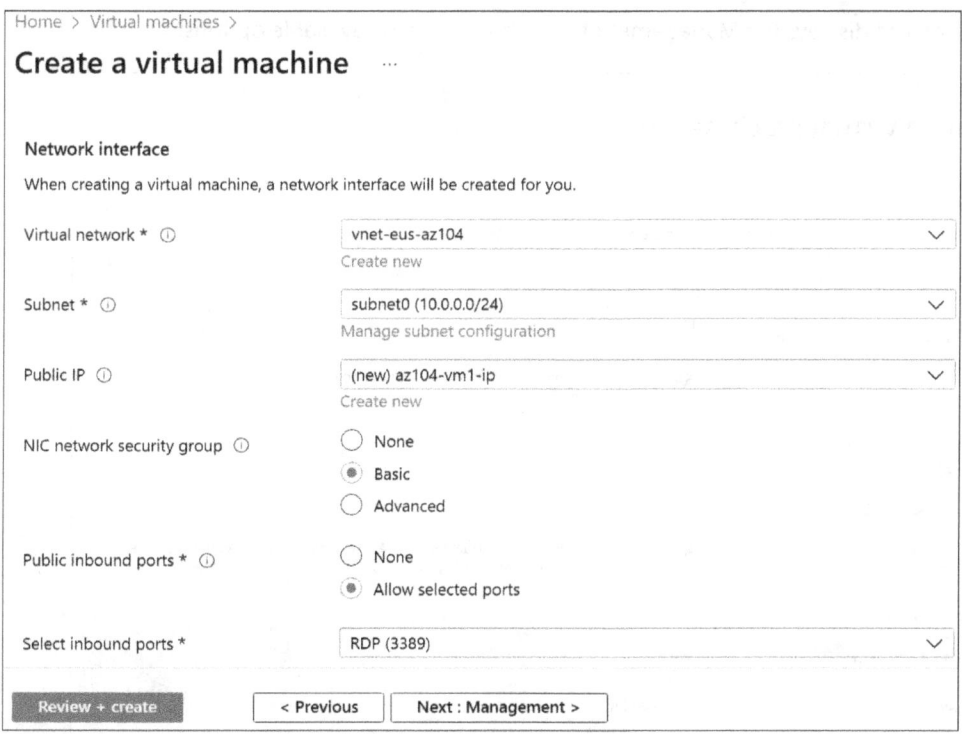

**FIGURE 3-13**   The Networking tab of the Create A Virtual Machine blade

## Management

The Management tab provides various options for administrative management of the VM in Azure. These are configuration items that either help you administer the VM or help you achieve other business goals your organization might have. The options on the Management tab include

- **Microsoft Defender For Cloud**   Whether the VM is monitored by Defender for Cloud.
- **Identity**   Whether the VM has a system-assigned managed identity associated with it.

- **Azure AD**   Whether you can log into the VM using Entra ID, previously Azure AD.
- **Auto-shutdown**   Schedule the VM to be deallocated automatically at a specific time.
- **Backup**   Configure backup options using Azure Backup Center for the VM.
- **Site Recovery**   Configure disaster recovery options using Azure Site Recovery.
- **Guest OS Updates**   Hot patching and update orchestration options for the VM.

> **NOTE   AZURE AD IS NOW ENTRA ID**
>
> **Azure AD was rebranded as Microsoft Entra ID. However, as of this update, the Create A Virtual Machine blade still references Azure AD. Note that this should be Entra ID.**

Figure 3-14 displays the Management tab with some of the available options.

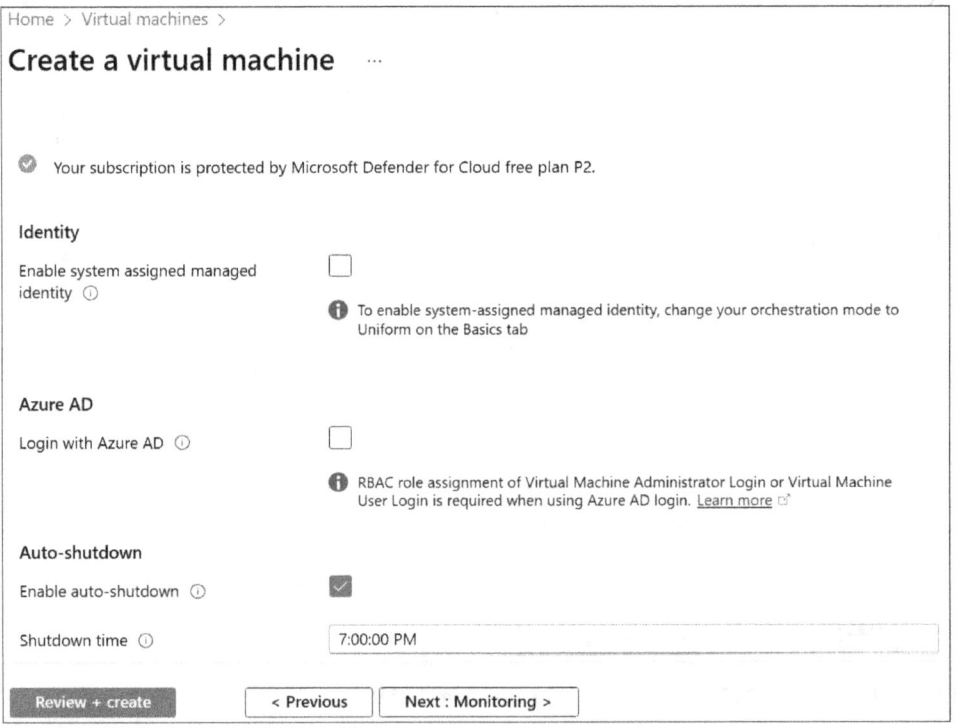

**FIGURE 3-14**   The Management tab of the Create A Virtual Machine blade

## Monitoring

The Monitoring tab is where you can configure any alerting and diagnostic settings for the VM during deployment. The options for VM monitoring include

- **Alerts**   The recommended alert rules for a VM.
- **Boot Diagnostics**   Whether these are enabled and where to store the logs if so.

- **OS Guest Diagnostics**   Whether to collect the events, logs, or other traces from the operating system and store them in a target location in Azure.

Figure 3-15 shows the Monitoring tab of the Create A Virtual Machine blade.

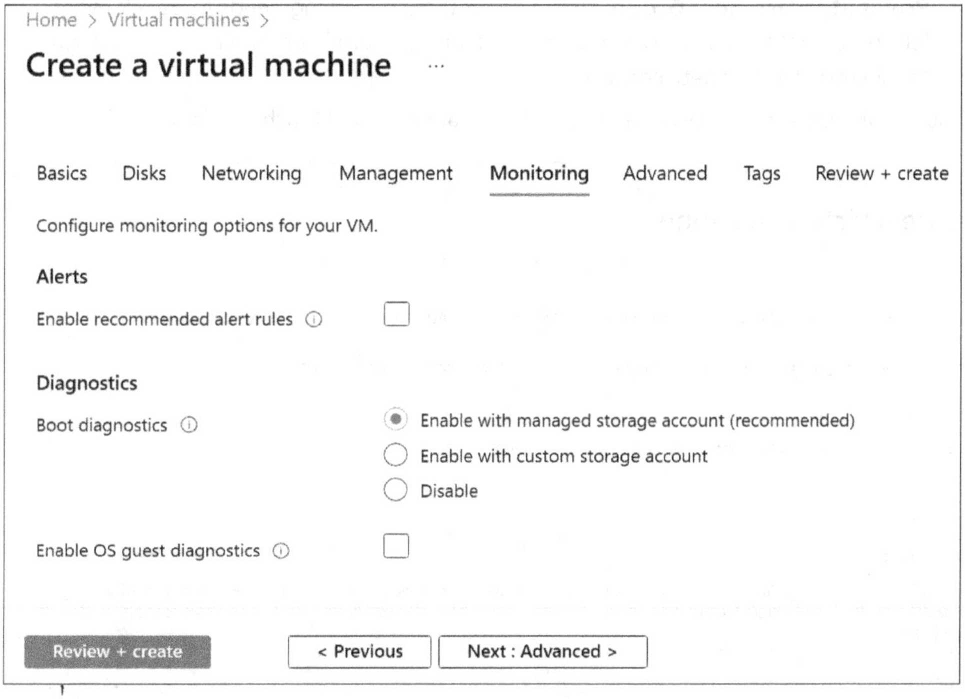

**FIGURE 3-15**   The Monitoring tab of the Create A Virtual Machine blade

## Advanced

The Advanced tab provides options for further customizing the VM by configuring extensions, applications, custom data, or VM proximity during the deployment. These options are convenient if you need the VM to immediately be available and working as part of your application after deployment, but it requires further customization. The options included on the Advanced tab are

- **Extensions**   Any Azure VM extensions that should be added to the VM.
- **VM Applications**   Applications that you have added to your gallery that should be installed with the VM.
- **Custom Data**   Scripts, configuration files, or other data that the VM might need during provisioning to complete application installs or customization.
- **User Data**   Scripts, configuration files, or other data that will be available to the VM *persistently*, including after deployment.
- **NVMe Performance**   Depending on the VM SKU size, additional disk performance options.

- **Host Group** Whether the VM will run on an Azure dedicated host.
- **Capacity Reservation Group** If you have reserved capacity in a specific Azure region, the capacity group that the VM should be associated with.
- **Proximity Placement Group** If the VM you are deploying needs to be relatively "close" to another VM in your environment, group them in proximity placement groups to minimize latency between VMs.

Figure 3-16 displays the Advanced tab of the Create A Virtual Machine blade.

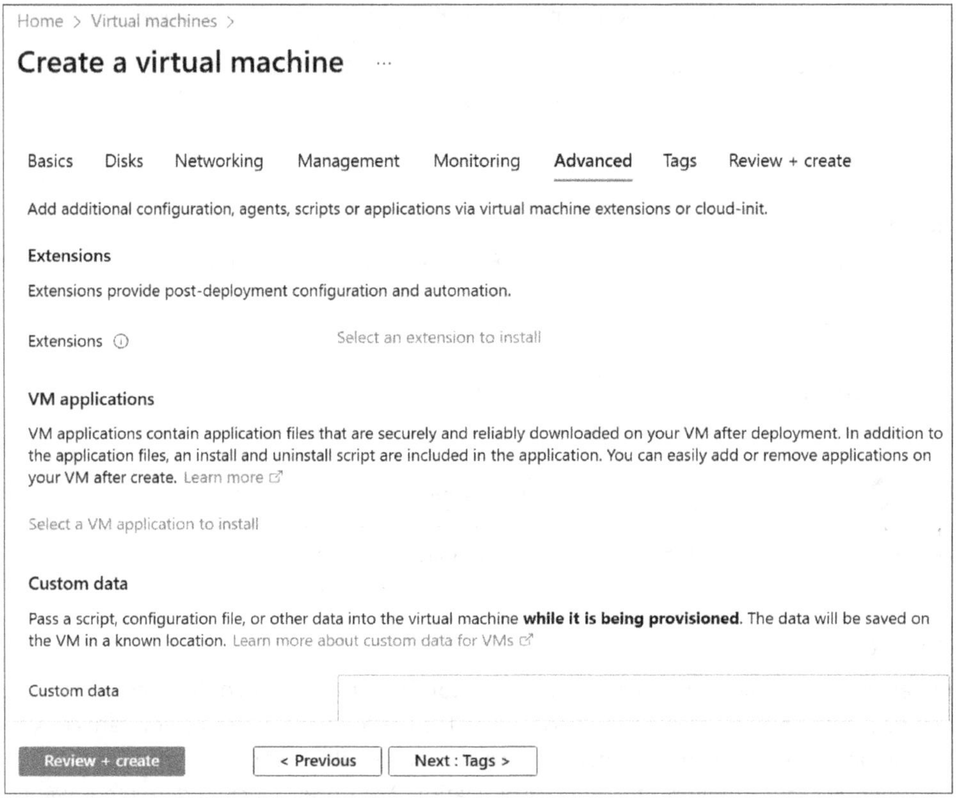

**FIGURE 3-16** The Advanced tab of the Create A Virtual Machine blade

## Configure Azure Disk Encryption

The disks of an Azure virtual machine are always encrypted. However, you have the option to configure how the disks are encrypted. By default, disks use platform-managed encryption, meaning that Microsoft manages the encryption key and key rotation for the disk. If you have a business or technical requirement to manage your own encryption keys, you can integrate the encryption with Azure Key Vault. In this section, you will learn how to manage Azure Disk Encryption with a few scenarios using the Azure portal. Please note that these steps can be performed using PowerShell or Azure CLI.

> **NOTE** **CHARGES FOR AZURE DISK ENCRYPTION**
>
> There is no charge for encrypting VM disks with Azure Disk Encryption, but there are charges associated with the use of Azure Key Vault. Key Vault pricing can be accessed at *https://azure.microsoft.com/en-in/pricing/details/key-vault/*.

## Enable encryption on an existing VM

Follow these steps to enable encryption on an existing VM:

1. Browse to the VM resource in the Azure portal and under Settings, select Disks (see Figure 3-17).

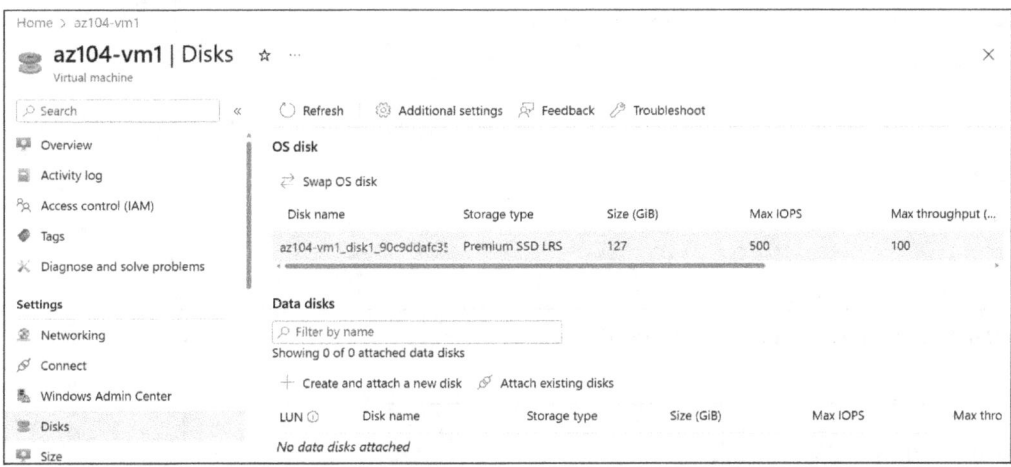

**FIGURE 3-17** Disks blade for an Azure VM

2. In the Command bar, click Additional Settings. Then, click Encryption. Under Disks To Encrypt, choose None, OS Disk, or OS And Data Disks, as shown in Figure 3-18. Select either OS Disk or OS And Data Disks to be prompted for the Azure Key Vault.

3. When you select the disks to encrypt, options for Azure Key Vault appear. Select the key vault where you have created a key to be used for Azure Disk Encryption. If you do not have a key vault or key, you can also create them directly from this page, as shown in Figure 3-19.

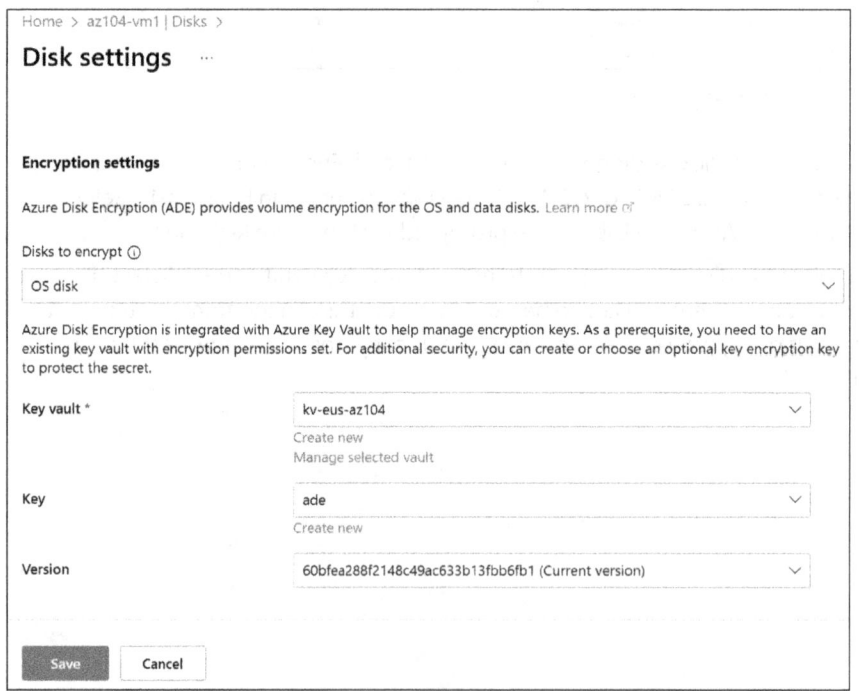

**FIGURE 3-18**  Encryption options for Azure VM disks

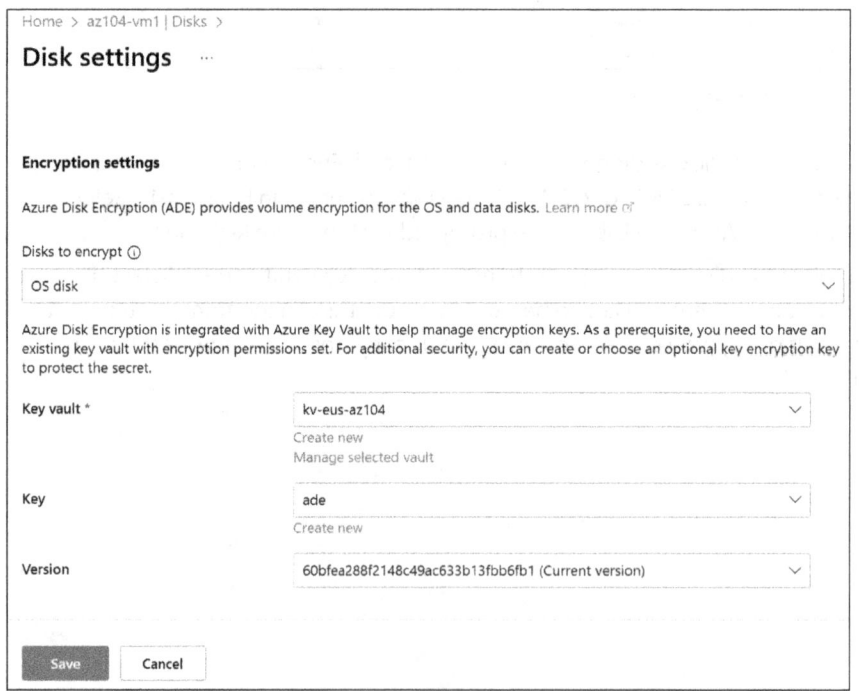

**FIGURE 3-19**  Encryption options for Azure VM disks

4. Click Save.

5. Click Review + Create. After the key vault has passed validation, click Create. This will return you to the Select Key From Azure Key Vault blade.

## Disable encryption

To disable encryption for operating system and data disks for an existing VM, select None from the Disks To Encrypt menu, as shown in Figure 3-20.

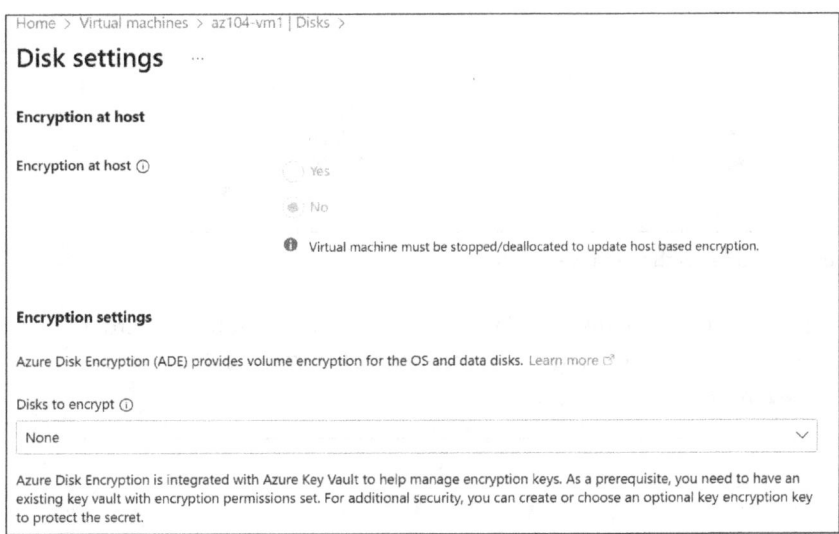

FIGURE 3-20 Disable disk encryption

# Move VMs from one resource group or subscription to another

Azure provides the ability to move some resources from one subscription to another or from resource group to resource group. You could choose to do this for ongoing governance, mergers and acquisitions, changing in billing chargebacks, or other reasons. Depending on the resource type, you might be able to move a single resource. For some resources, such as a virtual machine, you need to move the related resources, such as the NIC, disk, etc., with the VM. Net new resources are not created in the target resource group or subscription.

Follow these steps to move a resource using the Azure portal:

1. From the Azure portal, navigate to the resource group where the resource is located.

2. On the Overview blade of the resource group, select the resources you plan to move by selecting the checkmark for the desired resources.

3. From the command bar, choose Move, and then select the destination: Move To Another Resource Group, Move To Another Subscription, or Move To Another Region. For this example, select Move To Another Resource Group. Depending on your screen size or resolution, Move might be hidden behind the ellipses, as shown in Figure 3-21.

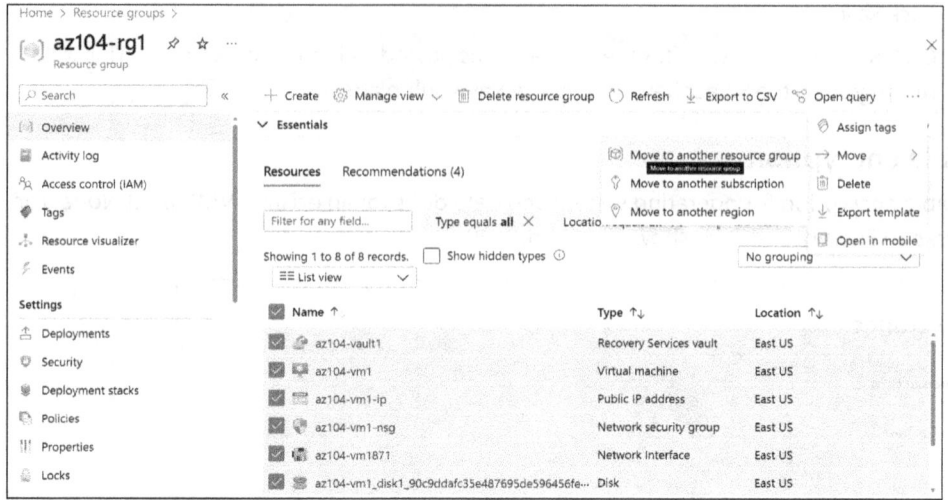

**FIGURE 3-21** Resources selected with the Move menu

4. The Move Resources blade opens in the Azure portal. Select the Source + Target tab, and then select the desired target resource group. Figure 3-22 shows az104-rg2 selected as the target.

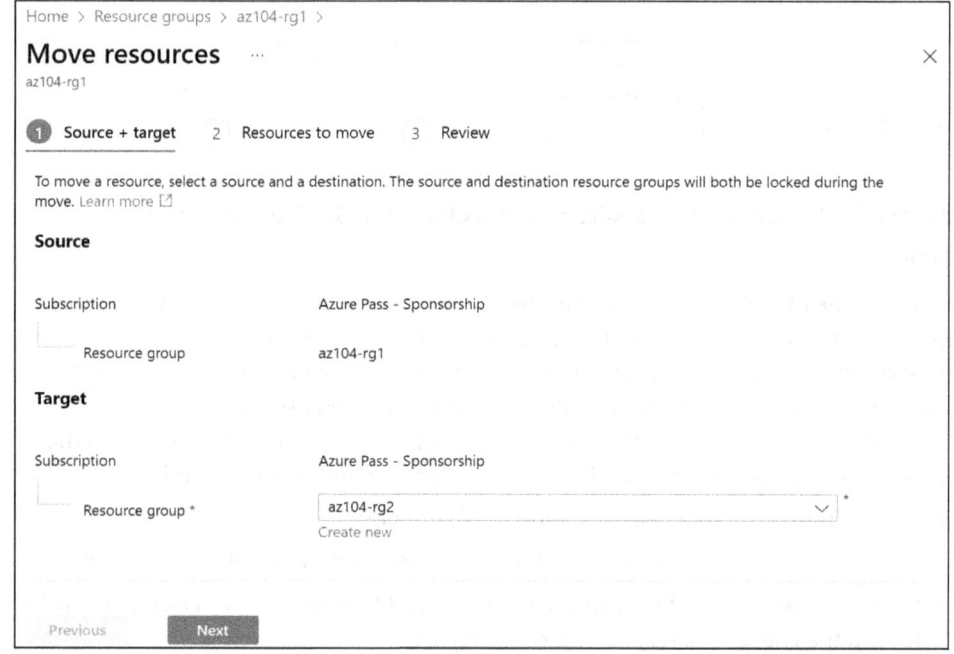

**FIGURE 3-22** The Source + Target tab of the Move Resources blade

5. Click Next. On the Resources To Move tab, the portal will validate whether the resources can be moved to the target resource group. Depending on the option you select, this

could fail because of quota (if moving to a new subscription), orphaned resources (if you did not select all required resources), or other factors. Figure 3-23 shows a successful validation.

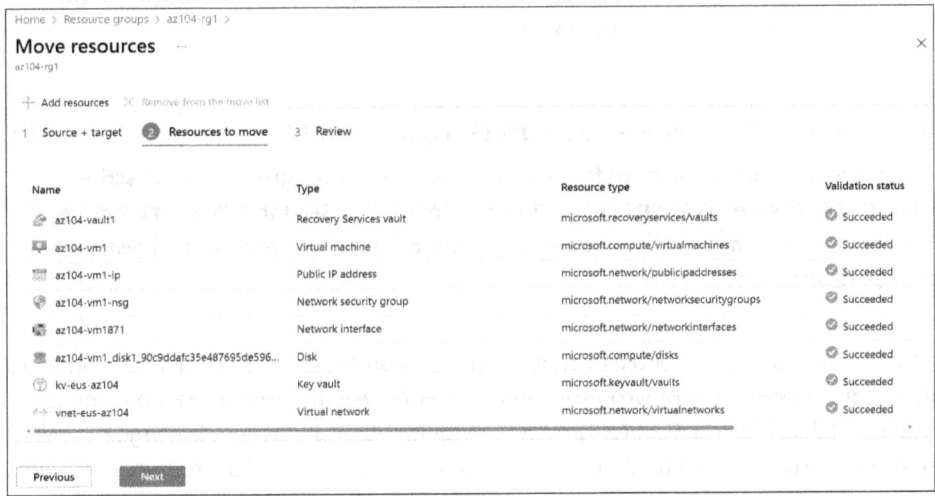

**FIGURE 3-23** The Resources To Move tab of the Move Resources blade

6. Click Next. Accept the terms, and click Move to start moving the resources, as shown in Figure 3-24.

7. Because the resource group will change, any existing scripts that target resources in this resource group will no longer work until they have been updated. The Azure portal prompts you to confirm that you are aware of this change before you can continue with the move.

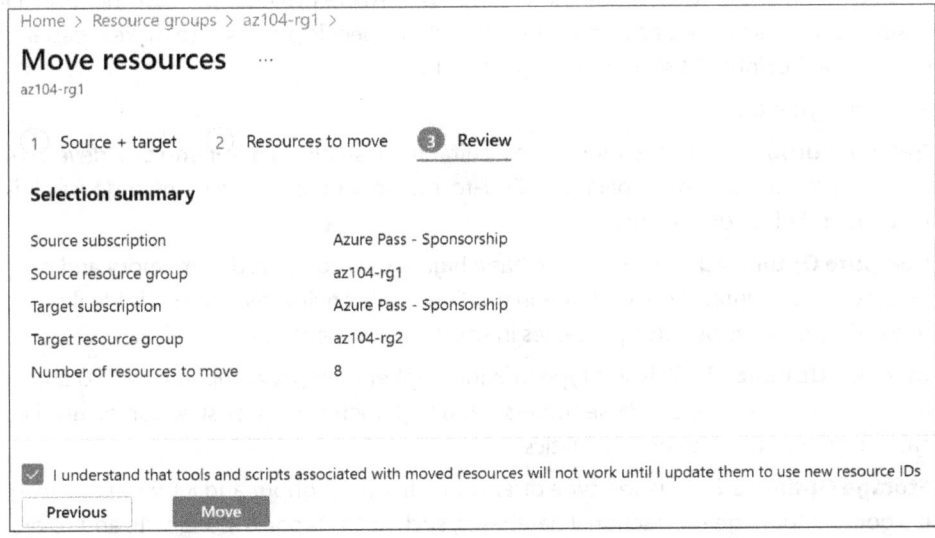

**FIGURE 3-24** The Review tab of the Move Resources blade

The process for moving a resource to another subscription or region is similar and progresses through the same steps on the Move Resources blade. However, this could have more impact on the resources than moving resource groups. Changing subscriptions could change the billing associated with the resource. Changing regions could impact the price of the resource, as well as connectivity, as routing or other network changes might also need to be made.

## Manage VM sizes

There are many situations where the amount of compute processing your workload needs varies dramatically from day to day or even hour to hour. For example, in many organizations, line of business (LOB) applications are used heavily during the workweek, but on the weekends, they see little actual usage. Other examples are workloads that require more processing time due to scheduled events such as backups or maintenance windows where having more compute time may make it faster to complete these tasks. Azure provides purpose-built virtual machine sizes. This means that each family is designed for specific purposes to make it easier for you to choose the right VM size for the right workload.

The different types are

- **General Purpose**   This size type is most suitable for small- to medium-scale development environments. It has a balanced CPU-to-memory ratio. As the name suggests, it is recommended for general use.
- **Compute Optimized**   This size type has a higher CPU compared to memory and can be used for CPU-intensive workloads in medium-scale environments. This is ideal for network appliances or batch processes in small environments.
- **Memory Optimized**   This size type provides higher memory compared to CPU and is ideal for medium-scale database servers. With high memory, these sizes can be used for caches, or used with memory analytics.
- **Storage Optimized**   This size type offers high disk throughput and IO, which makes it a good fit for large transactional databases, such as Cassandra, MongoDB, and so on. Also, it can be used for Big Data and data warehousing.

- **GPU Optimized** This size type provides VMs with one or many NVIDIA GPUs. It provides high compute and graphics, which are ideal for visualization workloads.
- **High Performance Compute** This size type is capable of handling batch processing, molecular modeling, and fluid dynamics. This size type offers substantial CPU power and diverse options for low-latency RDMA networking using FDR InfiniBand and several memory configurations to support memory-intensive computational requirements.

Azure Virtual Machines make it relatively easy to change the size of a virtual machine, even after it has been deployed. There are a few things to consider with this approach.

First, ensure that the region your VM is deployed to supports the instance size that you want to change the VM to. In most cases, this is not an issue, but if you have a use case where the desired size isn't in the region to which the existing VM is deployed, your only options are to either wait for the size to be supported in the region or to move the existing VM to a region that already supports it. This can become complicated because then the new region must also have the networking and other resources required by the VM.

Second, ensure that the new size is supported in the current hardware cluster in which your VM is deployed. This can be determined by viewing the Size blade in the virtual machine configuration blade in the Azure portal of a running virtual machine, as shown in Figure 3-25. If the size is available, you can select it. Changing the size reboots the virtual machine.

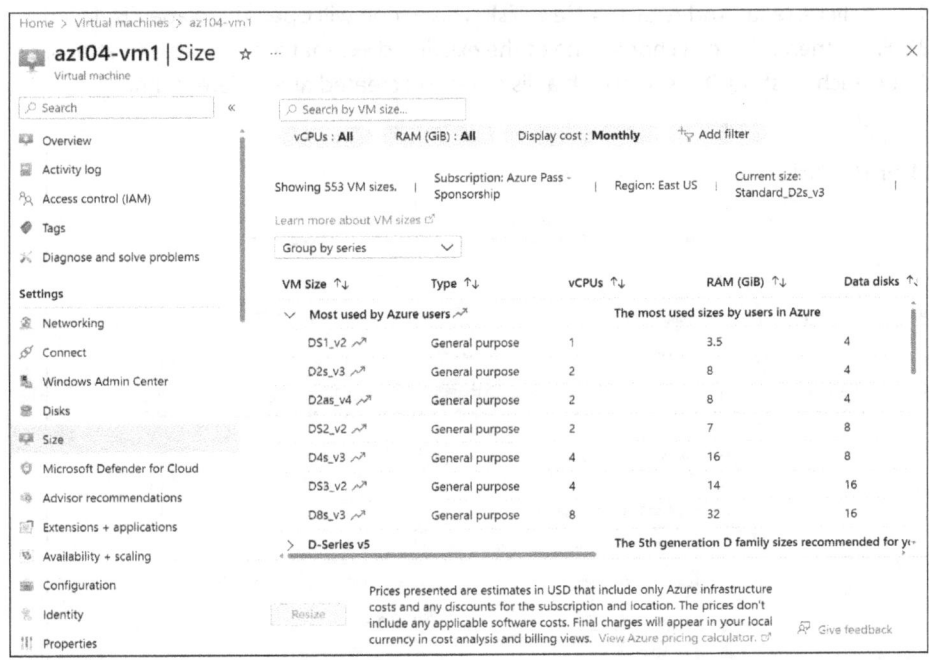

**FIGURE 3-25** Changing the size of an Azure virtual machine using the Azure portal

If a desired size is not available, it means either the size is not available in the region or on the current hardware cluster. You can view the available sizes by region at *https://azure. microsoft.com/regions/services/*. If you need to change to a different hardware cluster, you must first deallocate the virtual machine, and if it is part of an availability set, you must stop all instances of the availability set at the same time. After all the VMs are deallocated, you can then change the size, which moves all the VMs to the new hardware cluster as they are resized and started. All VMs in the availability set must be stopped before performing the resize operation to a size that requires different hardware because all running VMs in the availability set must use the same physical hardware cluster. Therefore, if you are required to change a physical hardware cluster in order to change the VM size, all VMs must be stopped and then restarted one by one to a different physical hardware cluster.

> ***NEED MORE REVIEW?*** **VIRTUAL MACHINE SIZES**
>
> **There are a lot of considerations when choosing the correct virtual machine size. For more information on sizes in the context of Windows-based virtual machines see** *https://learn. microsoft.com/en-us/azure/virtual-machines/sizes*. **For the Linux version of the article, see** *https://learn.microsoft.com/en-us/azure/virtual-machines/sizes*.

## Manage VM disks

Adding a data disk to an existing Azure virtual machine using the Azure portal is almost identical to the creation process. From within the virtual machine configuration blade, click Disks, and then click Create And Attach A New Disk. This action will open the blade displayed in Figure 3-26. From there, you can choose one of the existing disks that are available to attach, or you can click Attach Existing Disks to attach a disk that you created at a different time.

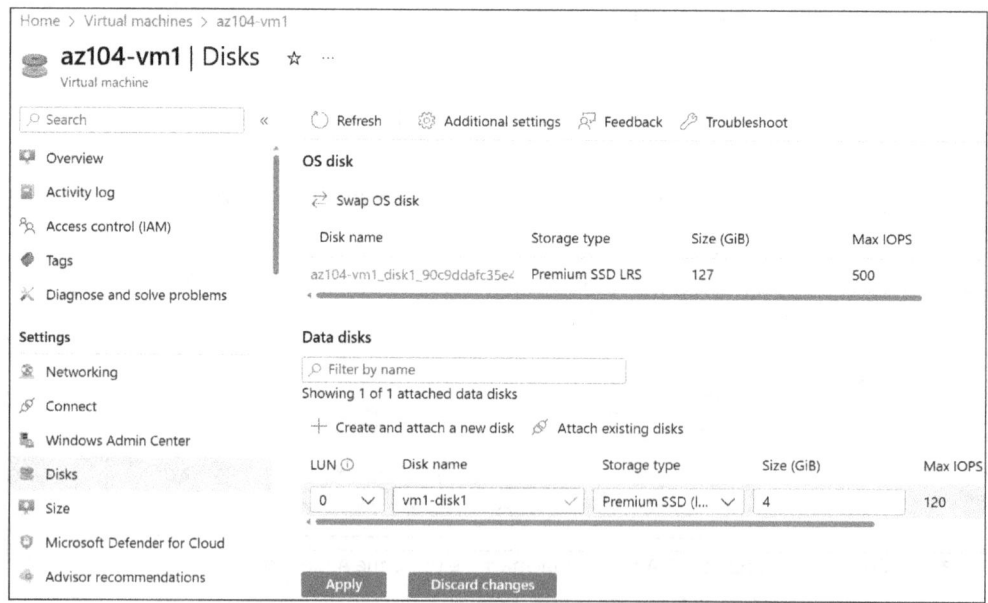

**FIGURE 3-26** Adding a data disk to an Azure virtual machine in the Azure portal

When you create the data disk, you will have options similar to those available for the operating system disk when creating a VM. The storage type can be configured with various options for Premium and Standard performance types, disk capacity, encryption options, and host caching options.

**EXAM TIP**

If the virtual machine is deployed into an availability zone and you are using PowerShell, use the `Zone` **parameter with the** `New-AzDiskConfig` **cmdlet to specify which availability zone to create the disk.**

**EXAM TIP**

If the virtual machine is deployed into an availability zone and you are using Azure CLI, the disk is automatically placed into the same zone as the virtual machine.

# Deploy VMs to availability sets and zones

Resiliency is a critical part of any application architecture. Azure provides several features and capabilities to make it easier to design resilient solutions. The platform helps you avoid a single point of failure at the physical hardware level and provides techniques to avoid downtime during host updates. Using features such as availability zones, availability sets, and load balancers provides you the capabilities to build highly resilient and available systems.

## Availability zones

Availability zones are separate units—each with its own power, cooling, and networking—which provide higher resiliency and protect applications and data from disruption in the data centers. To ensure resiliency, there is a minimum of three separate zones in all enabled regions. The physical and logical separation of availability zones within a region protects applications and data from zone-level failures. Availability zones provide a 99.99 percent SLA uptime when two or more VMs are deployed into two or more availability zones. Figure 3-27 illustrates how a three-tier application can be deployed with a virtual machine from each tier deployed in each of the three zones for increased availability.

When you create VMs in three availability zones, those will be automatically distributed across three zones. A zone represents one or more datacenters that all have shared power, cooling, and networking within that zone.

To deploy a VM to an availability zone, select the zone you want to use on the Basics tab of the Create A Virtual Machine blade, as shown in Figure 3-28.

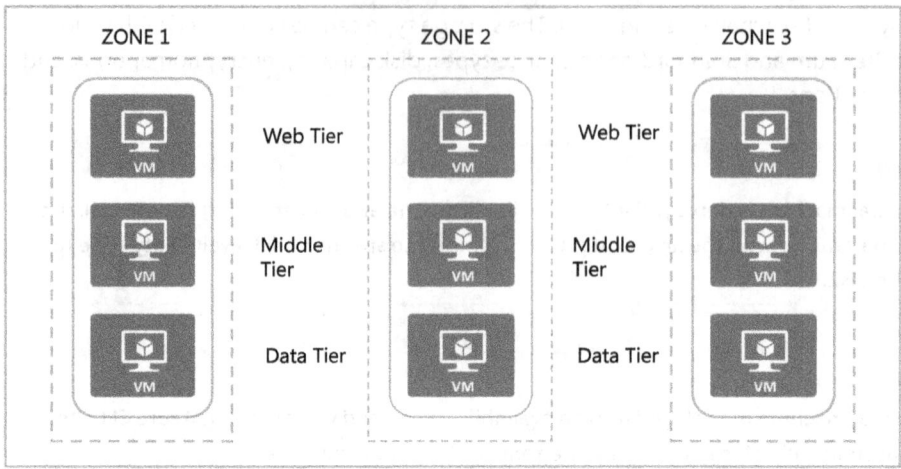

**FIGURE 3-27** Architectural view of an availability zone

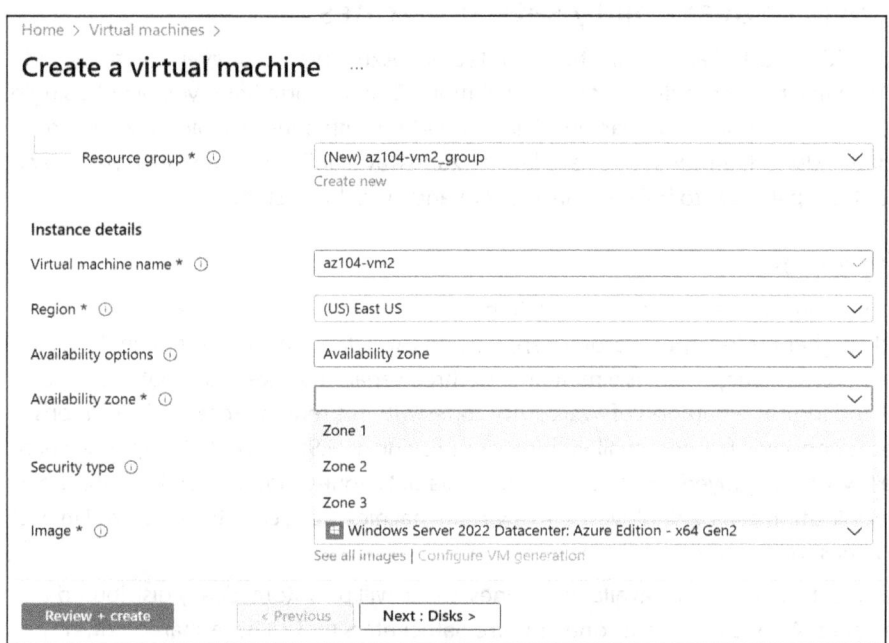

**FIGURE 3-28** Specifying the availability zone for a VM

> **NOTE** **AVAILABILITY ZONES AND AZURE REGION**
>
> If you are unable to set an availability zone, you have probably selected a region where availability zones are not available. For a list of all Azure regions and zone support, visit *https://azure.microsoft.com/en-us/explore/global-infrastructure/geographies.*

## Availability sets

Deploying a multitier application into an availability set can provide redundancy and high availability to the virtual machines. To provide redundancy for your virtual machines, you must place at least two virtual machines in an availability set. This configuration ensures that at least one virtual machine is available in the event of a host update, or a problem with the physical hardware the virtual machines are hosted on. Having at least two virtual machines in an availability set is a requirement for the service level agreement (SLA) for virtual machines of 99.95 percent.

You can place a single instance virtual machine in an availability set, too, but doing so provides comparatively lower SLAs. A Premium SSD provides an SLA of 99.9 percent, while Standard SSD and Standard HDD provide SLAs of 99.5 percent and 95 percent, respectively.

Virtual machines should be deployed into availability sets according to their workload or application tier. For instance, if you are deploying a three-tier solution that consists of web servers, a middle tier, and a database tier, each tier would have its own availability set, as Figure 3-29 illustrates.

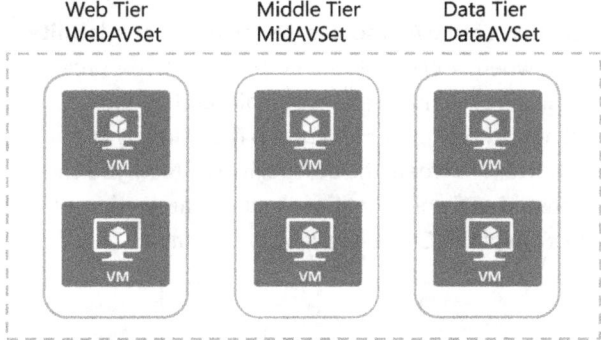

**FIGURE 3-29**   Availability set configurations for a multitier solution

Availability sets can be configured by assigning a fault domain and an update domain. A fault domain represents a group of servers that have shared power, cooling, and networking, while an update domain represents a group of servers that can be rebooted at the same time. Each availability set can have up to 20 update domains and 3 fault domains. This reduces the impact to VMs from physical hardware failures, such as server, network, or power interruptions on one of the physical racks. It is important to understand that the availability set must be set when the virtual machine is created.

## Create an availability set

To create an availability set, specify a name for the availability set that is not in use by any other availability sets within the resource group, along with the number of fault and update domains, as well as whether you will use managed disks with the availability set or not. In order to create the availability set at the time of virtual machine creation, go to the homepage, click Create A Resource, search for **virtual machine,** and click Create. The Basics tab opens, as shown in Figure 3-30. On the Create Availability Set page, you can create a new availability set. As shown in Figure 3-30, you can select the number of fault domains and update domains.

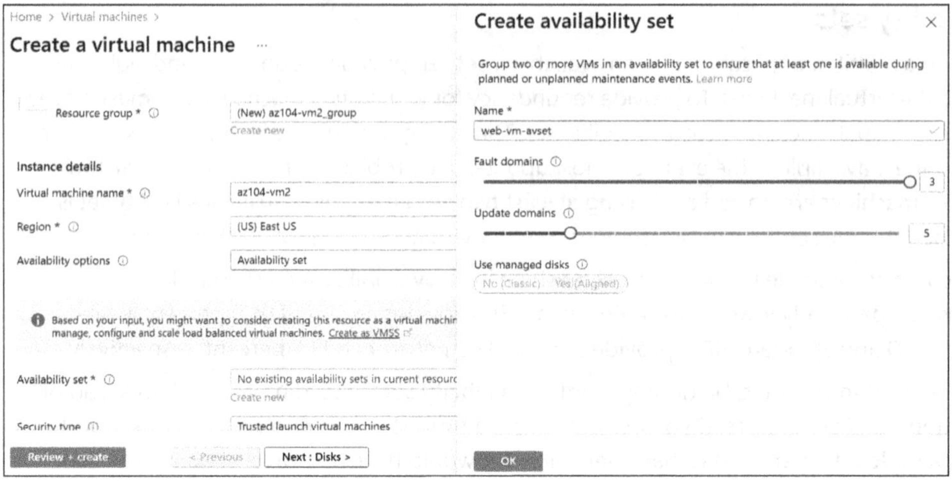

**FIGURE 3-30** Creating an availability set

You can also create an availability set by clicking Create A Resource, searching for **availability set**, and clicking Create. The Basics tab appears, where you can select a subscription, resource group, and region, and you can specify the availability set name, fault domain, and update domain. On the Advanced tab, you can select a proximity placement group if it's already created. Click the Review + Create button at the bottom to create the availability set. Now you can place resources such as VMs into this newly created availability set by selecting it at the time of resource creation. Figure 3-31 shows the Basics tab of the Create Availability Set blade.

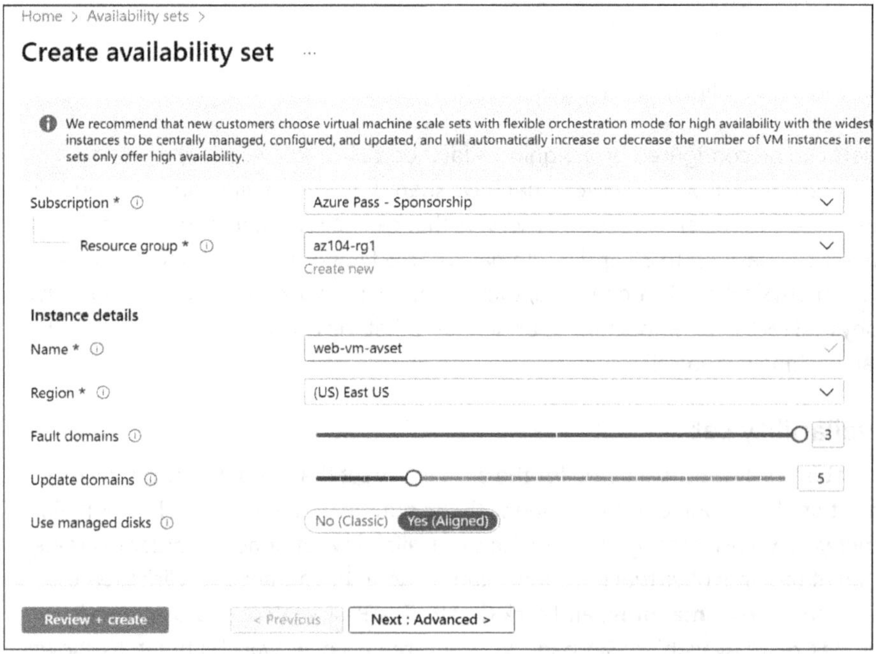

**FIGURE 3-31** Creating an availability set

## Availability sets and managed disks

Availability sets and managed disks complement each other. When the VM uses managed disks and is placed in an availability set (known as an aligned availability set), it ensures that the VM disks are placed in different storage fault domains, as shown in Figure 3-32. This alignment ensures that all the managed disks attached to a VM are within the same managed disk fault domain. The number of fault domains for an availability set depends on the region it belongs to, with either two or three fault domains per region.

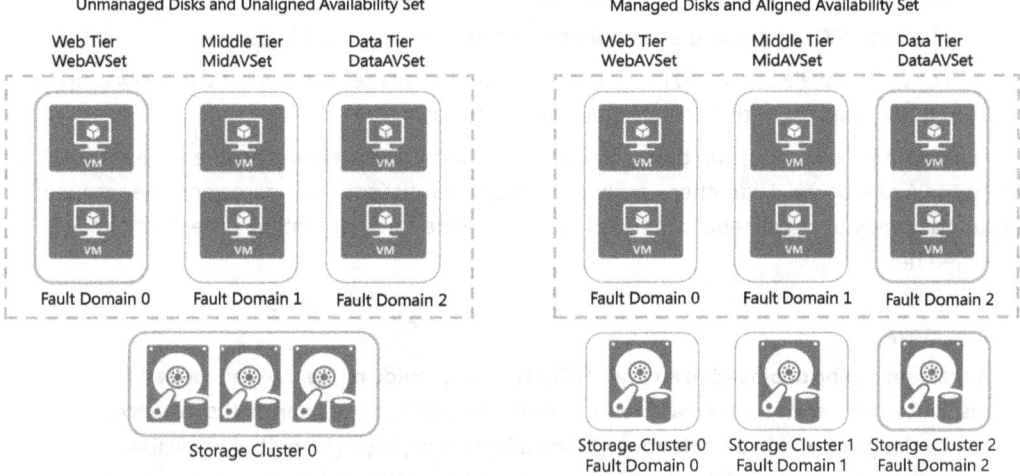

**FIGURE 3-32**   Aligning managed disks with an availability set

# Deploy and configure Virtual Machine Scale Sets

A Virtual Machine Scale Set (VMSS) is a compute resource that you can use to deploy and manage a set of identical virtual machines.

By default, a VMSS supports up to 100 instances. However, it is possible to create a scale set up to 1,000 instances by placing instances into multiple placement groups. A placement group is a construct, such as an Azure availability set, with its own fault domains and upgrade domains. If you define an instance count that is higher than 100 in the Azure portal when the scale set is created, the Azure portal will automatically enable the scale set for multiple placement groups. By default, a scale set consists of a single placement group with a maximum size of 100 VMs. If the scale set property called `singlePlacementGroup` is set to `false` or if you define an instance count higher than 100 in the Azure portal, the scale set can be composed of multiple placement groups and has a range of 0-1,000 VMs.

Using multiple placement groups is commonly referred to as a "large scale set." The `singlePlacementGroup` property can be set using ARM templates or the command-line tools. Be aware of the following conditions when working with large scale sets:

- If you are using a custom image (not a default available image from marketplace), your scale set supports up to 600 instances instead of 1,000.
- The basic SKU of the Azure Load Balancer can scale up to 300 instances.
- For a large scale set (> 100 instances), you should use the Standard SKU (supports up to 1,000 instances) or the Azure Application Gateway.

Figure 3-33 shows a portion of the blade for creating a new virtual machine scale set (VMSS) using the Azure portal. Like other Azure resources, you must specify a name and the resource group to deploy to. All instances of the VMSS will use the same operating system disk image specified here.

 **EXAM TIP**

A scale set can be deployed to an availability zone to provide higher redundancy and resiliency. If the scale set is created with a single availability zone, then all the instances will be deployed within a single zone. If the scale set is deployed in multiple availability zones (known as a zone-redundant scale set), based on scaling rules, the instances can be deployed to multiple zones if needed.

On the Scaling tab, you can configure the Scaling Policy as Manual or Custom. When you set the Scaling Policy to Custom, you see the configuration options for setting the default rules, as shown in Figure 3-34. Here, you can specify the minimum and maximum number of VMs in the set, and you can set the actions to scale out (add more) or to scale in (remove instances).

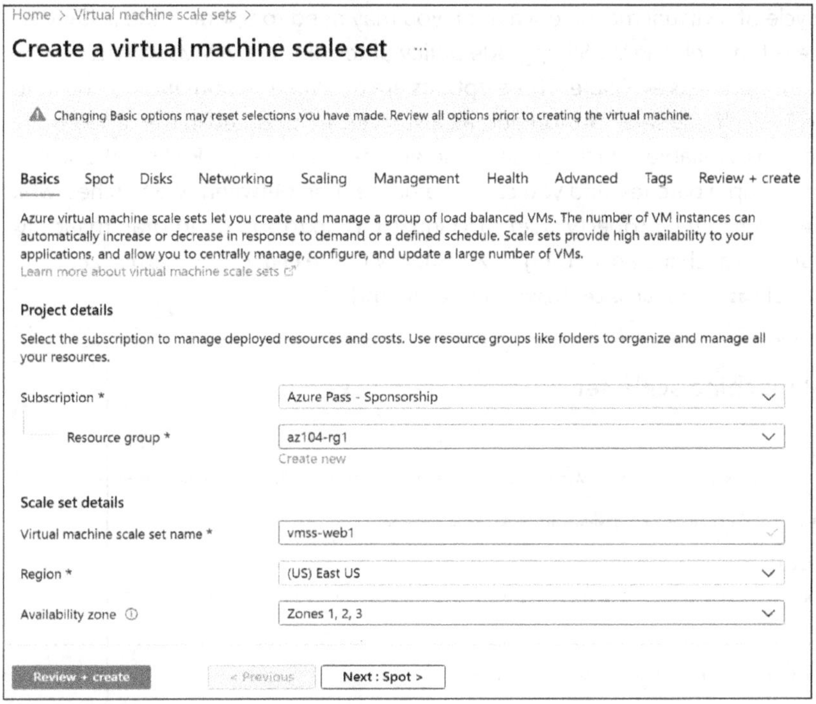

# Create a virtual machine scale set ···

⚠ Changing Basic options may reset selections you have made. Review all options prior to creating the virtual machine.

**Basics**  Spot  Disks  Networking  Scaling  Management  Health  Advanced  Tags  Review + create

Azure virtual machine scale sets let you create and manage a group of load balanced VMs. The number of VM instances can automatically increase or decrease in response to demand or a defined schedule. Scale sets provide high availability to your applications, and allow you to centrally manage, configure, and update a large number of VMs.
Learn more about virtual machine scale sets ⧉

### Project details

Select the subscription to manage deployed resources and costs. Use resource groups like folders to organize and manage all your resources.

| | |
|---|---|
| Subscription * | Azure Pass - Sponsorship ⌄ |
| Resource group * | az104-rg1 ⌄ |
| | Create new |

### Scale set details

| | |
|---|---|
| Virtual machine scale set name * | vmss-web1 |
| Region * | (US) East US ⌄ |
| Availability zone ⓘ | Zones 1, 2, 3 ⌄ |

[ Review + create ]  [ < Previous ]  [ Next : Spot > ]

**FIGURE 3-33** Creating a VM scale set

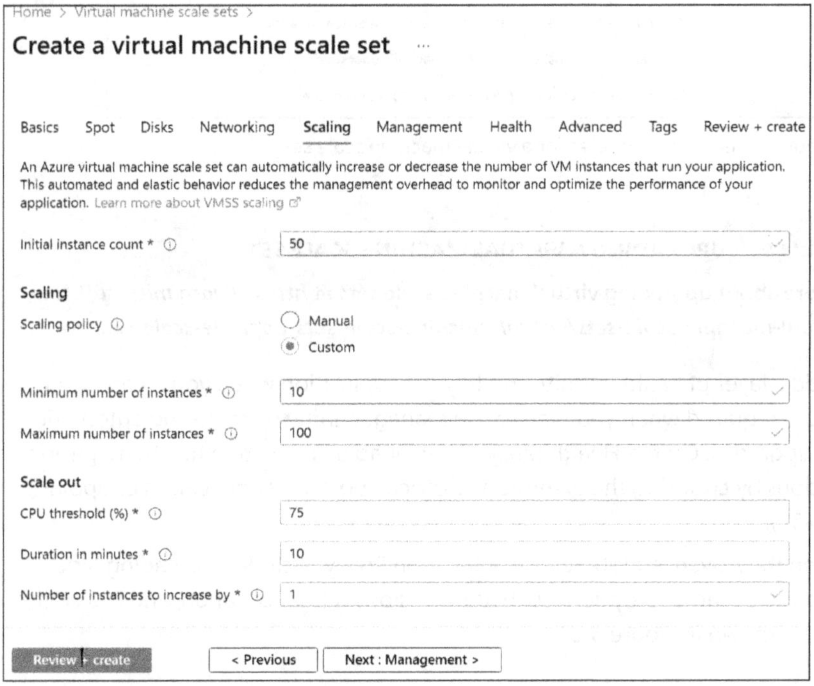

---

# Create a virtual machine scale set ···

Basics  Spot  Disks  Networking  **Scaling**  Management  Health  Advanced  Tags  Review + create

An Azure virtual machine scale set can automatically increase or decrease the number of VM instances that run your application. This automated and elastic behavior reduces the management overhead to monitor and optimize the performance of your application. Learn more about VMSS scaling ⧉

| | |
|---|---|
| Initial instance count * ⓘ | 50 |

### Scaling

| | |
|---|---|
| Scaling policy ⓘ | ◯ Manual |
| | ◉ Custom |
| Minimum number of instances * ⓘ | 10 |
| Maximum number of instances * ⓘ | 100 |

### Scale out

| | |
|---|---|
| CPU threshold (%) * ⓘ | 75 |
| Duration in minutes * ⓘ | 10 |
| Number of instances to increase by * ⓘ | 1 |

[ Review + create ]  [ < Previous ]  [ Next : Management > ]

**FIGURE 3-34** Configuring scaling rules for a virtual machine scale set

During the lifecycle of a virtual machine scale set, you may need to upgrade the instances with the latest scale set model. The VMSS upgrade policy determines how VMs will be upgraded once a new update is available. Three options are available: Automatic, Rolling, and Manual (see Figure 3-35). If you select Automatic, all instances are updated in the random order when an update is available, which can cause downtime. If you select Rolling, the scale set updates VMs in multiple batches, and you can set a pause time between two batches, which can avoid total downtime. If the property is set to Manual, it is up to you to programmatically step through and update each instance using PowerShell with the `Update-AzVmssInstance` cmdlet or the Azure CLI `az vmss update-instances` command.

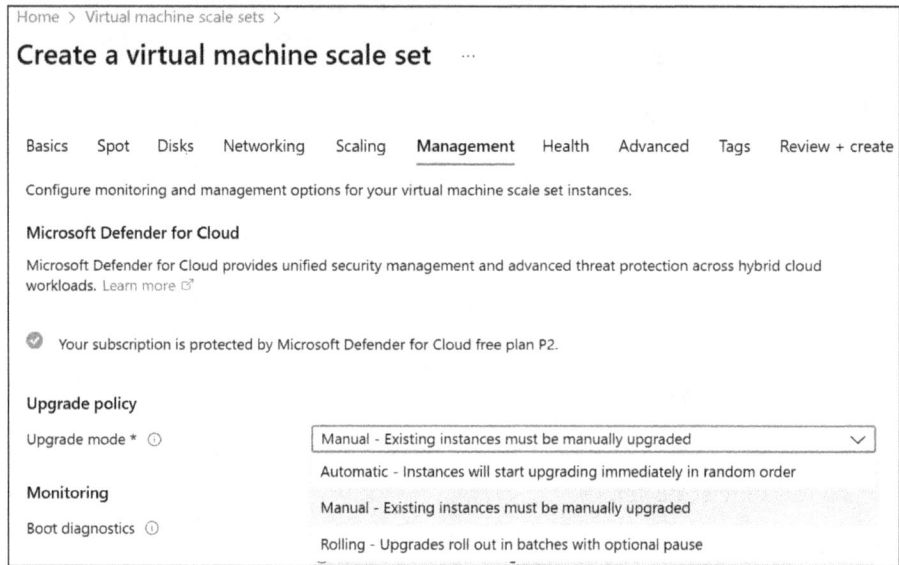

**FIGURE 3-35** Configuring management rules for a virtual machine scale set

> *NEED MORE REVIEW?* **UPGRADING A VIRTUAL MACHINE SCALE SET**
>
> You can learn more about upgrading virtual machine scale sets at *https://learn.microsoft.com/ en-us/azure/virtual-machine-scale-sets/virtual-machine-scale-sets-upgrade-scale-set*.

You can also add a layer of health monitoring to your application when you create VMSS. Health monitoring is required when you plan to use managed infrastructure and automatic operating system upgrades. On the Health tab, you can enable application health monitoring and configure options by choosing the extension, protocol, port, and application endpoint path (see Figure 3-36).

Some options on the Advanced tab, such as Allocation Policy, include a spreading algorithm. Also, you can select among options such as Extensions, VM application, and Proximity Placement Group, as shown in Figure 3-37.

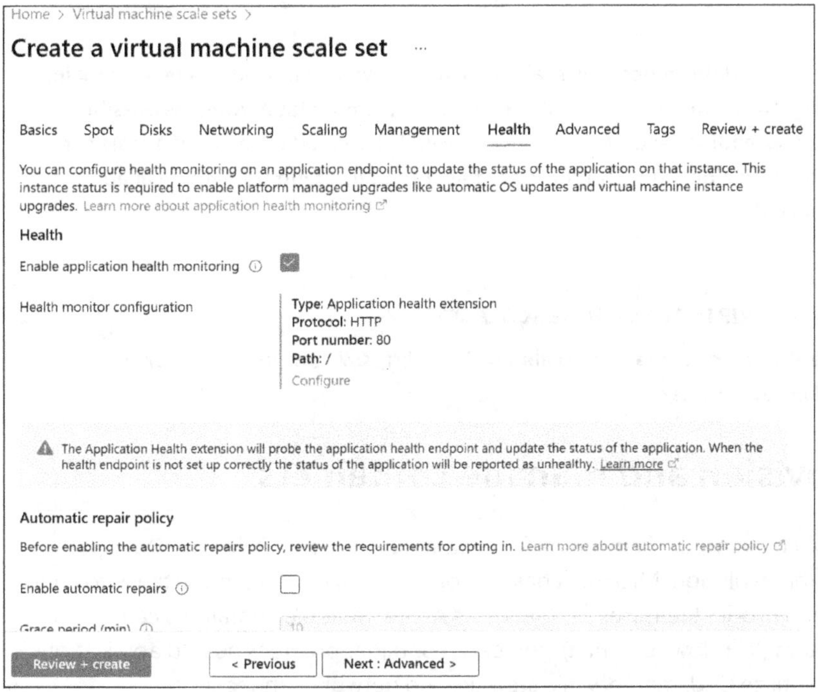

Home > Virtual machine scale sets >

**Create a virtual machine scale set** ...

Basics  Spot  Disks  Networking  Scaling  Management  **Health**  Advanced  Tags  Review + create

You can configure health monitoring on an application endpoint to update the status of the application on that instance. This instance status is required to enable platform managed upgrades like automatic OS updates and virtual machine instance upgrades. Learn more about application health monitoring ⧉

**Health**

Enable application health monitoring ⓘ ☑

Health monitor configuration

> **Type:** Application health extension
> **Protocol:** HTTP
> **Port number:** 80
> **Path:** /
> Configure

⚠ The Application Health extension will probe the application health endpoint and update the status of the application. When the health endpoint is not set up correctly the status of the application will be reported as unhealthy. Learn more ⧉

**Automatic repair policy**

Before enabling the automatic repairs policy, review the requirements for opting in. Learn more about automatic repair policy ⧉

Enable automatic repairs ⓘ ☐

Grace period (min) ⓘ  10

Review + create    < Previous    Next : Advanced >

**FIGURE 3-36**  Configuring health monitoring for a virtual machine scale set

Home > Virtual machine scale sets >

**Create a virtual machine scale set** ...

Basics  Spot  Disks  Networking  Scaling  Management  Health  **Advanced**  Tags  Review + create

Add additional configuration, agents, scripts or applications via virtual machine extensions or cloud-init.

**Allocation policy**

Enable scaling beyond 100 instances ⓘ ☑

Force strictly even balance across zones ☐
ⓘ

Spreading algorithm ⓘ    ⦿ Max spreading
                          ○ Fixed spreading

**Extensions**

Extensions provide post-deployment configuration and automation. Add new features, like configuration management or antivirus protection, to your virtual machine using extensions.

+ Add

Click 'Add' to get started with VM extensions.

**VM applications**

VM applications contain application files that are securely and reliably downloaded on your VM after deployment. In addition to

Review + create    < Previous    Next : Tags >

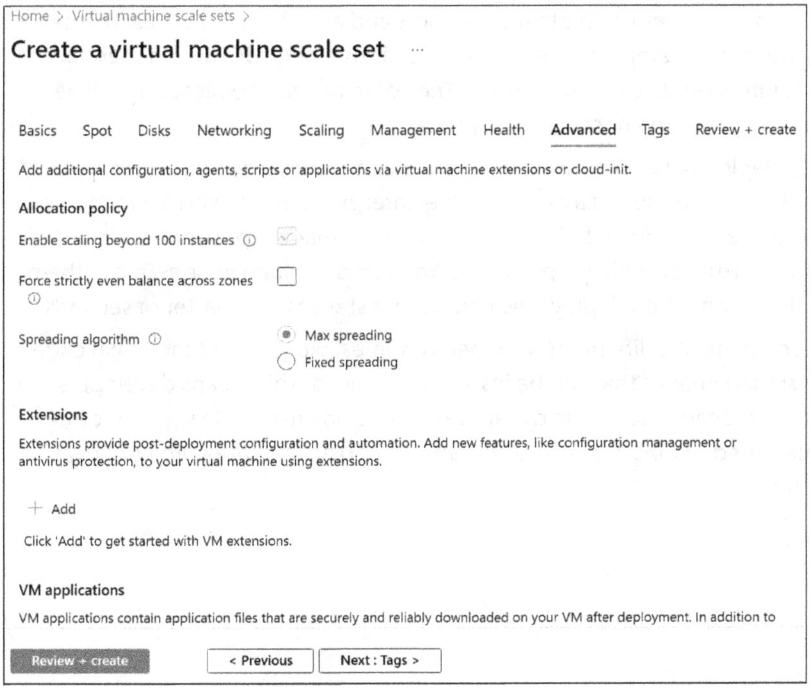

**FIGURE 3-37**  Configuring advanced rules for a virtual machine scale set

**EXAM TIP**

The spreading algorithm determines how scale set instances will be placed in a fault domain. With max spreading, the instances are distributed in the maximum fault domains possible for each zone. Fixed spreading restricts instances to exactly five fault domains. If a scale set is using a fixed spreading algorithm and there are fewer than five fault domains available, the deployment will fail.

**NEED MORE REVIEW?** **VIRTUAL MACHINE SCALE SETS**

You can learn more about virtual machine scale sets here: *https://learn.microsoft.com/en-us/azure/virtual-machine-scale-sets/.*

# Skill 3.3: Provision and manage containers

Today, containers are considered a first choice for instant deployments following DevOps methodology. With this evolution, Microsoft has also given equal or more importance to containers while designing product features for Microsoft Azure. Multiple containers can run on a host operating system (Windows or Linux), and each container provides hosted applications with a mechanism to utilize CPU, memory, file storage, and network connections.

Although containers do not have their own operating systems, so they can be deployed and run anywhere, most distributions include most user-mode operating system components. They share the host operating system with other containers and are isolated from each other. Containers are widely used to develop and package applications along with their dependencies and configurations into a single container image. The containerized applications will be deployed as container images on the host operating system.

By using containers, deployment times are shortened, maintenance is easier, and scale-out/in is possible in seconds. This is the sole reason why large enterprises are slowly moving to container technology. Containers can be built from your private image repository using Azure Container Registry, Docker Hub, or similar repositories. You can download images, make them available to container hosts, and then deploy new container instances in a matter of seconds.

Developers can even further modify them with their runtimes, plugins, and their own code, creating their own versioned images that can be instantly deployed. This means developers are more productive and can manage their code dynamically to deploy new code versions quickly. This can be further integrated to CI/CD tolls to automate the end-to-end deployment flow for containerized applications.

# Create and manage an Azure Container Registry

An Azure Container Registry (ACR) is a container image repository based on Docker Registry 2.0. You can create an ACR resource instance in your Azure subscription to store and manage container images and related components. An ACR can be used with your existing container pipelines, or you can use ACR Tasks to build new container images directly in Azure.

When you deploy an ACR to your subscription, there are three SKU options to choose from: Basic, Standard, and Premium. Table 3-1 outlines some of the primary differences between tiers.

**TABLE 3-1** Azure Container Registry tiers

| Feature | Basic | Standard | Premium |
|---|---|---|---|
| Included storage | 10 GiB | 100 GiB | 500 GiB |
| Webhooks | 2 | 10 | 500 |
| Download bandwidth | 30 Mbps | 60 Mbps | 100 Mbps |
| Geo-replication | Not supported | Not supported | Supported |
| Availability zones | Not supported | Not supported | Supported |
| Customer-managed keys | Not supported | Not supported | Supported |

> **NEED MORE REVIEW?** **AZURE CONTAINER REGISTRY TIERS**
>
> This is only a sample of the differences between ACR tiers. For a complete list of feature differences, visit *https://learn.microsoft.com/en-us/azure/container-registry/container-registry-skus*.

# Create an Azure container registry

You can use any of the available tools to deploy an ACR instance: the Azure portal, CLI, PowerShell, or REST API. You can also automate the deployment by using ARM templates or Bicep files. To create an instance from the Azure portal, follow these steps:

1. From the Azure portal, search for **Container registries**.

2. On the Container Registries blade, click Create.

3. On the Basics tab of the Create Container Registry blade, complete the fields and then click Next. The fields on the Basics tab, as shown in Figure 3-38, include

   - **Subscription**   The Azure subscription to deploy the resource to.
   - **Resource Group**   The logical resource group for the resource.
   - **Registry Name**   The name of the resource, which must be globally unique across all Azure customers and regions, and can only contain 5–50 alphanumeric characters.
   - **Location**   The Azure region to deploy the resource to.
   - **Use Availability Zones**   Whether the ACR is zone-redundant, only available with Premium tier.
   - **Pricing Plan**   The tier of service: Basic, Standard, or Premium.

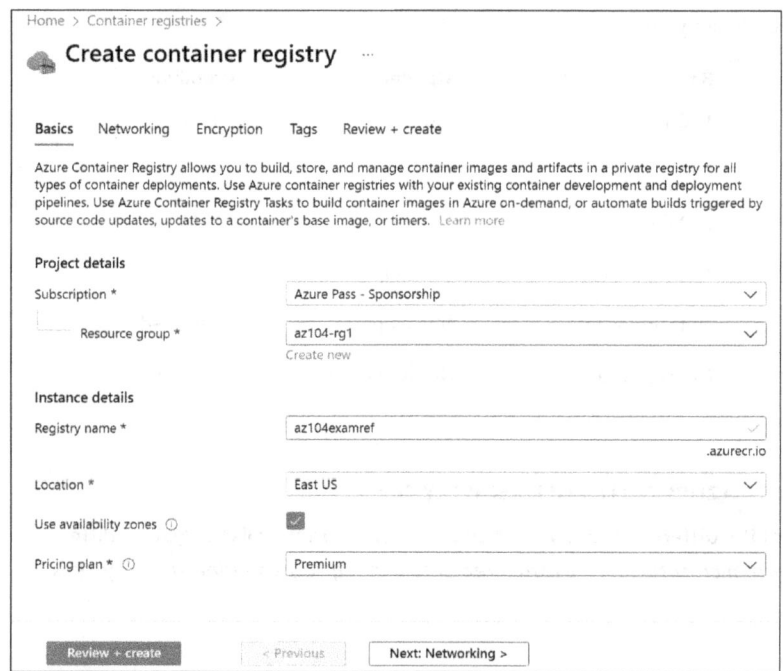

**FIGURE 3-38**   The Basics tab of the Create Container Registry blade

4. On the Networking tab, choose either Public Access or Private Access for the registry. Private access requires creating a private endpoint connection for the ACR but is only available for the Premium tier. Figure 3-39 displays the Networking tab.

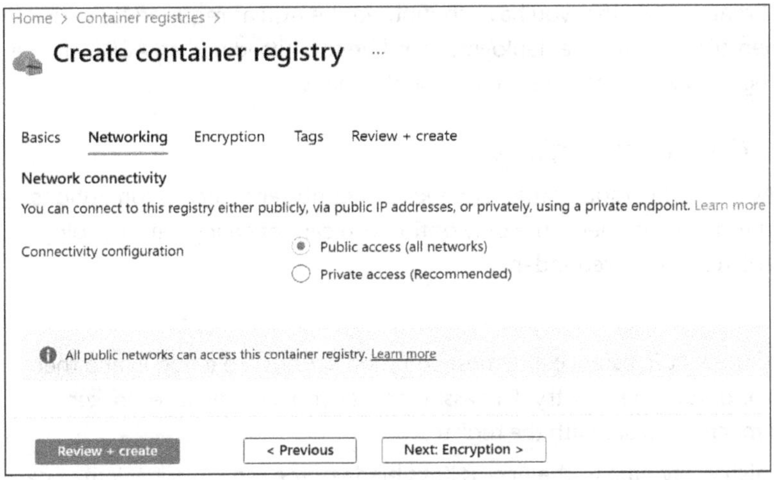

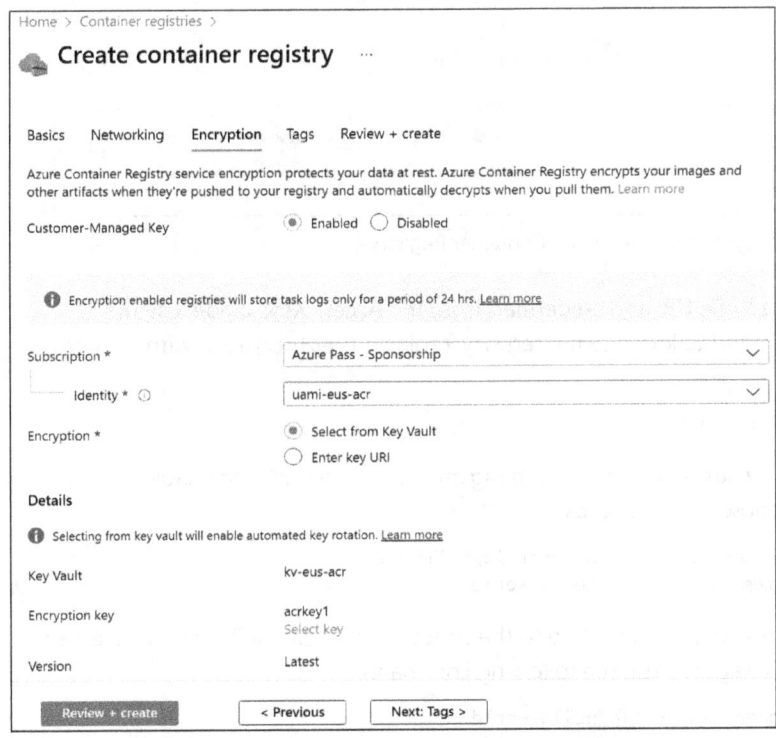

FIGURE 3-39   The Networking tab of the Create Container Registry blade

5. On the Encryption tab, choose whether to use a customer-managed encryption key or to accept the default Microsoft-managed platform key. Selecting Enabled for the Customer-Managed Key requires that a key be created in an Azure Key Vault and a managed identity has been created and given permissions to access the key from the vault. Figure 3-40 displays the Encryption tab with these settings selected.

FIGURE 3-40   The Encryption tab of the Create Container Registry blade

6. Click Review + Create. Verify that you have configured the ACR as required for your scenario and then click Create. The deployment will be submitted and an ACR instance created in the region that you selected during deployment.

## Manage an Azure Container Registry

After you deploy the ACR, you can modify the networking, identity, and encryption options that were configured during deployment. The only option you cannot change after deployment is whether the registry is zone-redundant.

### GETTING STARTED

To get started with your new ACR instance, the most common actions are to log in and then push or pull images to and from the registry. This assumes that you already have Docker installed on your local machine to use with the registry.

To log in to your registry, navigate to the Access Keys blade of the registry. Check the box next to Admin User to activate the admin account for the registry; it will provide two passwords to use with the account. The Access Keys blade is shown in Figure 3-41.

**FIGURE 3-41**  The Access Keys blade of an Azure Container Registry

After you obtain the login URI and credentials from the Access Keys blade, use the following Docker command to log in to the registry, replacing `registryname` with the name of your resource:

```
docker login registryname.azurecr.io
```

After you successfully authenticate, you can tag an image and push the image to your repository using the following commands:

```
docker tag hello-world registryname.azurecr.io/hello-world
docker push registryname.azurecr.io/hello-world
```

Finally, after the image has been pushed to the container, you can pull it when necessary. To pull an image from a registry, run the following command:

```
docker pull registryname.azurecr.io/hello-world
```

## WEBHOOKS

Webhooks can be used to trigger specific events when an action has occurred in the repository. This could be for all repository items or for specific tags. If you have enabled geo-replication to another Azure region, you can also configure the webhook to respond to specific region-based events.

To create a webhook from the Azure portal, navigate to the container registry and then select Webhooks. On the Webhooks blade, click Add. The configuration items for a webhook include

- **Webhook Name** The alphanumeric name of the webhook, including 5–50 characters.
- **Location** The Azure region where the webhook is deployed. If using a replica, select the region of the replica instance.
- **Service URI** The URI where the webhook sends POST notifications.
- **Custom Headers** Any custom header that should be included with the POST notification.
- **Trigger Actions** The actions on the repository that should trigger the webhook. These can include push and delete for both images and Helm charts.
- **Active Status** Whether the webhook is enabled or not.
- **Scope** Specific tags that the webhook should apply to or blank to apply to all items in the repository.

Figure 3-42 shows a sample webhook being created.

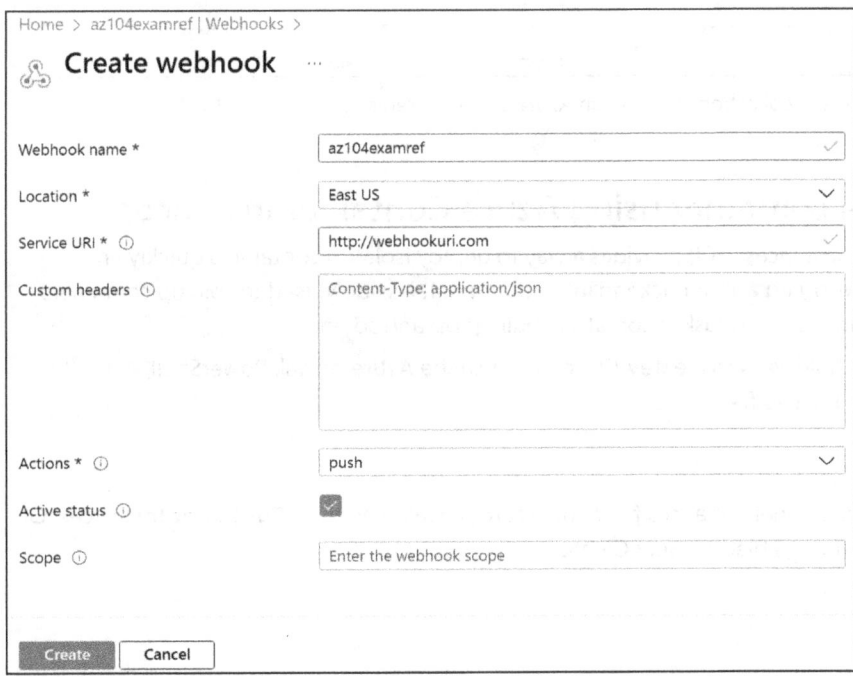

**FIGURE 3-42** The Create Webhook blade for an Azure container registry

### GEO-REPLICATIONS

If you have business or technical requirements that the images in the repository be replicated to another Azure region, you must deploy the ACR in the Premium tier. This will allow you to select additional Azure regions to replicate the data to.

To enable replication, navigate to the ACR instance in your subscription and select Geo-replications. A world map with selectable icons will be displayed, as shown in Figure 3-43. Select a region icon on the map to open the Create Replication blade.

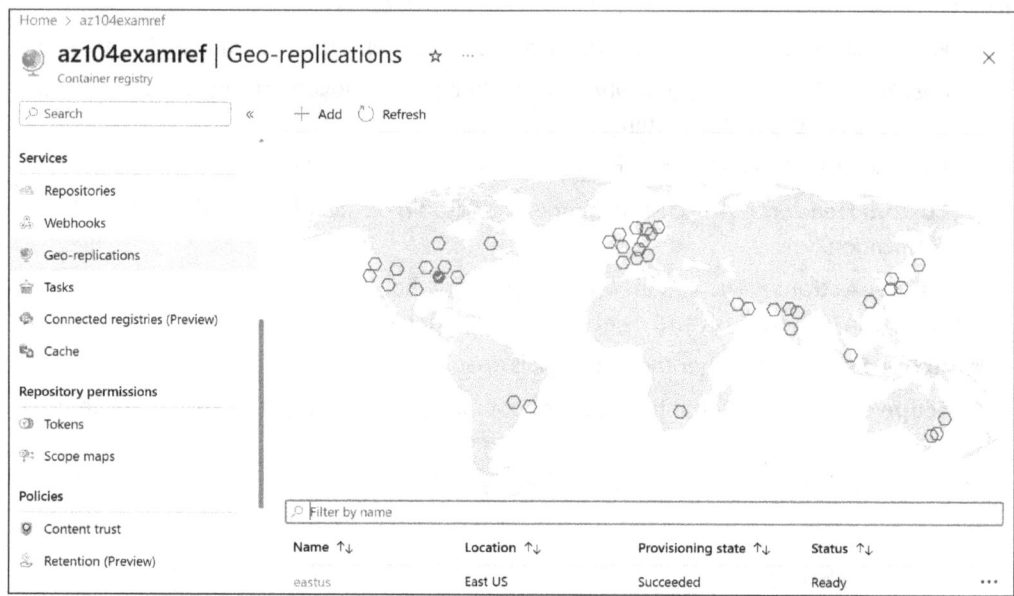

**FIGURE 3-43**  The Geo-replications blade of an Azure container registry

## Provision a container using Azure Container Instances

Azure Container Instances (ACI) provides a way to deploy isolated containers quickly and simply without worrying about backend infrastructure. It is widely used to spin up containers quickly for activities such as task automation, build jobs, and so on.

There are multiple ways to create ACI. You can use the Azure portal, PowerShell, Azure CLI, ARM templates, or Bicep files.

### Create ACI

To create Azure container instances from the Azure portal, search for **Container Instances**. On the Container Instances blade, select Create.

As shown in Figure 3-44, you can fill in the details, such as Subscription, Resource Group, Region, and so on. Choose the Image Source for your container instance. You can select the available image using QuickStart Images, Azure Container Registry (ACR), or Docker Hub registry.

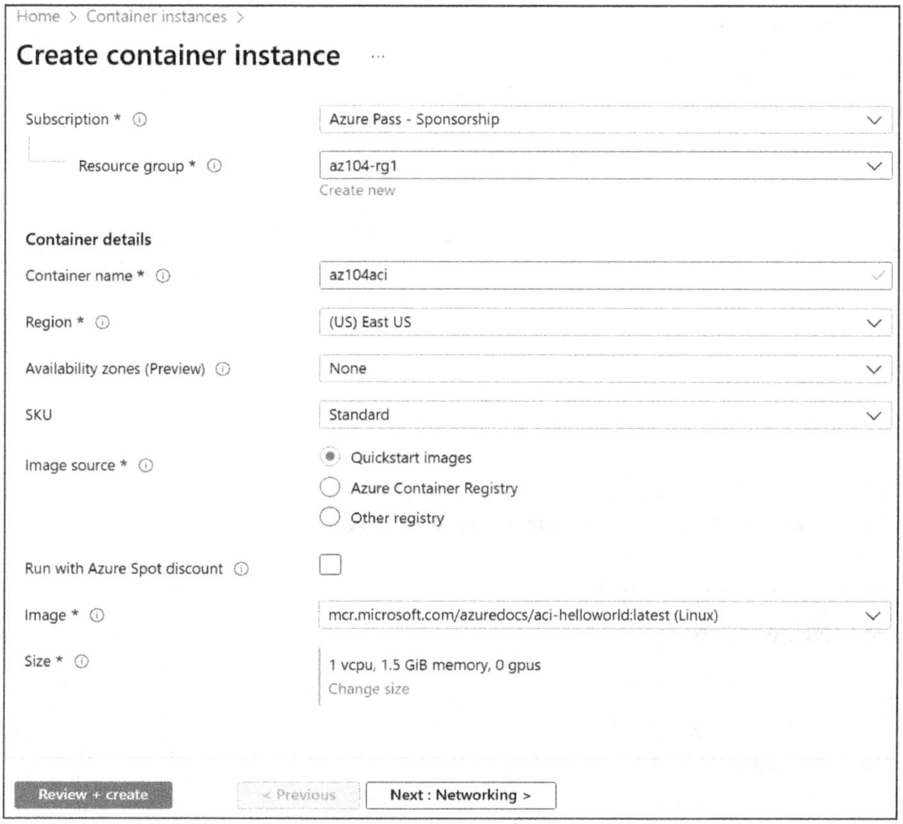

**FIGURE 3-44**   Basics tab of the Create Container Instance blade

On the Networking tab, set the Networking Type to Public, Private, or None (see Figure 3-45). If you choose Public, you will be asked to provide a DNS Name label for your public IP. You can also select the ports for your container instances.

On the Advanced tab, define the Restart Policy for your containers. The Restart Policy determines when the container would restart. You can choose On Failure, Always, or Never (see Figure 3-46). Choosing Always means your containers will restart irrespective of how they are stopped. Choosing On Failure means your containers restart only if they failed earlier, and choosing Never means your containers never restart.

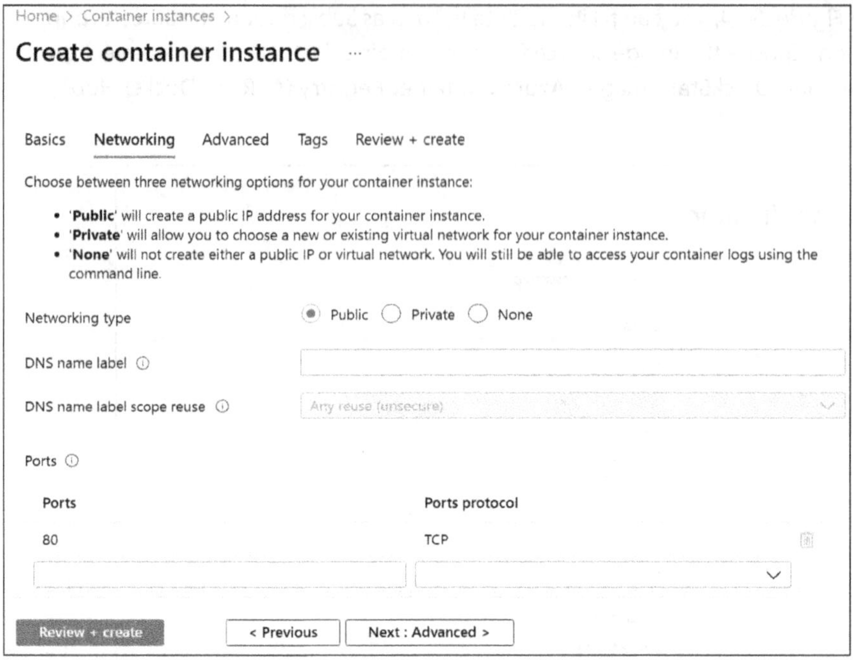

**FIGURE 3-45** Networking tab for the Create Container Instance blade

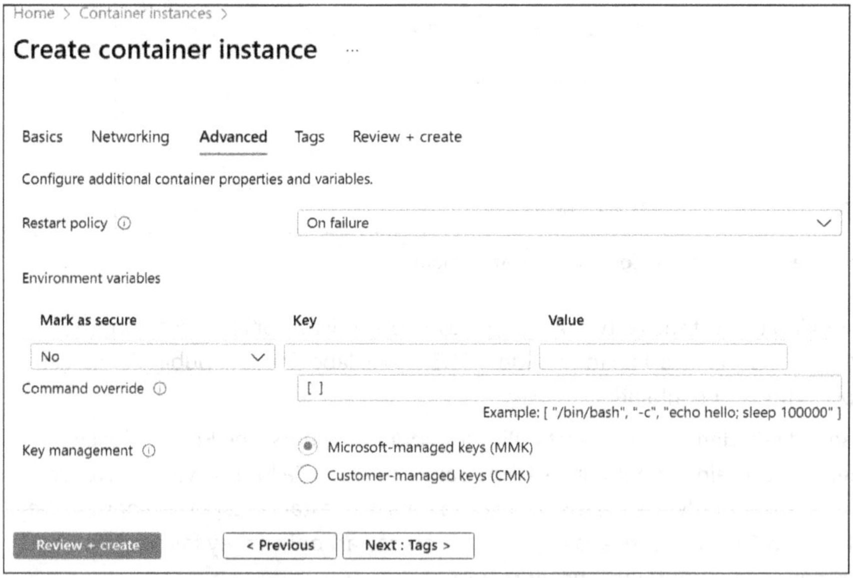

**FIGURE 3-46** Advanced tab of the Create Container Instance blade

Once the container instance is created, the Azure Container Instance overview page will look similar to the one shown in Figure 3-47.

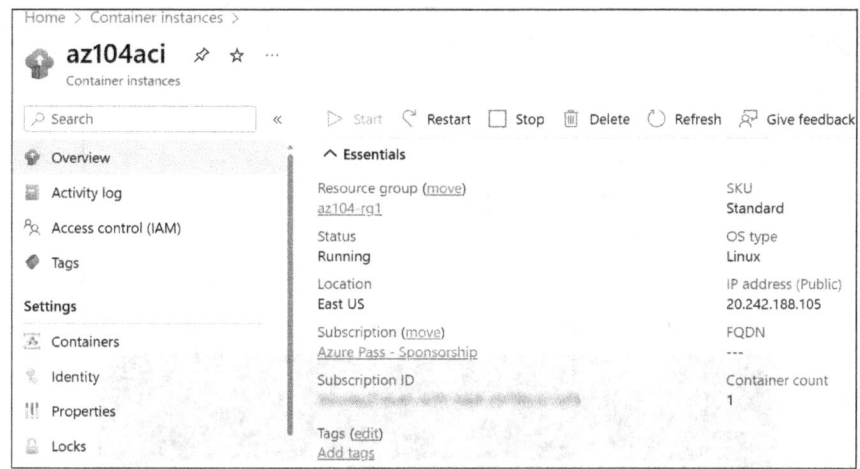

**FIGURE 3-47** Azure container instances in the Azure portal

## Connect to ACI

To review container events, properties, and logs, click Containers under Settings, as shown in Figure 3-48.

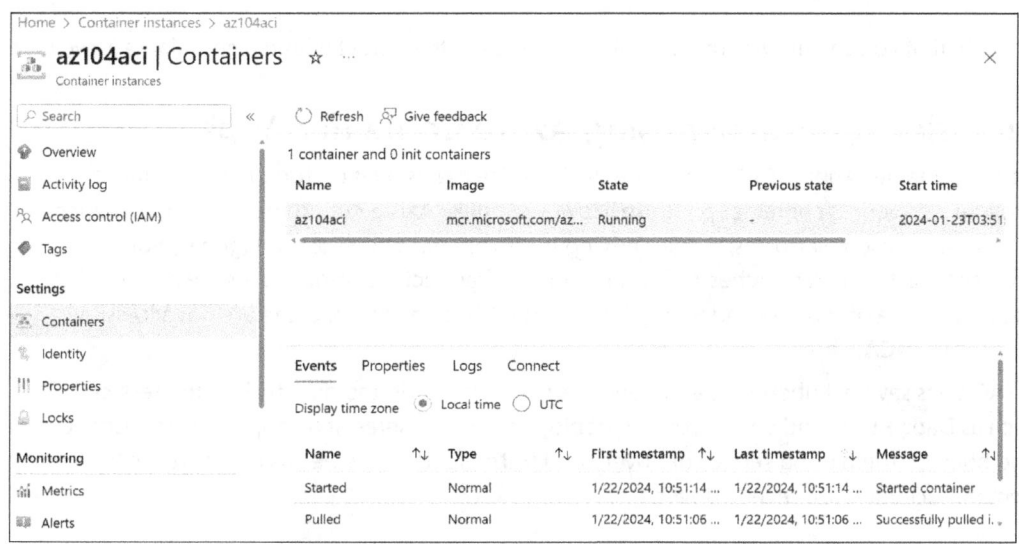

**FIGURE 3-48** Containers blade for ACI

You can connect to containers directly from the Azure portal, as shown in Figure 3-49. You can choose between shell environments (bash and sh) to start interacting with containers.

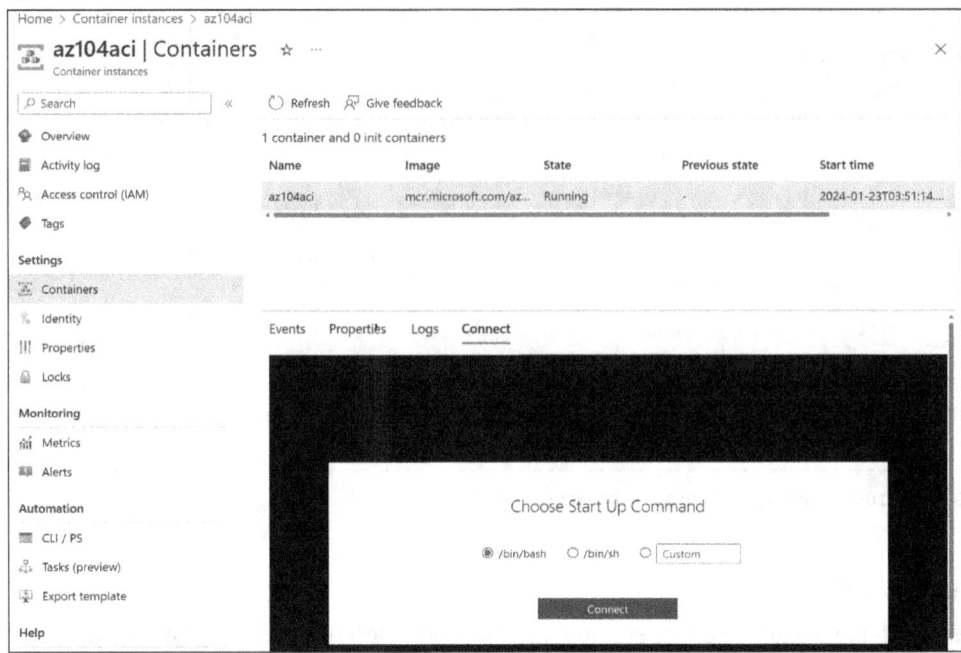

**FIGURE 3-49**  Azure container instances in the Azure portal

Note that you can use Azure PowerShell and the CLI to interact with the containers as well.

## Provision a container using Azure Container Apps

Azure Container Apps (ACA) is an Azure PaaS service that is built on top of Kubernetes to provide a serverless container platform. However, unlike Azure Kubernetes Services or other Kubernetes-based platforms, the underlying nodes are not visible or managed by you. As a PaaS service, the nodes, orchestration, and other infrastructure components are managed by Azure. Because of this, the underlying Kubernetes APIs and control plane are not accessible when using ACA.

ACA has several Kubernetes and open-source tools integrated directly into the service, such as Dapr, KEDA, and envoy. You can deploy your Kubernetes-style apps and microservices and build on them using service discovery and traffic splitting, as well as jobs that can be on demand, scheduled, or event-driven.

### Create an ACA instance

You can use any of the available tools to deploy an ACA instance, including the Azure portal, PowerShell, CLI, ARM Templates, and Bicep files. To create an ACA instance using the Azure portal, follow these steps:

1. From the Azure portal, search for **Azure Container Apps**.
2. On the Azure Container Apps blade, select Create.

3. On the Create Container App blade, complete the required fields:

- **Subscription** The Azure subscription to deploy the resource to.
- **Resource Group** The logical resource group that the resource should be deployed in.
- **Container App Name** The name of the ACA resource instance.
- **Region** The Azure region where the ACA instance is deployed.
- **Container Apps Environment** The managed compute environment for the ACA instance.

Figure 3-50 shows the Basics tab of the Create Container App blade containing these fields.

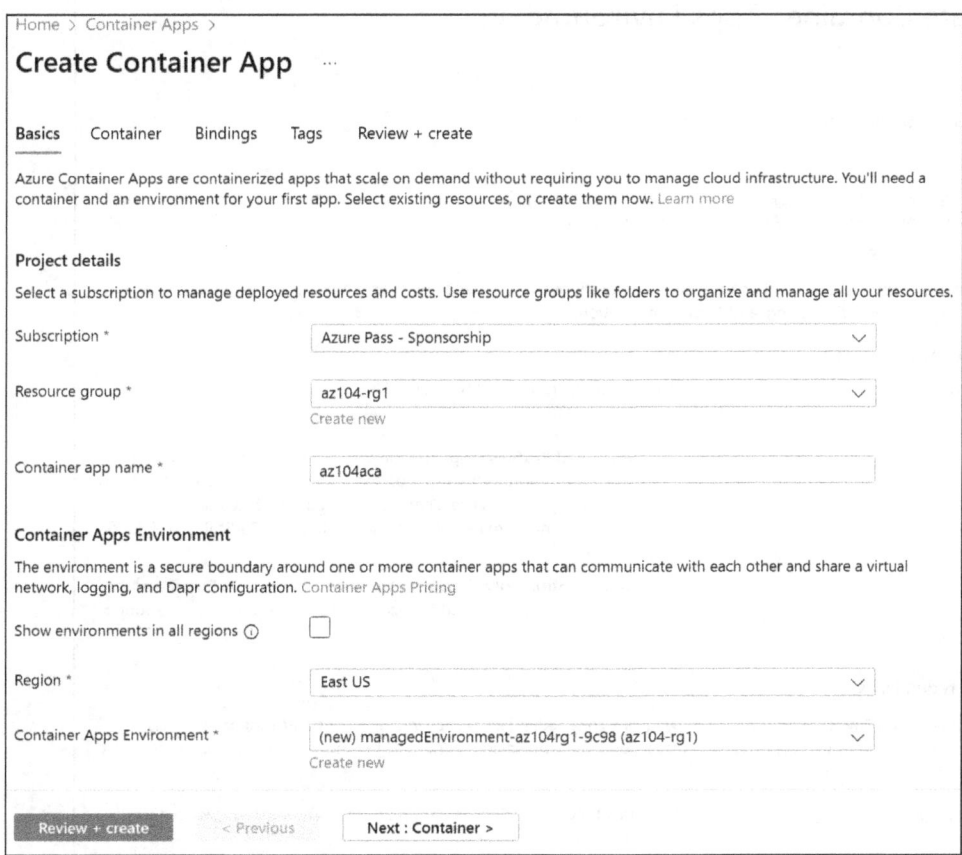

FIGURE 3-50 Create Container App blade

4. On the Basics tab, for Container Apps Environment, click Create New. The ACA instance requires an environment to run in. The environment defines the compute for the ACA instance, and whether the instance spans availability zones.

5. On the Create Container Apps Environment blade, the Environment Type and Zone Redundancy options can be configured. For Environment Type, choose Workload Profiles or Consumption Only.

- **Workload Profiles** Enables you to use both consumption-based compute, as well as defined hardware profiles that provides predictive costs and known performance.

- **Consumption Only** Provides a serverless app environment with scaling options where you pay only for what you use.

Figure 3-51 shows the Basics tab of the Create Container Apps Environment blade.

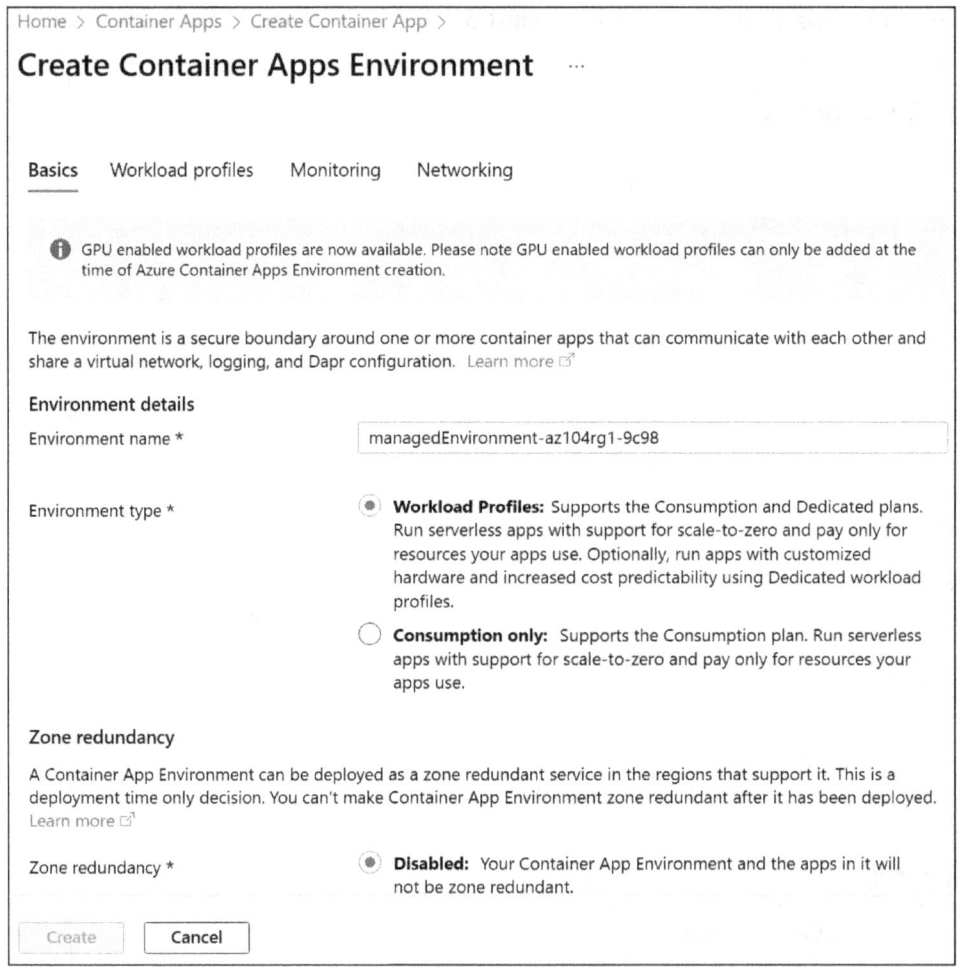

FIGURE 3-51   Create Container Apps Environment

6. On the Create Container Apps Environment blade, select the Workload Profiles tab, then click Add Workload Profile. Note that this tab is not available if you select Consumption Only on the Basics tab. Workload profiles are dedicated amounts of CPU and memory that get assigned to the ACA instance.

7. On the Add Workload Profile blade, select Choose A Size. Figure 3-52 displays the predefined workload profiles.

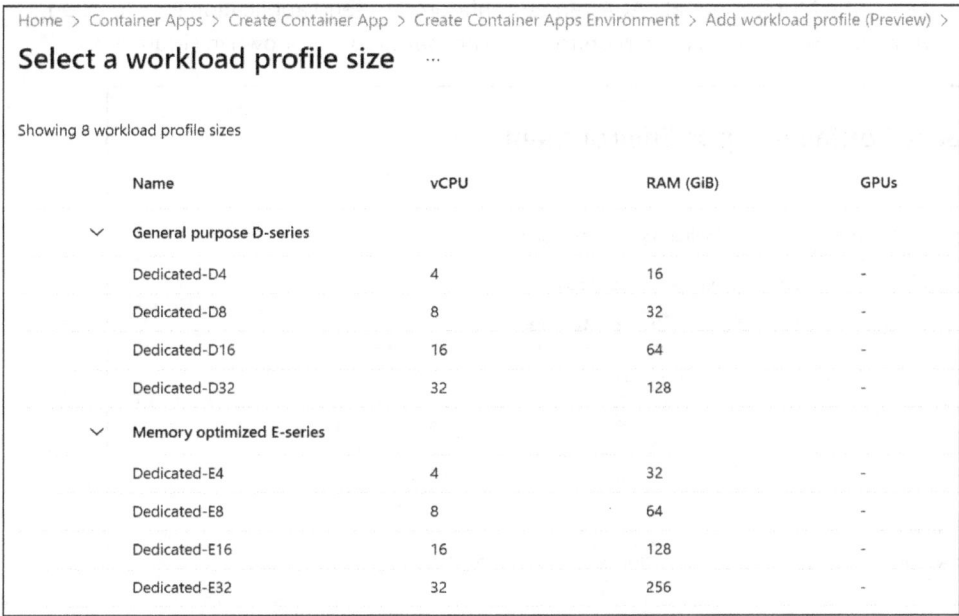

**FIGURE 3-52** Workload profile sizes

8. If necessary, adjust the autoscaling instances range for the workload profile. This can be adjusted from 0–20. However, setting the lower end of the range below 3 is not recommended for resiliency. Figure 3-53 displays the completed Add Workload Profile blade.

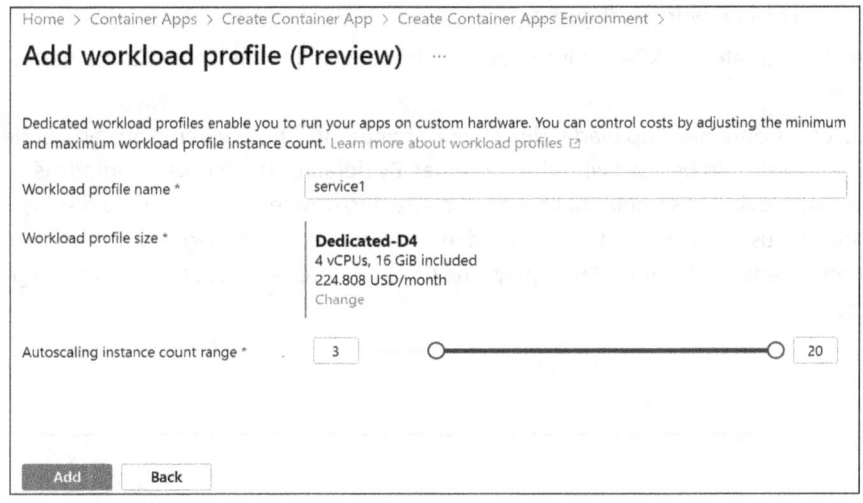

**FIGURE 3-53** Add Workload Profile

9. Click Add to add the profile to the container apps environment and be returned to the Create Container Apps Environment blade.

10. Select the Monitoring tab. ACA can be configured to send application logs to a Log Analytics workspace, Azure Monitor, or to not save logs, as shown in Figure 3-54.

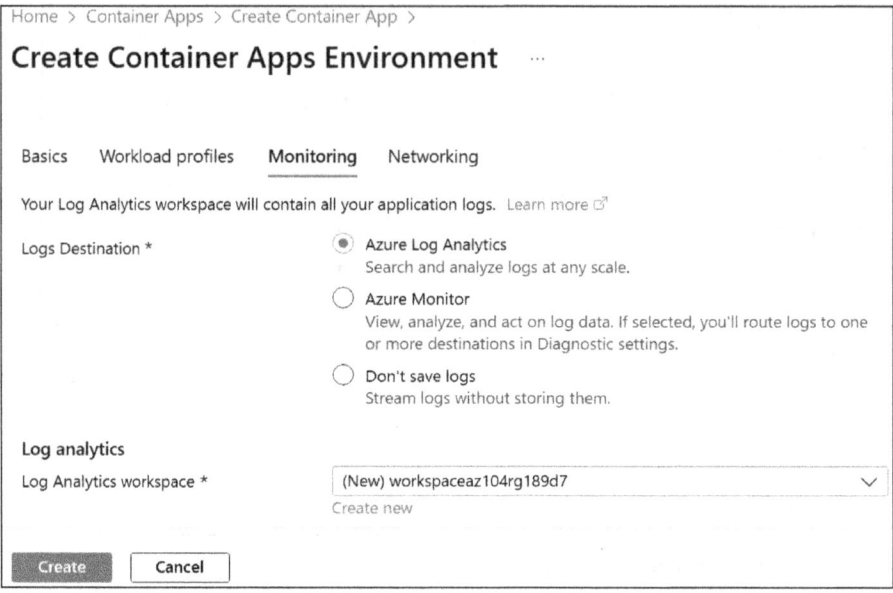

FIGURE 3-54  Monitoring tab for the Create Container Apps Environment blade

11. On the Create Container Apps Environment blade, click the Networking tab. You can configure the ACA instance either to be integrated with a virtual network, or to be a standalone instance with a public IP address. Figure 3-55 displays the Networking tab with a new virtual network configured.

12. Click Create to create the ACA environment and be returned to the Create Container App blade.

13. On the Create Container App blade, click the Container tab. This is where you can define the image that should be used with the container. By default, Use Quickstart Image is selected and provides a sample "hello world" container. Deselect this option to display the options for using an image that is stored in an Azure Container Registry, Docker Hub, or other registry. Figure 3-56 displays the Container tab with Use Quickstart Image deselected.

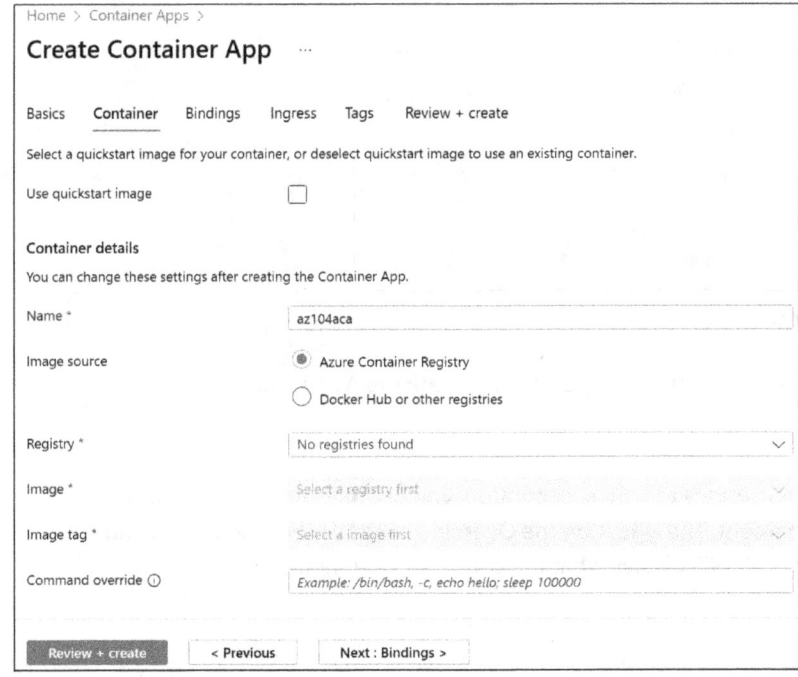

# Create Container Apps Environment   ···

Basics     Workload profiles     Monitoring     **Networking**

Selecting your own virtual network allows you to connect your application to other Azure resources or on-premises systems through the same network. Learn more ⬚

**Virtual network**

Use your own virtual network *     ○ No   ◉ Yes

Virtual network *     | (New) aca-vnet     ∨ |
Create new

Infrastructure subnet * ○     | (New) aca-infra     ∨ |
Create new

Virtual IP     ○ Internal: The endpoint is an internal load balancer

          ◉ External: Exposes the hosted apps on an internet-accessible IP address

Infrastructure resource group     | |

       ⚠ Azure recommends using the default name unless it does not conform with your company's Azure Policies. Learn more ⬚

[ Create ]     [ Cancel ]

**FIGURE 3-55** Container Apps Environment networking options

# Create Container App   ···

Basics     **Container**     Bindings     Ingress     Tags     Review + create

Select a quickstart image for your container, or deselect quickstart image to use an existing container.

Use quickstart image     ☐

**Container details**

You can change these settings after creating the Container App.

Name *     | az104aca |

Image source     ◉ Azure Container Registry

          ○ Docker Hub or other registries

Registry *     | No registries found     ∨ |

Image *     | Select a registry first     ∨ |

Image tag *     | Select a image first     ∨ |

Command override ⓘ     | Example: /bin/bash, -c, echo hello; sleep 100000 |

[ Review + create ]     [ < Previous ]     [ Next : Bindings > ]

**FIGURE 3-56** Containers

The Bindings tab can be used to connect add-ons to run in the same environment as your container. As of this writing, the available add-ons are shown in Figure 3-57 and include

- Redis
- MariaDB
- PostgreSQL
- Kafka
- Qdrant
- Milvus
- Weavitae

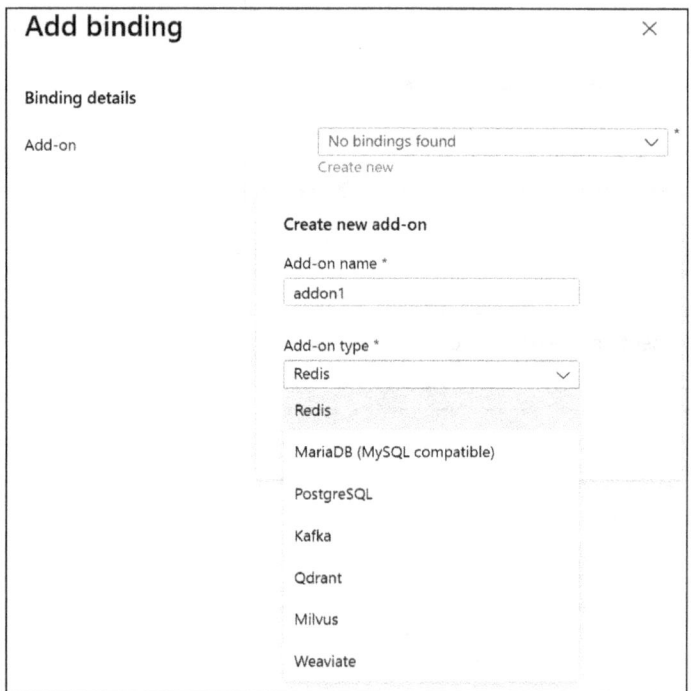

**FIGURE 3-57**   Bindings for containers

14.   Click Review + Create and then click Create to create the ACA instance.

## Connect to ACA

After deploying the ACA instance to your subscription, you will receive an Application URL for the container environment. The URL is on the Overview blade of the resource, as shown in Figure 3-58. Clicking the URL will display what the container is running.

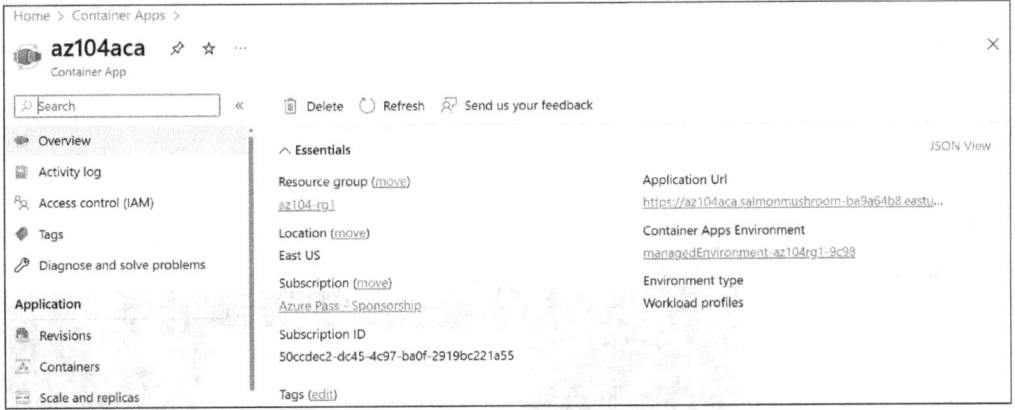

**FIGURE 3-58** Azure Container App overview

From the Azure portal and other tools, you can manage several aspects of the instance, including

- **Authentication**   Add a Microsoft or third-party identity provider to manage the authentication flow for the app.
- **Secrets**   Key/value pair secrets that the app needs access to within the container.
- **Ingress**   Determine whether ingress traffic is accepted by the container and the type and certificate mode to use if so.
- **Continuous Deployment**   Integrate the instance with GitHub and/or an Azure Container Registry for automatic provisioning.
- **Dapr**   Enable Dapr in the environment for service-to-service calls in your container environment.
- **Console**   Direct console access to a container using *bash* or *sh* shell options.
- **Log Stream**   View the log stream of the container directly from the Azure portal.

The ACA instance also supports common Azure components such as tagging, locks, metrics, alerts, diagnostic settings, and more.

To connect to the console of a container, from the ACA instance, select Console. You can choose to connect to a specific replica and container, as well as whether to use *sh* or *bash*, as shown in Figure 3-59.

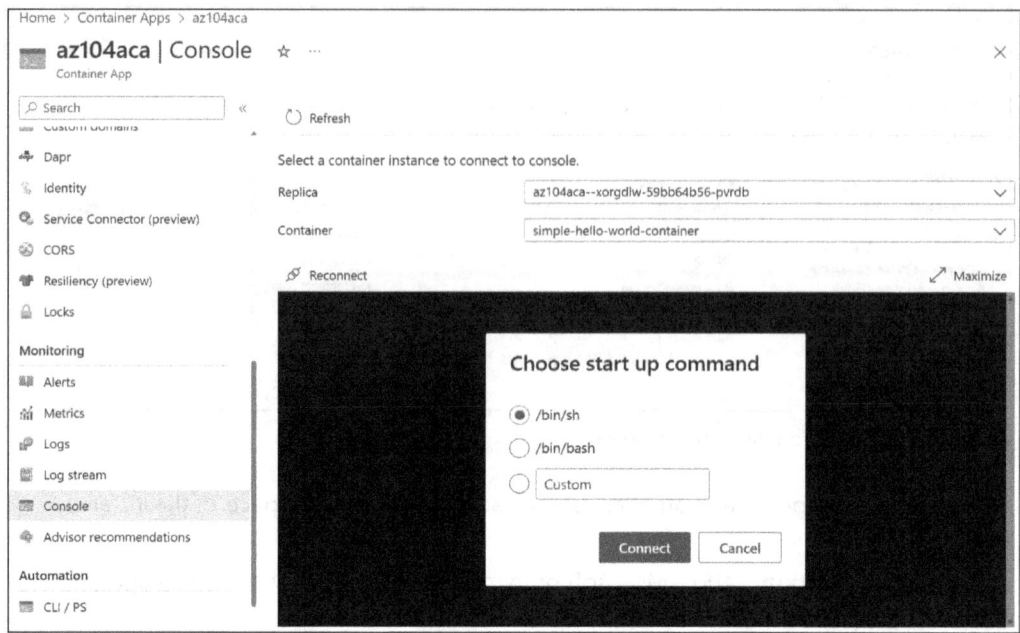

**FIGURE 3-59**  Azure container app console

## Manage sizing and scaling for containers

The sizing and scaling of your containers will depend on the service or platform you choose for deployment.

### Azure Container Instances

ACI does not provide any additional orchestration, scaling, or extensive add-ons that are available in other container environments. ACI is intended to be used for quick, lightweight container needs where you can easily create and deallocate containers.

Sizing an ACI instance occurs during deployment when you select the compute, memory, and GPU options. If you need more compute, simply deploy more instances using automation tools such as ARM templates or Bicep files. Figure 3-60 displays the size options for an ACI instance during deployment. The available values might depend on your subscription quota, Azure region, and container operating system type.

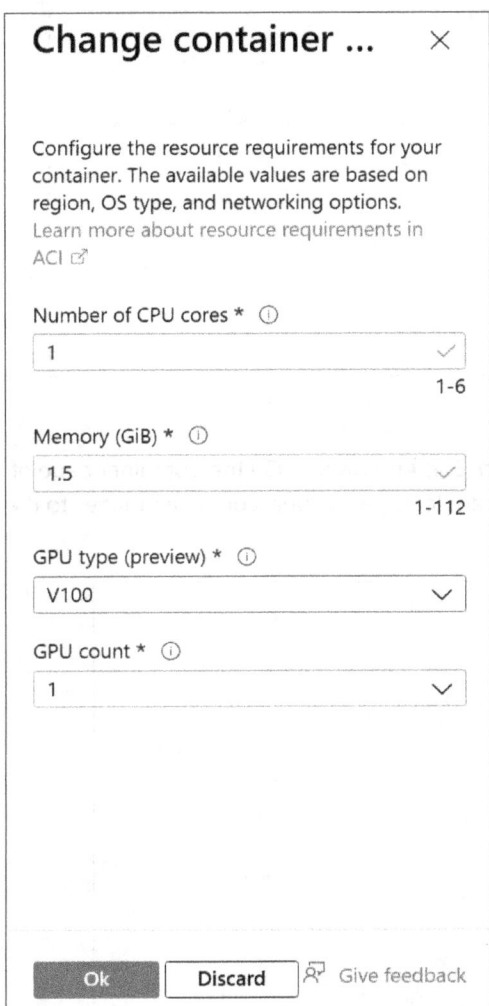

**Change container ...**  ✕

Configure the resource requirements for your container. The available values are based on region, OS type, and networking options. Learn more about resource requirements in ACI ☑

Number of CPU cores * ⓘ

| 1 | ✓ |

1-6

Memory (GiB) * ⓘ

| 1.5 | ✓ |

1-112

GPU type (preview) * ⓘ

| V100 | ⌄ |

GPU count * ⓘ

| 1 | ⌄ |

| Ok | Discard | 🗩 Give feedback |

**FIGURE 3-60** Azure container instance sizing

## Azure Container Apps

Compared to ACI, Azure Container Apps provides more scaling and sizing options. First, when creating the ACA environment, you choose between using a consumption-only model or both consumption and workload profiles. The consumption-only option bills only for what is used by the container app, while workload profiles require dedicated CPU and memory that is allocated to a container, so you pay for it even if it is not fully used.

After an ACA is deployed, you can scale the replica that a container is associated with to provide it with more resources. To configure this, navigate to the ACA instance, and then select the Scale And Replicas blade, as shown in Figure 3-61.

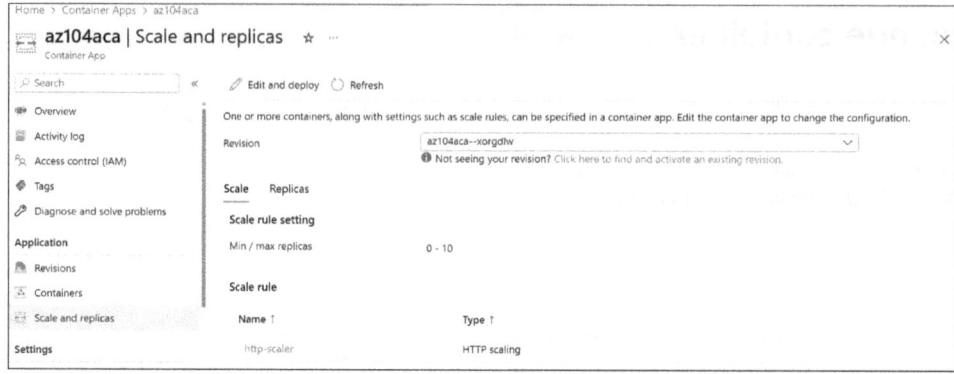

FIGURE 3-61   Azure Container App Scale And Replicas

To modify the scale settings of the replica, click Edit And Deploy. On the Container page of the Create And Deploy New Revision page, you can deploy additional container images to the replica, as shown in Figure 3-62.

Home > Container Apps > az104aca | Scale and replicas >

# Create and deploy new revision   …

**Container**   Scale   Secrets Volumes   Bindings

**Revision details**

| | |
|---|---|
| Based on revision * | az104aca--xorgdfw |
| Name / suffix | Enter name / suffix |
| Termination grace period (seconds) ⓘ | |
| Workload profile type | Consumption |

**Container image**

A revision needs one main app container. Add sidecar containers that support the main app container, or init containers that run before app containers are started.

✏ Edit   🗑 Delete   ➕ Add ⌄

| Name | Source | Type | Tag | CPU cores | Memory (... |
|---|---|---|---|---|---|
| simple-hello-w... | Azure Containe... | App container | latest | 0.25 | 0.5Gi |

Create   < Previous   **Next : Scale >**

FIGURE 3-62   Container tab of the Create And Deploy New Revision blade

When you add a new *app* or *init* container, part of the configuration options includes the CPU cores and memory that will be assigned to the container. These columns are also displayed in Figure 3-62.

To modify the minimum and maximum number of replicas used to power the containers, select the Scale tab. On the Scale tab, you can control the automatic scaling limits as well as additional scale rules to determine when or how the replica scales. The replicas can be set in a range from 0 to 300, as shown in Figure 3-63.

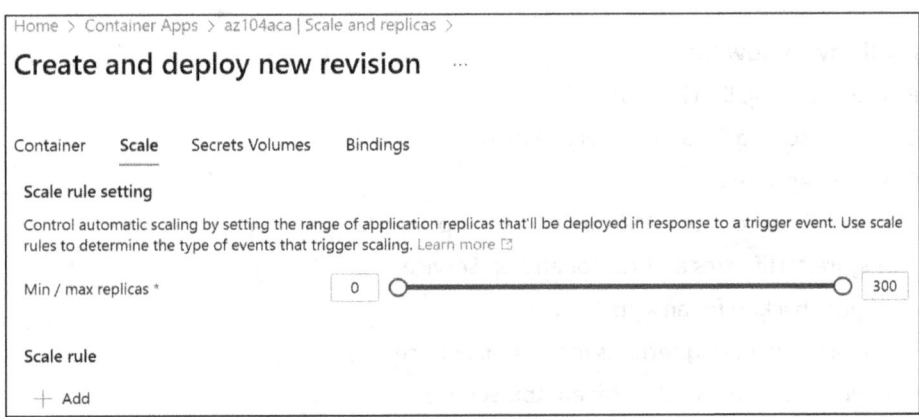

FIGURE 3-63 Scale tab of the Create And Deploy New Revision blade

To add scale rules, click Add. On the Add Scale Rule blade, you can create scaling rules based on the following options, as shown in Figure 3-64:

- **HTTP scaling** Scales the instance when the number of concurrent requests hits the defined threshold.
- **Azure Queue** Scales the instance when a queue reaches the defined length threshold.
- **Custom** Authenticate and configure additional metadata for a custom scale metric.

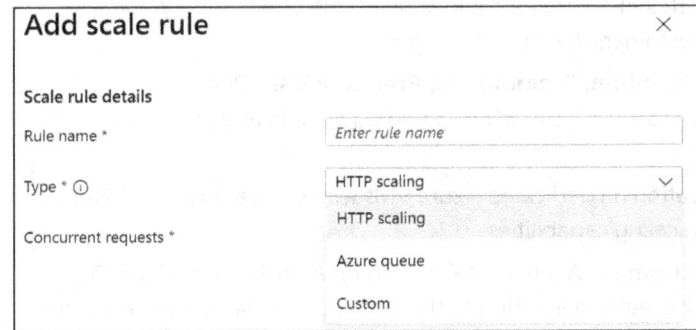

FIGURE 3-64 Add Scale Rule

# Skill 3.4: Create and configure Azure App Service

Azure App Service is a platform to develop an application in Azure without worrying about the required back-end infrastructure. You do not have to create, configure, and maintain a VM to host your applications if you are using a web app created with App Service. Moreover, Azure provides you with flexibility to choose the code language, such as ASP.NET, PHP, Node.js, Python, and so on. For creating and maintaining your code, you can still leverage common development environments, such as Visual Studio or GitHub.

---

**This skill covers how to:**

- Provision an App Service plan
- Configure scaling for an App Service plan
- Create an App Service
- Map an existing custom DNS name to an App Service
- Configure certificates and TLS for an App Service
- Configure backup for an App Service
- Configure networking settings for an App Service
- Configure deployment slots for an App Service

---

## Provision an App Service plan

When you deploy an App Service, it actually runs in an App Service plan. An App Service plan offers compute resources to the app for its execution. This App Service plan can be shared with multiple web apps, too.

App Service plans have a few pricing tiers that define the features, along with the cost for each one of them. Mainly, there are three major categories:

- **Dev/Test (Free, Shared, Basic)**  Runs an application on the same Azure VM as applications from other customers. It cannot scale out.
- **Production (Standard, Premium, Premium V2, Premium V3)**  Runs an application on allocated hardware. It can scale up and down or in and out based on the defined tier.
- **Isolated**  Runs an application on dedicated Azure VMs with dedicated Azure VNets. It provides the maximum scaling capabilities.

To create an App Service plan from the Azure portal, search for **App Service plans**. On the App Service Plans blade, click Create. In addition to the Subscription Name and Resource Group, you need to specify an App Service Plan Name along with the Operating System (see Figure 3-65). You also need to select an appropriate Pricing Plan (SKU) based on the requirements of the application.

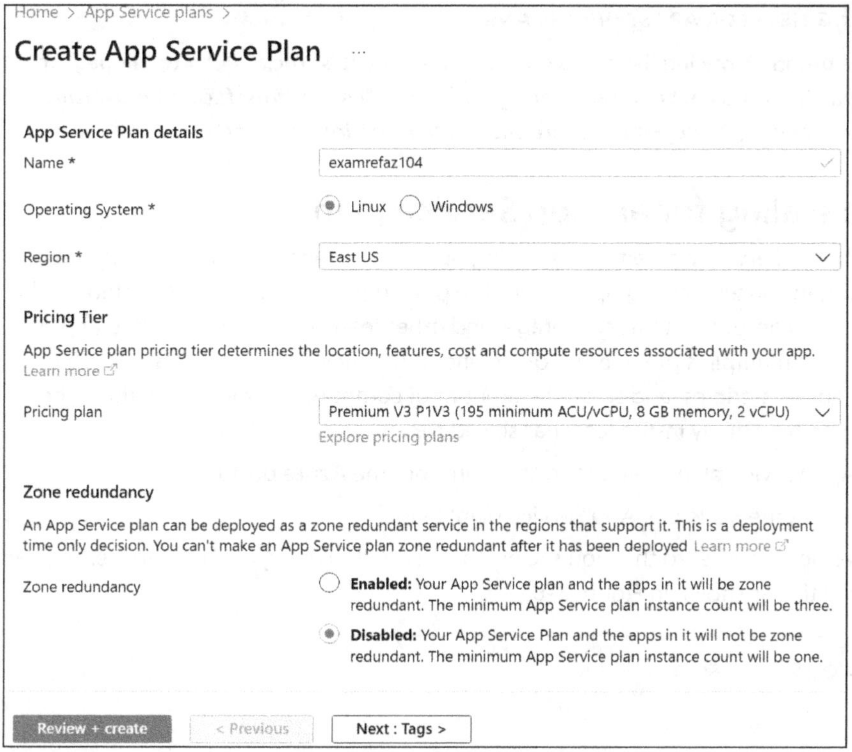

# Create App Service Plan  ⋯

**App Service Plan details**

Name *  | examrefaz104 | ✓

Operating System *  ◉ Linux  ◯ Windows

Region *  | East US | ⌄

**Pricing Tier**

App Service plan pricing tier determines the location, features, cost and compute resources associated with your app.
Learn more ⌕

Pricing plan  | Premium V3 P1V3 (195 minimum ACU/vCPU, 8 GB memory, 2 vCPU) | ⌄
Explore pricing plans

**Zone redundancy**

An App Service plan can be deployed as a zone redundant service in the regions that support it. This is a deployment time only decision. You can't make an App Service plan zone redundant after it has been deployed Learn more ⌕

Zone redundancy  ◯ **Enabled:** Your App Service plan and the apps in it will be zone redundant. The minimum App Service plan instance count will be three.

◉ **Disabled:** Your App Service Plan and the apps in it will not be zone redundant. The minimum App Service plan instance count will be one.

[ Review + create ]  [ < Previous ]  [ Next : Tags > ]

**FIGURE 3-65** Basics tab for the Create App Service Plan blade

Click Explore Pricing Plans to see pricing tiers, as shown in Figure 3-66. It is important to review the features and hardware provided before selecting the tiers.

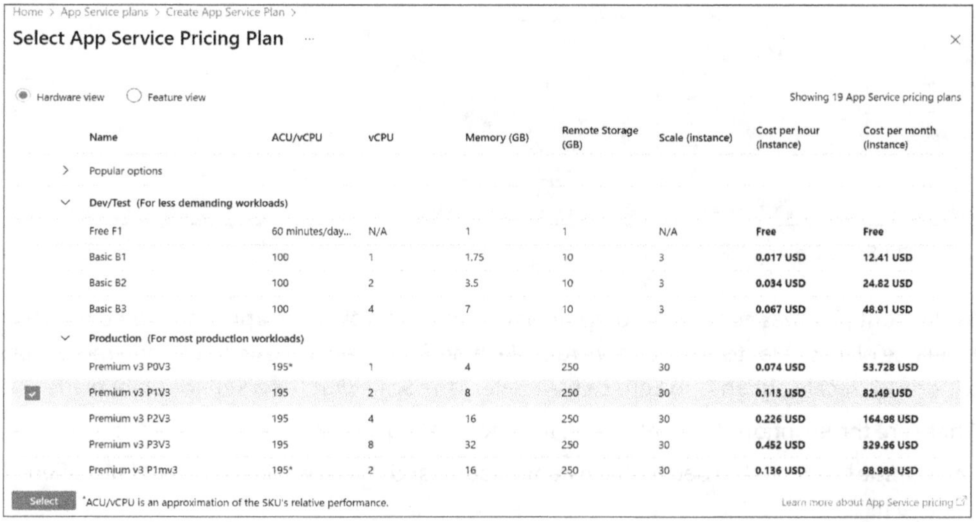

## Select App Service Pricing Plan  ⋯                                         ✕

◉ Hardware view    ◯ Feature view                                Showing 19 App Service pricing plans

| Name | ACU/vCPU | vCPU | Memory (GB) | Remote Storage (GB) | Scale (instance) | Cost per hour (instance) | Cost per month (instance) |
|------|----------|------|-------------|---------------------|------------------|--------------------------|---------------------------|
| > Popular options | | | | | | | |
| ∨ Dev/Test (For less demanding workloads) | | | | | | | |
| Free F1 | 60 minutes/day... | N/A | 1 | 1 | N/A | Free | Free |
| Basic B1 | 100 | 1 | 1.75 | 10 | 3 | 0.017 USD | 12.41 USD |
| Basic B2 | 100 | 2 | 3.5 | 10 | 3 | 0.034 USD | 24.82 USD |
| Basic B3 | 100 | 4 | 7 | 10 | 3 | 0.067 USD | 48.91 USD |
| ∨ Production (For most production workloads) | | | | | | | |
| Premium v3 P0V3 | 195* | 1 | 4 | 250 | 30 | 0.074 USD | 53.728 USD |
| ☑ Premium v3 P1V3 | 195 | 2 | 8 | 250 | 30 | 0.113 USD | 82.49 USD |
| Premium v3 P2V3 | 195 | 4 | 16 | 250 | 30 | 0.226 USD | 164.98 USD |
| Premium v3 P3V3 | 195 | 8 | 32 | 250 | 30 | 0.452 USD | 329.96 USD |
| Premium v3 P1mv3 | 195* | 2 | 16 | 250 | 30 | 0.136 USD | 98.988 USD |

[ Select ]  *ACU/vCPU is an approximation of the SKU's relative performance.          Learn more about App Service pricing ⌕

**FIGURE 3-66** Select App Service Pricing Plan

## Configure scaling for an App Service plan

You can scale an App Service plan either by scaling up or down the tier, or scaling out (or in) to more (or fewer) instances. Scaling up changes the pricing tier that you have selected for the plan, along with the compute, memory, storage, and other features that are included at the new tier. Scaling out multiplies the number of instances that the plan has associated with it. This effectively doubles, triples, and so on the amount of compute and memory behind the plan depending on how many instances it has scaled to.

To scale an App Service plan, navigate to the plan from the Azure portal.

To scale up, select the Scale Up (App Service plan) blade.

The user interface is similar to that you see when you are deploying an App Service. You can select the desired tier, as shown in Figure 3-67.

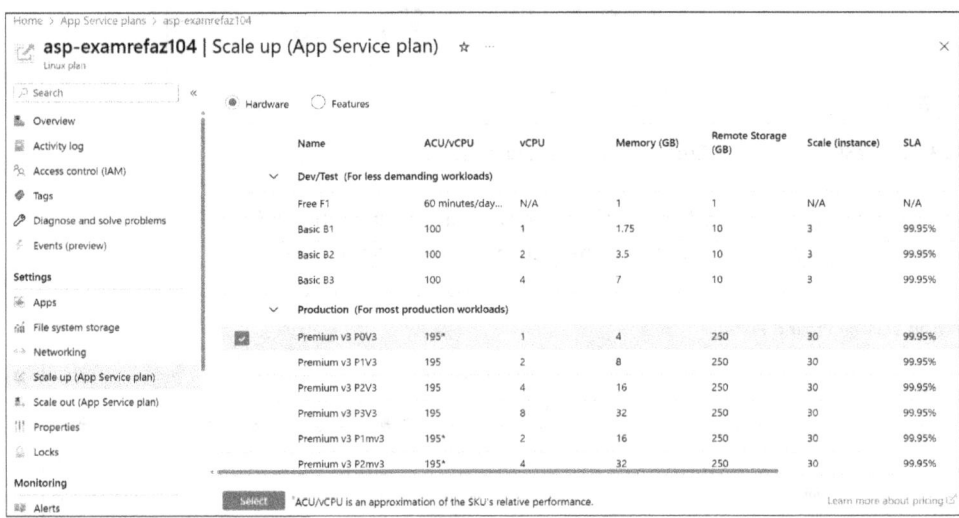

**FIGURE 3-67**    Scale Up App Service Plan

Scaling out provides additional compute and memory for the web apps associated with the App Service plan, similar to the way a Virtual Machine Scale Set can scale to provide more compute nodes. To scale out an App Service plan, select the Scale Out (App Service plan) blade.

There are three options for configuring instance scaling, as shown in Figure 3-68:

- **Manual**    Manually specify the instance count, and it will remain static until modified again.

- **Automatic**   Allow Azure to manage scaling based on traffic, but you can set a maximum instance count for cost management.
- **Rules Based**   Requires you to create rules to determine when and how to scale out (more capacity) or in (less capacity). Rules can be created based on common App Service metrics, such as bandwidth, requests, errors, and more.

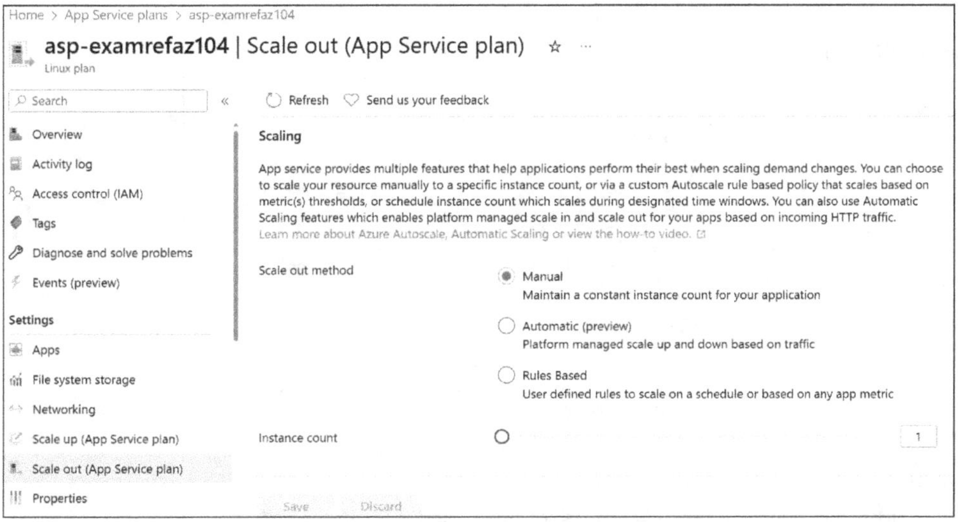

**FIGURE 3-68**   Scale Out App Service Plan

> **NOTE   SCALE OUT METHODS**
>
> As of this writing, the Automatic scale method is in preview. This topic might or might not be included on the exam, and services in preview could change before becoming generally available.

## Create an App Service

App Service is a hosting service that lets you build and deploy your applications to run on Windows, Linux, or your Docker container in a managed infrastructure. You can also choose the programming language with which you would like to build your code. App Service is currently supported on both Windows and Linux environments. Also, App Service can host your Docker containers running on either Linux or Windows, and it brings additional capabilities, such as autoscaling, load balancing, security, monitoring, and the like.

With App Service, it is easy to acquire DevOps capabilities, such as CI/CD integration with various platforms, such as Azure DevOps, GitHub, Docker Hub, or other sources. Additionally, App Service has extensive monitoring capabilities by using Application Insights and Azure Monitor.

To create a web app, from the Azure portal homepage, search for **App Services**. On the App Services page, select Create, Web App. In addition to selecting the Subscription Name and Resource Group, select Web App-Specific Details. You can choose either to publish your code or publish a Docker container, as shown in Figure 3-69. You will also need to select or create a new App Service plan, which is discussed earlier in this Skill.

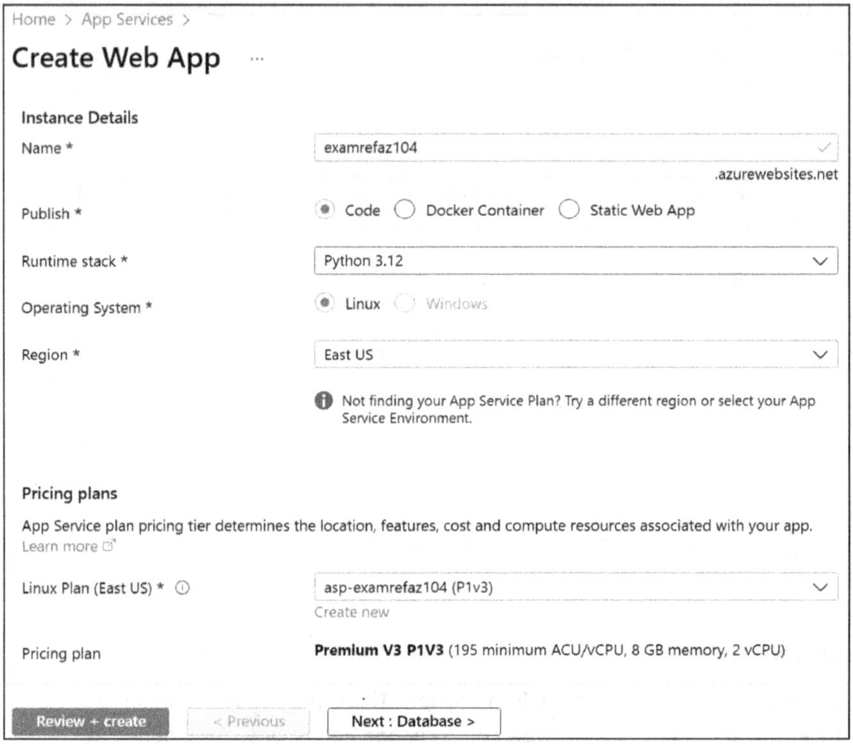

Home > App Services >

## Create Web App ...

**Instance Details**

Name *                                examrefaz104                                                        ✓

                                                                                               .azurewebsites.net

Publish *                     ● Code    ◯ Docker Container    ◯ Static Web App

Runtime stack *               Python 3.12                                                          ⌄

Operating System *          ● Linux   ◯ Windows

Region *                      East US                                                              ⌄

                            ℹ Not finding your App Service Plan? Try a different region or select your App
                              Service Environment.

**Pricing plans**

App Service plan pricing tier determines the location, features, cost and compute resources associated with your app.
Learn more ⌕

Linux Plan (East US) *  ℹ        asp-examrefaz104 (P1v3)                                             ⌄
                                 Create new

Pricing plan                    **Premium V3 P1V3** (195 minimum ACU/vCPU, 8 GB memory, 2 vCPU)

[ Review + create ]    [ < Previous ]    [ Next : Database > ]

**FIGURE 3-69**   Basics tab on the Create Web App blade

If you choose to publish your code, you need to select an option from the Runtime Stack drop-down menu and select the Operating System (either Windows or Linux) to run your application code. As of this writing, the supported runtime stacks for an App Service include

- .NET
- Java
- Node
- PHP
- Python

If you choose to publish a Docker Container, select the supported operating system (either Windows or Linux) to run your container, as shown in Figure 3-70. You will need to select the Container Image on the Docker page.

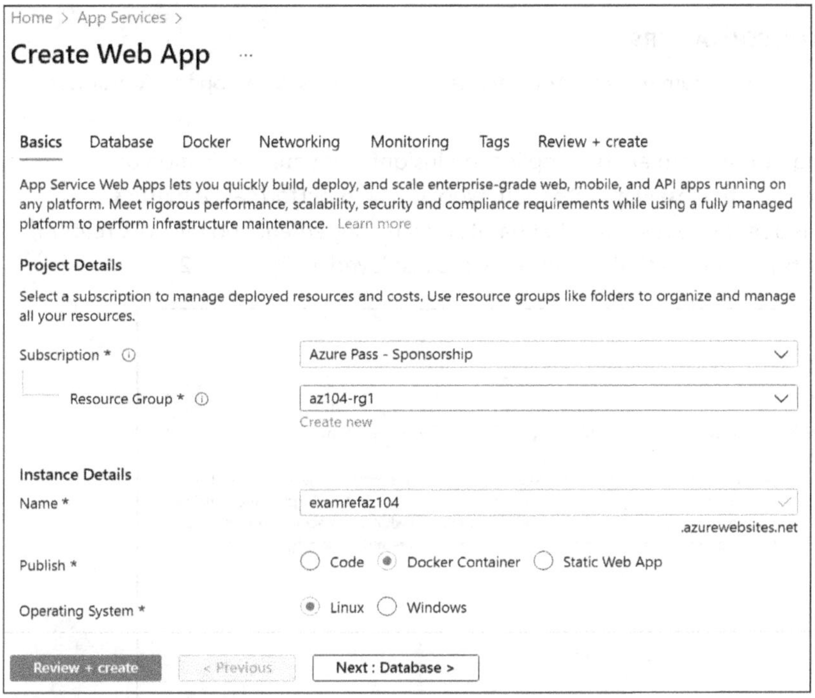

# Create Web App  ...

Basics    Database    Docker    Networking    Monitoring    Tags    Review + create

App Service Web Apps lets you quickly build, deploy, and scale enterprise-grade web, mobile, and API apps running on any platform. Meet rigorous performance, scalability, security and compliance requirements while using a fully managed platform to perform infrastructure maintenance. Learn more

### Project Details

Select a subscription to manage deployed resources and costs. Use resource groups like folders to organize and manage all your resources.

Subscription *  ⓘ        Azure Pass - Sponsorship                          ⌄

Resource Group *  ⓘ      az104-rg1                                         ⌄
                         Create new

### Instance Details

Name *                   examrefaz104                                      ⌄
                                                              .azurewebsites.net

Publish *                ○ Code   ◉ Docker Container   ○ Static Web App

Operating System *       ◉ Linux   ○ Windows

[ Review + create ]   [ < Previous ]   [ Next : Database > ]

**FIGURE 3-70**   Docker Container options for the Create Web App blade

You must choose either Single Container or Docker Compose (Preview). From the Image Source setting, you choose where to pull the images from (see Figure 3-71). Your choices are QuickStart Templates (only applicable for the Single Container option), Azure Container Registry, Docker Hub, or any other private registry.

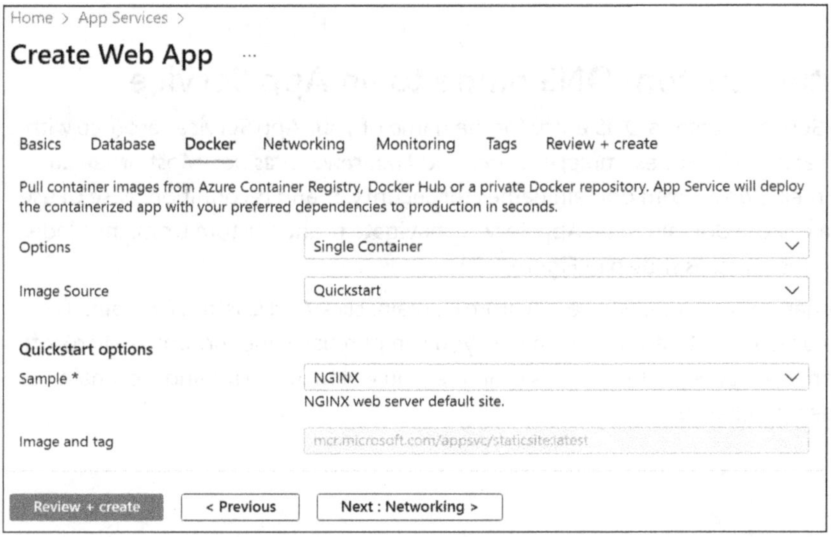

# Create Web App  ...

Basics    Database    **Docker**    Networking    Monitoring    Tags    Review + create

Pull container images from Azure Container Registry, Docker Hub or a private Docker repository. App Service will deploy the containerized app with your preferred dependencies to production in seconds.

Options                  Single Container                                  ⌄

Image Source             Quickstart                                        ⌄

### Quickstart options

Sample *                 NGINX                                             ⌄
                         NGINX web server default site.

Image and tag            mcr.microsoft.com/appsvc/staticsite:latest

[ Review + create ]   [ < Previous ]   [ Next : Networking > ]

**FIGURE 3-71**   Docker options for creating a web app

On the Monitoring tab, you can enable Application Insights for your application or containers. This allows you to closely monitor your web app with performance metrics. Application Insights is a separate resource that has its own management and billing. Creating an Application Insights resource with the App Service is displayed in Figure 3-72.

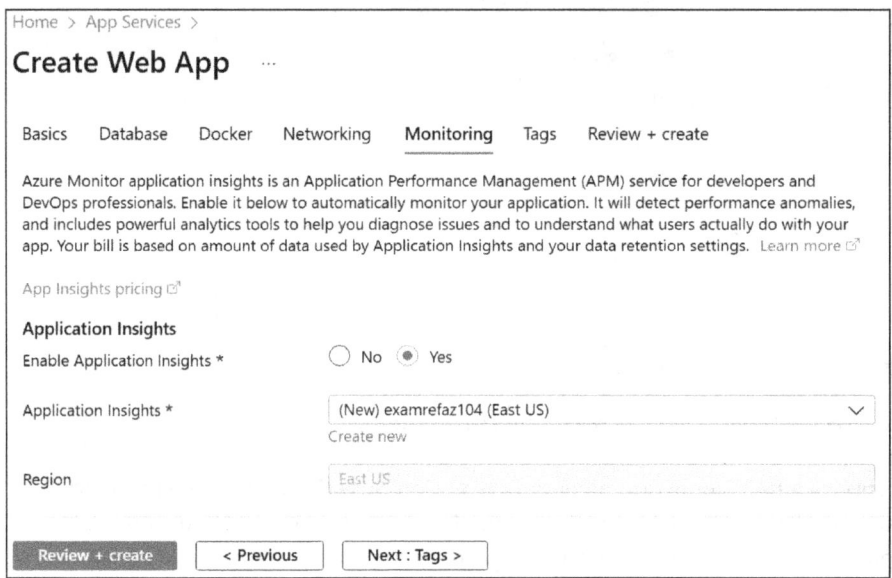

**FIGURE 3-72** Monitoring options for a web app

## Map an existing custom DNS name to an App Service

By default, an App Service creates a DNS entry for the name of your App Service resource with the *azurewebsites.net* domain. For example, *examrefaz104.azurewebsites.net*. Most organizations prefer to use their own custom domain names for security, brand recognition, ease of use, and more. To add a custom domain to an App Service, navigate to the Custom Domains blade of your App Service resource, as shown in Figure 3-73.

If you or your organization already have a custom domain, click Add Custom Domain. If you do not already own a registered domain name, you can purchase one through a Microsoft partner directly from the Azure portal. For this example, you will focus on adding a domain that has already been purchased.

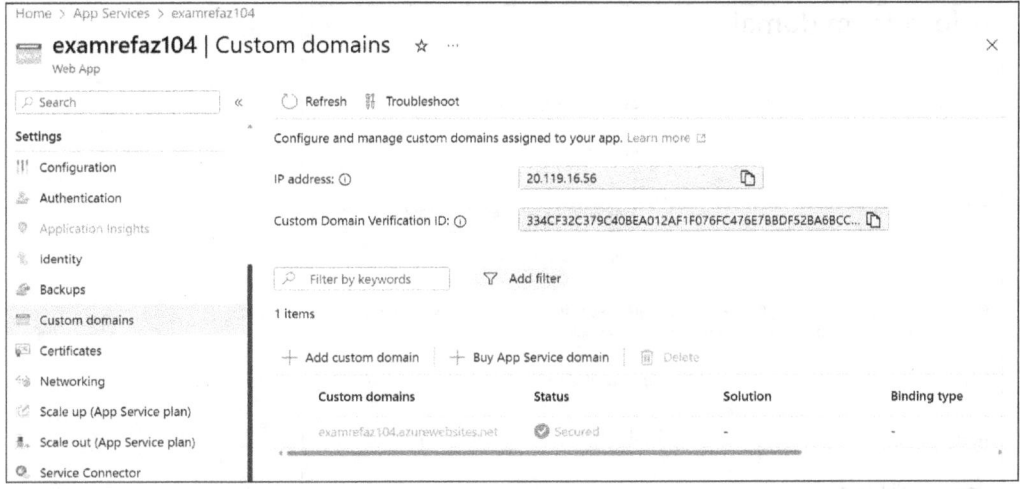

**examrefaz104 | Custom domains** ☆ ⋯
Web App

🔍 Search

**Settings**

Configuration

Authentication

Application Insights

Identity

Backups

Custom domains

Certificates

Networking

Scale up (App Service plan)

Scale out (App Service plan)

Service Connector

○ Refresh   ⫶ Troubleshoot

Configure and manage custom domains assigned to your app. Learn more ↗

IP address: ⓘ                               20.119.16.56

Custom Domain Verification ID: ⓘ       334CF32C379C40BEA012AF1F076FC476E7BBDF52BA6BCC...

🔍 Filter by keywords        ▽ Add filter

1 items

➕ Add custom domain   ➕ Buy App Service domain   🗑 Delete

| Custom domains | Status | Solution | Binding type |
|---|---|---|---|
| examrefaz104.azurewebsites.net | ✓ Secured | - | - |

**FIGURE 3-73**   App Service Custom Domains

To add an existing domain to an App Service, you first need to verify that you actually own the domain you are trying to add. This security measure prevents other App Services from trying to impersonate your web app.

The Add Custom Domain blade, shown in Figure 3-74, provides these options:

- **Domain Provider**   Specify whether you purchased the domain through the App Service or through a third party.
- **TLS/SSL Certificate**   Specify whether to use an App Service Managed Certificate or add a certificate later. We discuss certificates later in this skill.
- **Domain**   The custom domain you are adding to the App Service.
- **Domain Validation**   The DNS records required to validate that you own the domain.

To complete the domain validation, add the DNS records that are presented to your public DNS records. In the case shown in Figure 3-74, the App Service requires that a CNAME for app1. hugelab.net points to examrefaz104.azurewebsites.net. Additionally, a TXT record named asuid.app1 requires a specific key string be added for the text value. The type of records you are required to create depends on the domain, but they're typically CNAME, A, and TXT record types.

After you have added the DNS records, click Validate. The service will check the public DNS to verify you have added the records. If successful, the required rows will show checkmarks, and you will receive a note that it has passed validation, as shown in Figure 3-75.

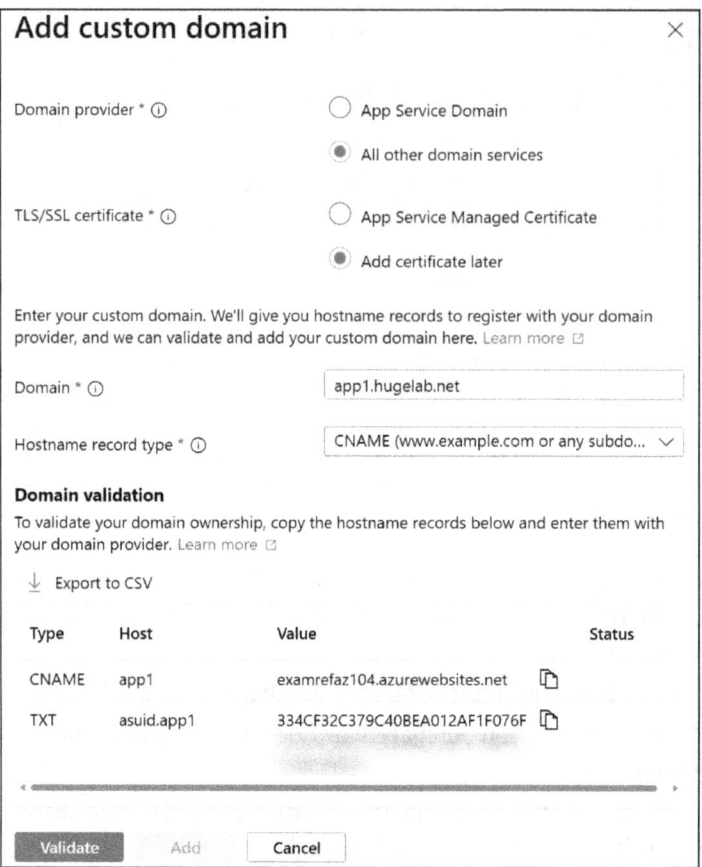

**FIGURE 3-74**   Add Custom Domain

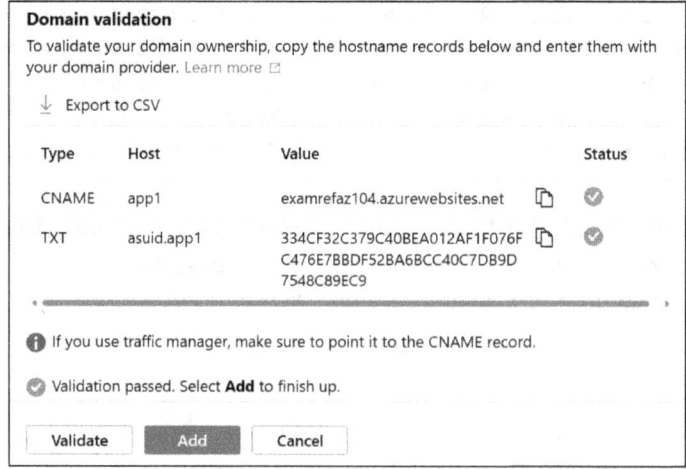

**FIGURE 3-75**   Domain validation

To complete the process, click Add to associate the custom domain with the App Service. The custom domain will be added to the App Service and will respond when accessing the service using the new URL. By adding the domain without configuring TLS/SSL, you will receive an error that the domain is not secured and does not have any bindings, as shown in Figure 3-76. This is expected. Certificates for App Service apps are discussed later in this skill.

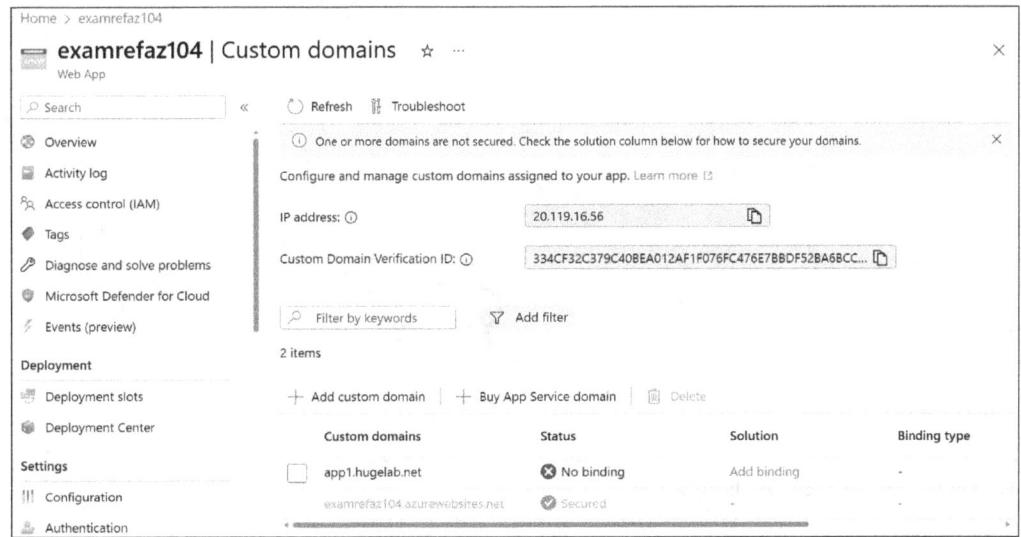

**FIGURE 3-76** Newly added custom domain

## Configure certificates and TLS for an App Service

App Services can use certificates along with custom domains to help secure and provide trust to the users and services that will access the resource. There are three supported methods of using certificates with an App Service:

- **Managed certificates**   Managed certificates are included with an App Service and managed by Digicert but have additional restrictions.
- **Private key certificates**   If you already have your own certificate, you can upload the *.pfx* private key certificate to be used by the App Service.
- **Public key certificates**   Upload *.cer* public certificates to be used by the App Service.

To manage the certificates of an App Service, navigate to the App Service and open the Certificates blade, as shown in Figure 3-77.

**FIGURE 3-77** App Service Certificates

## Managed certificates

Managed certificates are free certificates available for an App Service and are fully managed by Microsoft, but they're issued by Digicert. Managed certificates are easy to obtain, but do have some prerequisites and limitations to be aware of. To create a free certificate, the App Service plan that the web app is associated with cannot be in the Free tier.

If you plan to use a custom domain with a managed certificate, the domain must already be added to the App Service. Adding a custom domain to an App Service was discussed earlier in this skill. Additionally, if you plan to use a root domain, the App Service configuration cannot include any IP restrictions. The app must be accessible from the internet to process the creation and renewal of the certificate.

Free certificates also have the following restrictions:

- Cannot use wildcards in the certificate.
- Cannot be used as a client certificate.
- Cannot be used with private DNS.
- Cannot be exported from the App Service.
- Only support alphanumeric characters, dashes, and periods.
- Custom domain names cannot exceed 64 characters.

To add a managed certificate to your App Service, navigate to the Certificates blade of the App Service. On the Managed Certificates tab, shown in Figure 3-77, click Add.

On the Add App Service Managed Certificate blade, select the custom domain you previously added from the Custom Domain drop-down menu. Specify a friendly name for the certificate and then click Validate.

If the App Service passes validation, as shown in Figure 3-78, click Add to generate the certificate and add it to the App Service. This process could take up to 10 minutes to complete.

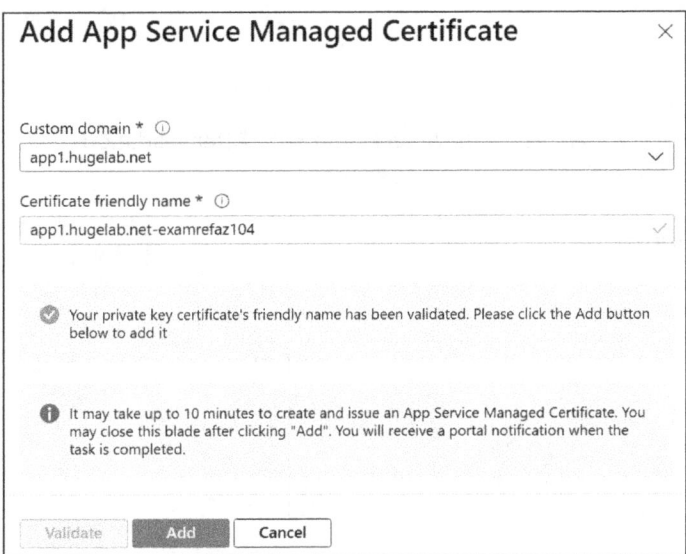

**FIGURE 3-78** Add a managed certificate

## Private key certificates

Private key certificates are .pfx certificates you have received from your Certificate Authority (CA) or trusted third-party provider. You can upload these certificates directly to an App Service or upload them to an Azure key vault that can be integrated with an App Service.

Private certificates must be password-protected, be encrypted using triple DES, be at least 2048 bits, and contain all intermediate and root certificate in the certificate chain. If you plan to use the certificate with a custom domain, the certificate must contain an Extended Key Usage attribute for server authentication.

To add a private key certificate to an App Service, navigate to the Certificates blade of the App Service and select the Bring Your Own Certificate (.pfx) tab. On this tab, click Add Certificate.

On the Add Private Key Certificate blade, there are three options in the Source drop-down menu:

- **Import From App Service**   This can be used if you previously uploaded a certificate.
- **Upload Certificate (.pfx)**   Select this option if you plan to upload the certificate directly to the App Service.

- **Import From Azure Key Vault**    Select this option if you have uploaded the certificate to an Azure Key Vault and plan to point the App Service to the Key Vault to obtain the certificate.

If you select Upload Certificate (.pfx), you will be prompted for information to upload the certificate, including

- PFX Certificate File
- Certificate Password
- Certificate Friendly Name

Complete the fields and then click Validate, as shown in Figure 3-79. After validation completes, click Add to upload the certificate to the App Service.

**FIGURE 3-79**    Add a private key certificate

To manage the domain and the certificate separately, you must associate the certificate with the domain by adding bindings to the domain. To add bindings, navigate to the App Service and then select Custom Domains. On the Custom Domains blade, select Add Bindings for the domain that you added and have a certificate for (refer to Figure 3-76).

On the Add TLS/SSL Binding blade, displayed in Figure 3-80, in the Certificate drop-down menu, select the certificate you uploaded to the App Service; then click Add. The certificate will be associated with the custom domain.

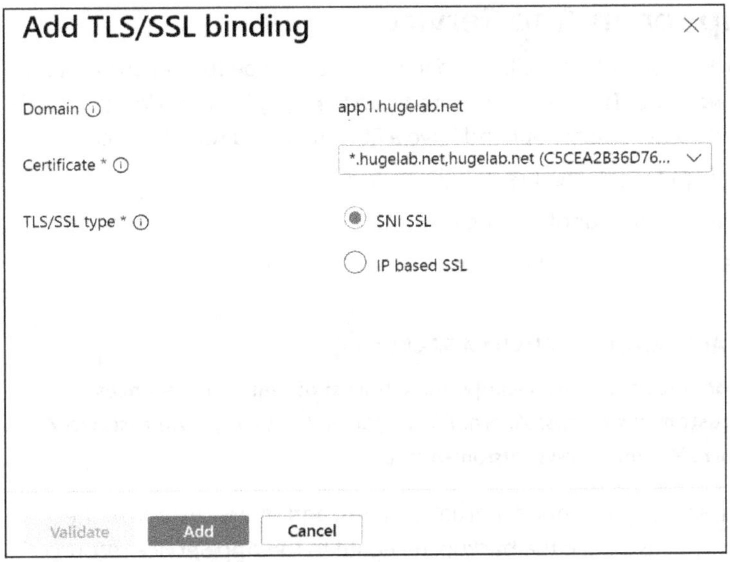

## Add TLS/SSL binding        ✕

| | |
|---|---|
| Domain ⓘ | app1.hugelab.net |
| Certificate * ⓘ | *.hugelab.net,hugelab.net (C5CEA2B36D76...  ⌄ |
| TLS/SSL type * ⓘ | ⦿ SNI SSL |
| | ◯ IP based SSL |

Validate    **Add**    Cancel

**FIGURE 3-80**   Add TLS/SSL Binding

## Public key certificates

Public key certificates can be uploaded directly to an App Service and must be in a .cer file for-mat without a private key. On the Public Key Certificates tab of the Certificates blade, click Add Certificate. You will be prompted to upload the .cer file and provide a friendly name, as shown in Figure 3-81. Click Add to upload the certificate.

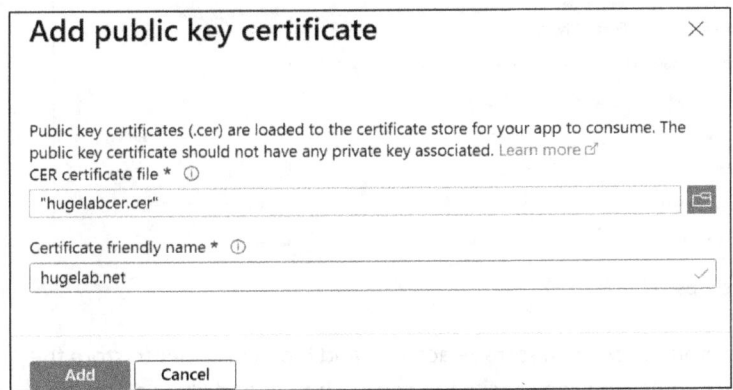

**FIGURE 3-81**   Add a public key certificate

# Configure backup for an App Service

Most App Service plans include an automatic backup feature that can be further customized to create backups of your web app. The Free and Shared App Service plan tiers do not include backups. Automatic backups occur every hour and have a 30-day sliding retention scale:

- Hourly backups are kept for the most recent three days.
- Days 4–14 retain every third hour of backups.
- Days 15–30 retain every sixth hour of backups.

> **NEED MORE REVIEW?** **AUTOMATIC VS CUSTOM BACKUPS**
>
> App Services can use automatic or custom backups. For a full list of feature differences between automatic and custom, visit *https://learn.microsoft.com/en-us/azure/app-service/manage-backup?tabs=portal#automatic-vs-custom-backups*.

If you need to modify the timing, retention, partial backups, and more, customize the backup for the App Service. To customize the backup, navigate to the App Service resource and select the Backups blade. The automatic hourly backups will be displayed, as shown in Figure 3-82, with an option on each row to restore from that backup. Alternatively, you can click Configure Custom Backups on the top menu.

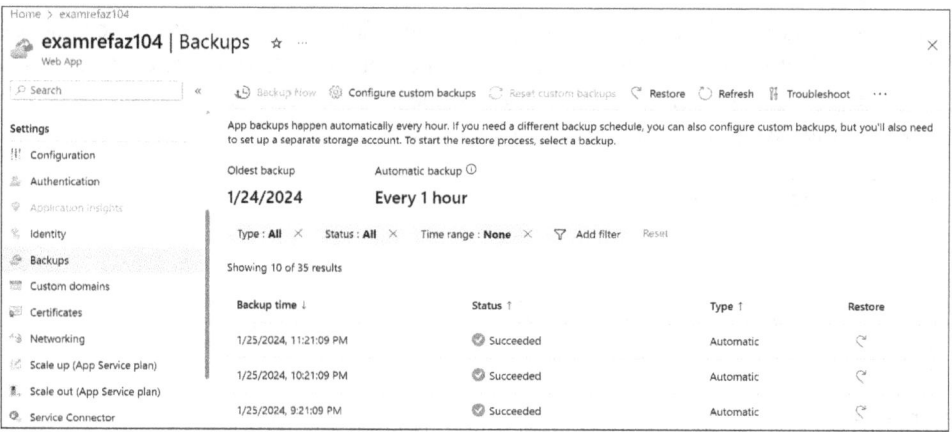

**FIGURE 3-82**  App Service backups

Custom backups require you to specify a storage account and blob container to store the backups in. As part of the schedule, you can specify the backup hourly or daily at a specific time, as well as retain the backup. Retention can be configured as a value from 0–60, with zero representing storing the backup indefinitely. Figure 3-83 displays the custom backup configuration page.

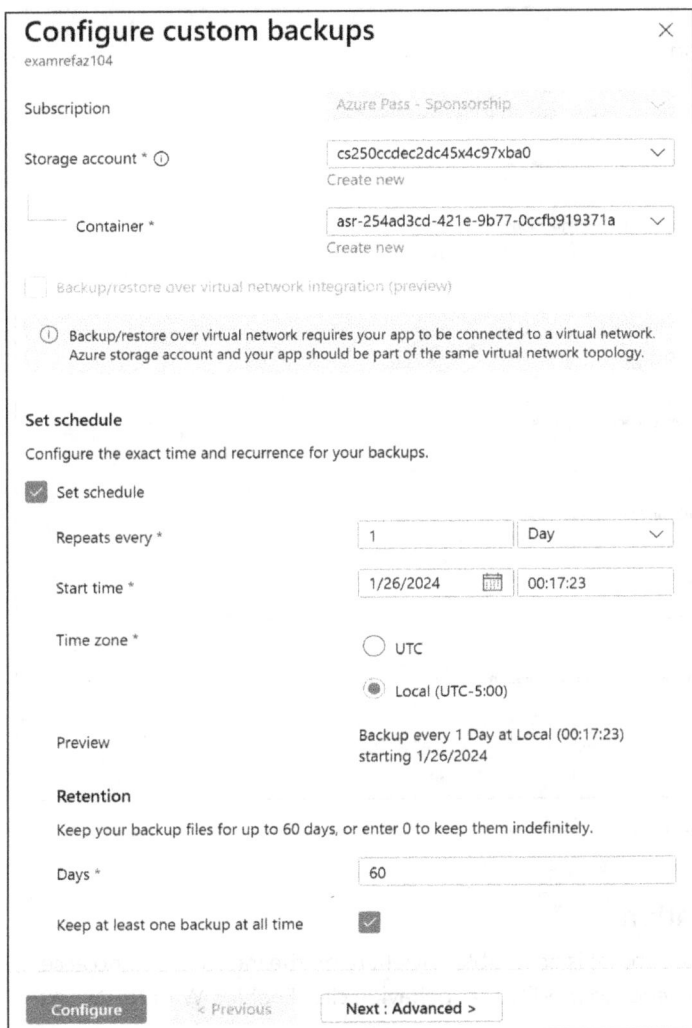

**Configure custom backups**

examrefaz104

| | |
|---|---|
| Subscription | Azure Pass - Sponsorship |
| Storage account * ⓘ | cs250ccdec2dc45x4c97xba0 |
| | Create new |
| Container * | asr-254ad3cd-421e-9b77-0ccfb919371a |
| | Create new |

☐ Backup/restore over virtual network integration (preview)

ⓘ Backup/restore over virtual network requires your app to be connected to a virtual network. Azure storage account and your app should be part of the same virtual network topology.

**Set schedule**

Configure the exact time and recurrence for your backups.

☑ Set schedule

| | | |
|---|---|---|
| Repeats every * | 1 | Day |
| Start time * | 1/26/2024 | 00:17:23 |
| Time zone * | ○ UTC | |
| | ◉ Local (UTC-5:00) | |
| Preview | Backup every 1 Day at Local (00:17:23) starting 1/26/2024 | |

**Retention**

Keep your backup files for up to 60 days, or enter 0 to keep them indefinitely.

| | |
|---|---|
| Days * | 60 |
| Keep at least one backup at all time | ☑ |

[ Configure ]   < Previous   [ Next : Advanced > ]

**FIGURE 3-83**   App Service custom backups

# Configure networking settings for an App Service

By default, App Services include a publicly accessible URL, and public network access is enabled with no restrictions. However, most organizations will have a security or compliance policy that requires all inbound traffic to first be inspected by a Web Application Firewall (WAF), or might require that all inbound traffic originate from a specific network.

To achieve this, modify the network settings of the App Service. From the Azure portal, navigate to the App Service and then select the Networking blade, as shown in Figure 3-84.

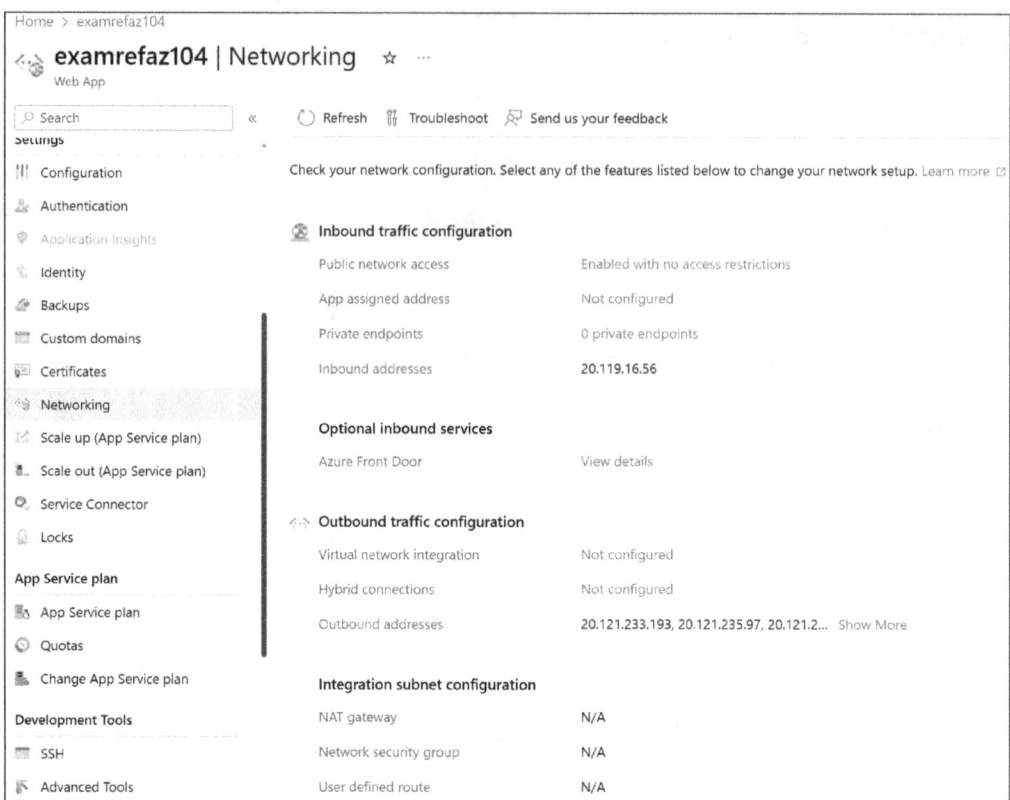

**examrefaz104** | Networking ☆ ...
Web App

🔍 Search «    🔄 Refresh    ⚙ Troubleshoot    🗨 Send us your feedback

Check your network configuration. Select any of the features listed below to change your network setup. Learn more ☐

**Settings**

||| Configuration
👤 Authentication
📊 Application Insights
🔑 Identity
💾 Backups
🖥 Custom domains
📜 Certificates
🔗 Networking
📈 Scale up (App Service plan)
📊 Scale out (App Service plan)
⚙ Service Connector
🔒 Locks

**App Service plan**

📊 App Service plan
🔄 Quotas
📊 Change App Service plan

**Development Tools**

💻 SSH
📄 Advanced Tools

🖥 **Inbound traffic configuration**

| | |
|---|---|
| Public network access | Enabled with no access restrictions |
| App assigned address | Not configured |
| Private endpoints | 0 private endpoints |
| Inbound addresses | 20.119.16.56 |

**Optional inbound services**

| | |
|---|---|
| Azure Front Door | View details |

↙ **Outbound traffic configuration**

| | |
|---|---|
| Virtual network integration | Not configured |
| Hybrid connections | Not configured |
| Outbound addresses | 20.121.233.193, 20.121.235.97, 20.121.2... Show More |

**Integration subnet configuration**

| | |
|---|---|
| NAT gateway | N/A |
| Network security group | N/A |
| User defined route | N/A |

**FIGURE 3-84**   App Service networking

## Inbound traffic configuration

By default, a public App Service endpoint is accessible anywhere on the internet with no access restrictions. If your organization needs to modify that behavior, click Enabled With No Access Restrictions.

On the Access Restrictions page, you can modify the inbound network to respond only from specific virtual networks and/or IP addresses, or to have inbound access disabled completely, as shown in Figure 3-85.

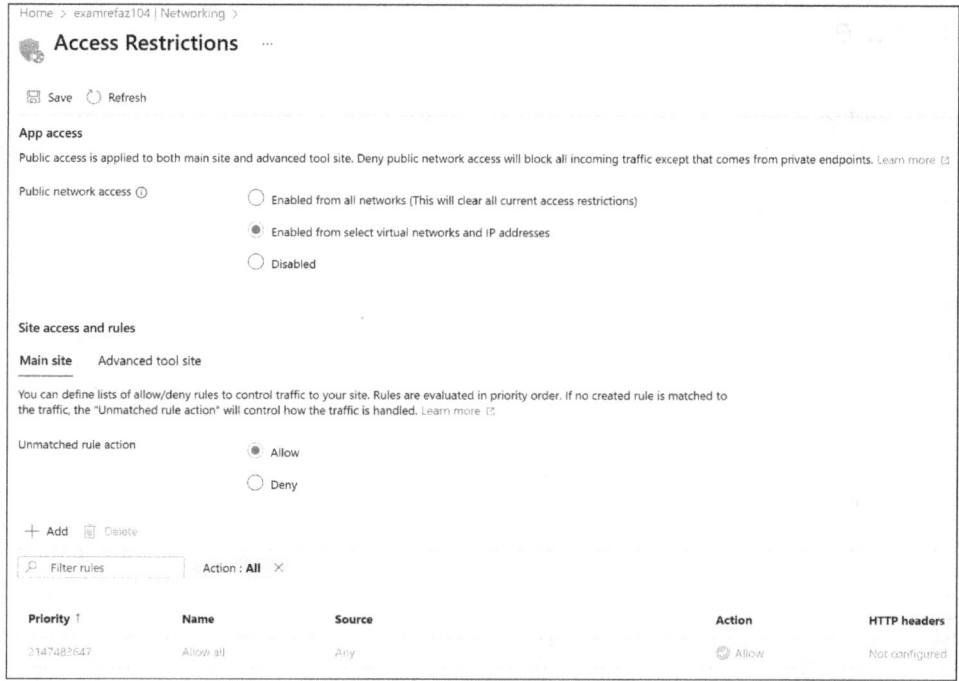

**Access Restrictions** ...

💾 Save ○ Refresh

**App access**

Public access is applied to both main site and advanced tool site. Deny public network access will block all incoming traffic except that comes from private endpoints. Learn more ☐

Public network access ⓘ
- ○ Enabled from all networks (This will clear all current access restrictions)
- ● Enabled from select virtual networks and IP addresses
- ○ Disabled

**Site access and rules**

**Main site**    Advanced tool site

You can define lists of allow/deny rules to control traffic to your site. Rules are evaluated in priority order. If no created rule is matched to the traffic, the "Unmatched rule action" will control how the traffic is handled. Learn more ☐

Unmatched rule action
- ● Allow
- ○ Deny

＋ Add   🗑 Delete

| Priority ↑ | Name | Source | | Action | HTTP headers |
|---|---|---|---|---|---|
| 🔍 Filter rules | | Action : **All** ✕ | | | |
| 2147483647 | Allow all | Any | | ⊘ Allow | Not configured |

**FIGURE 3-85**   App Service networking

If you select Enabled From Select Virtual Networks And IP Addresses, you can create individual rules that allow or deny specific IP addresses. There is also an Unmatched Rule Action field to either allow or deny traffic that does not explicitly match any of the rules that you configure.

On the Access Restrictions page, click Add. The Add Rule page has the following fields:

- **Name**   The display name of the rule
- **Action**   Whether to allow or deny the traffic that matches
- **Priority**   The priority of the rule for processing
- **Description**   A description of the rule for internal documentation
- **Type**   The source of the traffic: IPv4, IPv6, a virtual network, or Azure service tag
- **IP Address Block**   When IPv4 is selected, the CIDR notation of the address(es) to configure
- **X-Forwarded-Host**   Any hostnames that should be forwarded to the App Service
- **X-Forwarded-For**   The IP address to forward to communicate
- **X-Azure-FDID**   An optional Front Door or reverse proxy ID
- **X-FD-HealthProbe**   The health probe ID to use with a reverse proxy

Figure 3-86 displays the Add Rule page configured to allow traffic from the 10.0.0.0/16 network.

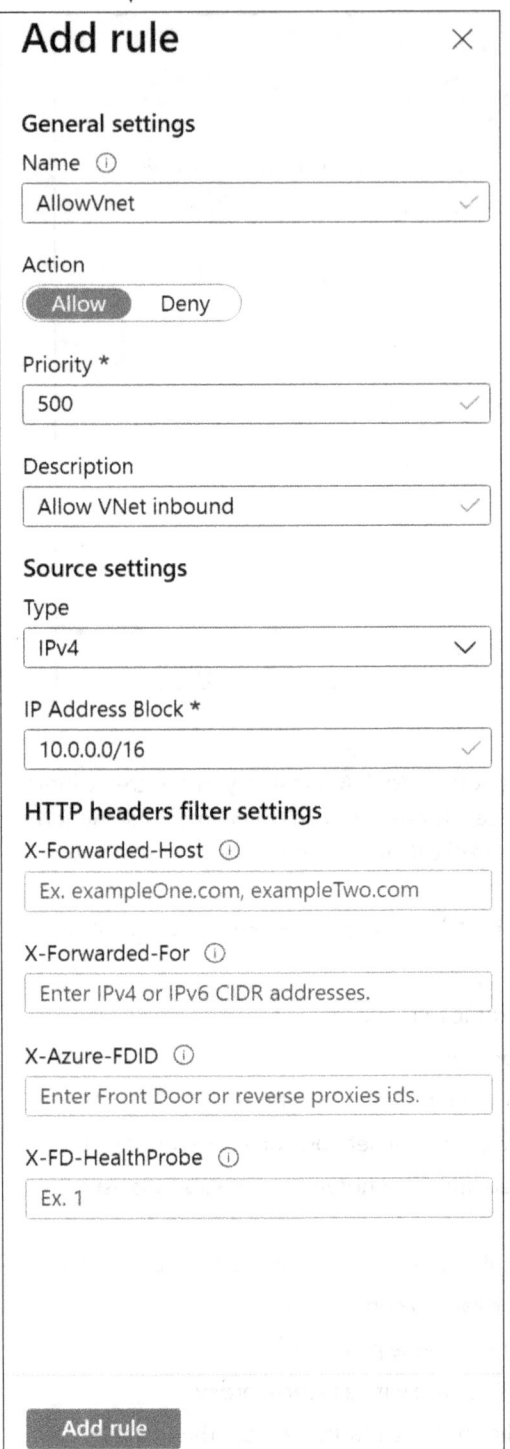

**FIGURE 3-86** Add inbound traffic rule

## Outbound traffic configuration

An App Service can be integrated with an Azure virtual network to allow the app to make outbound connections to VMs or other endpoints in a virtual network. The virtual network and the App Service must be in the same Azure region to be integrated. Integrating an App Service with a virtual network enables the App Service to

- Access resources in the integrated virtual network and other virtual networks that are peered with the integrated network.
- Access resources that use service or private endpoints.
- Access resources across ExpressRoute circuits or VPN tunnels.
- Force tunnel outbound traffic through a firewall or other network appliance.

To configure virtual network integration for an App Service, navigate to the Networking blade of the App Service. Under Outbound Traffic Configuration, click Not Configured next to Virtual Network Integration (refer to Figure 3-84).

The Virtual Network Integration blade will be displayed. Select Add Virtual Network Integration. On the Add Virtual Network Integration blade, select the subscription, virtual network, and specific subnet to integrate the App Service with, as shown in Figure 3-87.

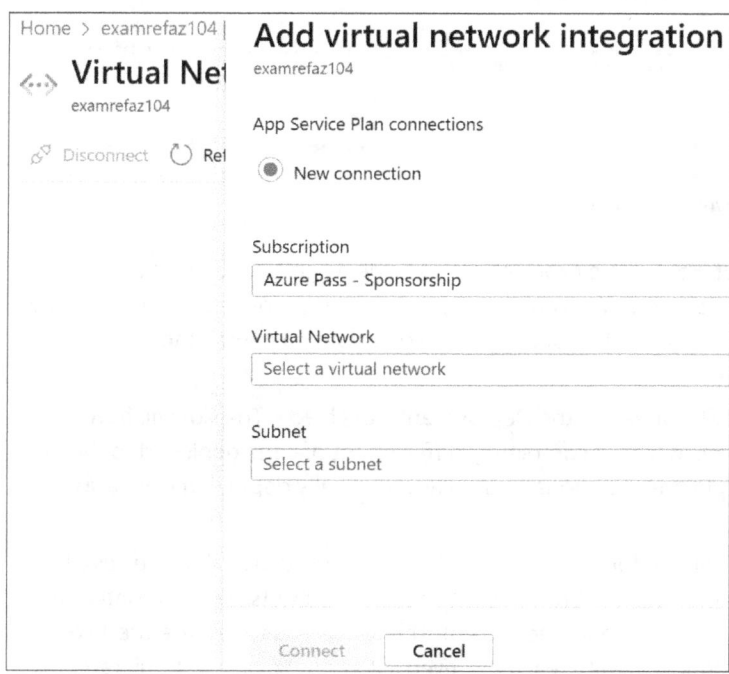

**FIGURE 3-87**  Add Virtual Network Integration

# Configure deployment slots for an App Service

Deployment slots are live, accessible versions of your web app that run in the same App Service plan, but each has its own hostname and can be configured separately. You can configure a production slot where active users and connections are hosted, as well as dev, canary, user acceptance, or other slots. Each slot can have its own configuration items, such as connection strings and environment variables. This makes it easy to perform A/B or blue/green testing for new versions of your application.

The number of slots and the specific features depend on the tier of App Service plan that you configured. To add a slot to your App Service, navigate to the resource and then select the Deployment Slots blade, as shown in Figure 3-88.

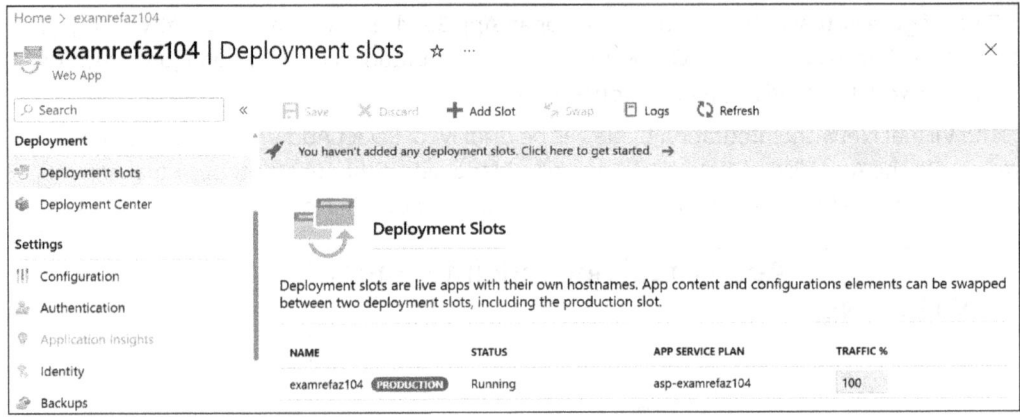

**FIGURE 3-88**   App Service Deployment Slots blade

To add a slot, click Add Slot. You will be prompted to provide a name for the slot and specify whether you want to clone the settings from an existing slot. Cloning provides an easy method of copying environment variables and other settings from one slot to a new slot to minimize changes that need to be made.

When you add the slot, it will appear on the Deployment Slots blade. The slot will have its own URL that is accessible and can have a completely different set of code deployed to the slot. Each slot can have its own deployment options, such as automatically deploying from a specific branch in a repository.

Figure 3-88 also displays a column for Traffic %. By default, the production slot receives all traffic that is destined for the App Service. When you add deployment slots, you can configure a specific percentage of new requests to be sent to the slot. This allows you to test the latest version of your application "in production" with real users, even if it's only 1% of your total traffic.

When you are ready to move a version of an application from one slot to another slot, such as from staging to production, you can easily swap the slots. This will prompt you to copy some configuration changes between environments, or to leave them as-is. Figure 3-89 displays the swap page with the source, target, and configuration changes.

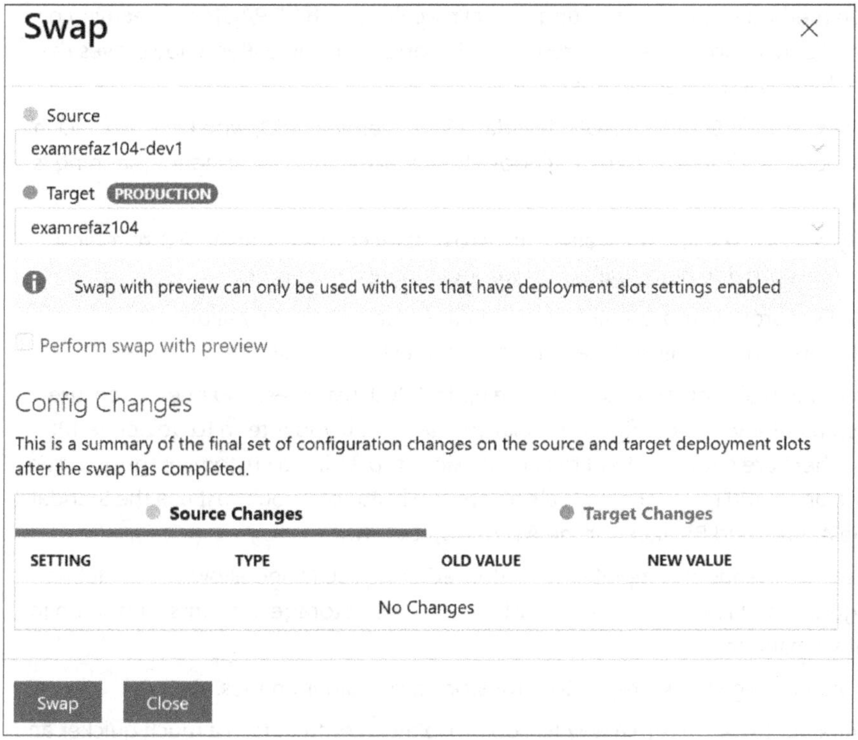

FIGURE 3-89   App Service swap slots

# Chapter summary

This chapter focused heavily on creating and configuring virtual machines in Azure as well as automated deployments using Azure Resource Manager templates and even the command-line tools. The chapter wrapped up focusing on container services, such as ACI and ACA, followed by Azure App Service and App Service Plans. Here are some of the key takeaways.

- Each compute family is optimized for either general or specific workloads. You should optimize your VM by choosing the most appropriate size.
- You can create VMs from the Azure portal, PowerShell, the CLI, Azure Resource Manager templates, or Bicep files. You should understand when to use which tool and how to configure the virtual machine resource during provisioning and after provisioning. For example, availability sets can only be set at provisioning time, but data disks can be added at any time.
- You can connect to Azure VMs using a public IP address or a private IP address with RDP, SSH, or even PowerShell. To connect to a VM using a private IP, you must also enable connectivity such as site-to-site, point-to-site, or ExpressRoute.

- A common method of troubleshooting virtual machines with RDP/SSH connectivity or unexplained application issues is to redeploy the virtual machine. Redeploy moves the virtual machine to a different Azure node.

- VM storage comes in Standard HDD, Standard SSD, Premium SSD, with Ultimate SSD in preview. Understanding which tier to choose for capacity and performance planning is important.

- Availability zones provide high availability across datacenters within an Azure region. Availability sets provide high availability within a single datacenter.

- Managed disks provide more availability than unmanaged disks by aligning with availability sets and providing storage in redundant storage units.

- Virtual machine scale sets (VMSS) can scale up to 1,000 instances. You need to ensure that you create the VMSS configured for large scale sets if you intend to go above 100 instances. There are several other limits to consider too. Using a custom image, you can only create up to 300 instances. To scale above 100 instances, you must use the Standard SKU of the Azure Load Balancer or the Azure App Gateway.

- Azure Resource Manager templates are authored using JSON and allow you to define the configuration of resources, such as virtual machines, storage accounts, and so on in a declarative manner.

- Azure Container Registry is a repository for storing container images.

- Azure Container Instances are a way to deploy isolated containers in a much quicker and simpler way without worrying about backend infrastructure.

- Azure Container Apps is a PaaS service built on Kubernetes but does not provide access to the Kubernetes API or to the control nodes.

- App Service is a hosting service that lets you build and deploy your applications to run on Windows, Linux, or a Docker container in a managed infrastructure. You can also choose the programming language with which you would like to build your code.

- An App Service plan offers computer resources to the web application for its execution. This App Service plan can be shared with multiple web apps, too.

# Thought experiment

In this thought experiment, apply what you have learned in this chapter. You can find answers to these questions in the next section, "Though experiment answers."

### Scenario 1

You are the IT administrator for Contoso, and you are tasked with migrating an existing web farm and database to Microsoft Azure. The web application is written in PHP and is deployed across 20 physical servers running RedHat for the operating system and Apache for the web server. The backend consists of two physical servers running MySQL in an active/passive configuration.

The solution must provide the ability to scale to at least as many web servers as the existing solution and ideally the number of web server instances should automatically scaling adjust based on the demand. All the servers must be reachable on the same network, so the administrator can easily connect to them using SSH from a jumpbox (a VM that is exposed to public IP and used to connect to other VMs in the network using private IPs internally) to administer the VMs.

Answer the following questions for your manager:

1. Which compute option would be ideal for the web servers?
2. How should the servers be configured for high availability?
3. What would be the recommended storage configuration for the web servers? What about the database servers?

**Scenario 2**

You are the solution architect for Contoso, and you must design a Python-based solution for hosting a web application in Microsoft Azure. Users must be able to access this application from multiple locations, and the application must be available around the clock. Also, the application should be implemented with DevOps capabilities, such as continuous deployment, package management, and the like.

Moreover, you don't want to manage the infrastructure, and you want to avoid the administration as much as possible. You do not want to manage the Windows and software updates on your own.

Answer the following questions about your solution:

1. Which compute option would be ideal for hosting this web application?
2. How will you avoid managing the infrastructure?
3. How will you make sure your application is highly available?

# Thought experiment answers

This section contains the solution to the thought experiment for the chapter.

**Scenario 1**

1. The web servers would be best served by deploying them into a virtual machine scale set (VMSS), because of the requirements for SSH connectivity. Scaling should be configured on the VMSS to address the requirement of automatically scaling up/down the number of instances based on the demand (CPU) used on the web servers.
2. The web servers should be deployed into their own availability set or availability zone if it is available within the region. The database tier should also be deployed into its own availability set or availability zone.
3. The web servers will likely not be I/O intensive so Standard SSD should be appropriate. The database servers will likely be I/O intensive so Premium SSD is the recommended approach. To minimize management overhead and to ensure that storage capacity planning is done correctly, managed disks should be used in both cases.

**Scenario 2**

1.  Azure App Service will be best suited for this set of requirements. You should deploy a Python web app to App Service on Linux.

2.  Azure App Service is a PaaS solution, so a managed infrastructure will host your application. You do not need to worry about the VMs on which the app is deployed. No additional administration efforts are required to manage Windows and software updates.

3.  Azure App Service provides 99.95 percent availability for the Basic tier (or higher). You should also consider using autoscaling per previous use in chapter so that it can still accept traffic during peak workload times.

# CHAPTER 4

# Configure and manage virtual networking

An Azure virtual network (or VNet) provides the foundation of the Azure networking infra-structure. Virtual machines are connected to virtual networks. This connection provides inbound and outbound connectivity to other virtual machines, to on-premises networks, and to the internet. Azure provides many networking features that will be familiar to those already experienced in networking, such as the abilities to control which network flows are permitted and to control network routing. These features allow Azure deployments to imple-ment familiar network architectures, such as network segmentation between layers of an N-tier application.

This chapter focuses on the core capabilities that allow you to connect your Azure virtual machines—flexibly and securely.

## Skills covered in this chapter:

- Skill 4.1: Configure and manage virtual networks in Azure
- Skill 4.2: Configure secure access to virtual networks
- Skill 4.3: Configure name resolution and load balancing

> **NOTE   MICROSOFT EXAM OBJECTIVES**
>
> The sections in this chapter align with the objectives that are listed in the AZ-104 study guide from Microsoft. However, the sections are presented in an order that is designed to help you learn, and do not directly match the order that is presented in the study guide. On the exam, questions will appear from different sections in a random order. For the full list of objectives, visit *https://learn.microsoft.com/en-us/credentials/certifications/resources/study-guides/az-104*.

## Skill 4.1: Configure and manage virtual networks in Azure

Azure Virtual Networks (VNets) form the foundation of the Azure networking infrastructure. Each VNet defines a network address space, comprising one or more IP address ranges. This network space is then carved into subnets. IP addresses for virtual machines, as well as some other services such as an internal Azure load balancer, are assigned from these subnets.

For each subnet, you define which network flows are permitted (using network security groups), and what network routes should be taken (using user-defined routes). You can use these features together to implement many common network topologies, such as a DMZ containing a network security appliance or a multitier application architecture with restricted communications between application tiers.

> **This skill covers how to:**
> - Create and configure virtual networks and subnets
> - Create and configure virtual network peering
> - Configure public IP addresses
> - Configure user-defined network routes
> - Troubleshoot network connectivity

## Create and configure virtual networks and subnets

A VNet is an Azure resource that defines address space, subnets, and connectivity for Azure resources. When you create a VNet, the most important setting is the IP range (or ranges) the VNet will use.

IP ranges are defined using classless inter-domain routing (CIDR) notation. For example, the range 10.5.0.0/16 represents all IP ranges starting with 10.5. The /16 represents the bitmask and indicates that the first 16 bits are the same for every IP in the address range. Each virtual network can use either a single IP range or multiple disjointed IP ranges.

> **NOTE  CIDR NOTATION**
>
> You need to understand CIDR notation to work effectively with virtual networks in Azure. There are many good explanations to be found online. For example, see *https://devblogs. microsoft.com/premier-developer/understanding-cidr-notation-when-designing-azure- virtual-networks-and-subnets/.*

> **NOTE  VIRTUAL NETWORK IP RANGES**
>
> It is normally a good idea to plan your network space in advance. Typically, you want to avoid creating overlaps with other virtual networks or with on-premises environments because any overlap will prevent you from connecting these networks later.

It is recommended that your VNet IP ranges be taken from the private address ranges defined in RFC 1918:

- 10.0.0.0–10.255.255.255 (10.0.0.0/8)
- 172.16.0.0–172.31.255.255 (172.16.0.0/12)
- 192.168.0.0– 192.168.255.255 (192.168.0.0/16)

You can also use public, internet-addressable IP ranges in your VNet. However, this is not recommended because the addresses within your VNet will take priority, and virtual machines in your VNet will no longer be able to access the corresponding internet addresses.

In addition, there are a small number of IP ranges you can't use because they are reserved by the Azure platform:

- 169.254.0.0/16 (Link-local)
- 168.63.129.16/32 (Azure-provided DNS)

## Subnets

Integrating Azure resources into a virtual network requires a subnet. Subnets are used to divide the VNet IP space. Different subnets can have different network security and routing rules, so applications and application tiers can be isolated and network flows between them can be controlled. For example, consider a typical three-tier application architecture comprising a web tier, an application tier, and a database tier. By implementing each tier as a separate subnet, you can control precisely which network flows are permitted between tiers and from the internet.

> **IMPORTANT   SUBNET REQUIREMENTS**
>
> The name of a subnet must be unique within that VNet. You cannot change the subnet name after it has been created. Each subnet must also define a single network range (in CIDR format). This range must be contained within the IP ranges defined by the VNet. Only IP addresses from within the subnets can be assigned to virtual machines and other resources. Subnets do not have to span the entire VNet address space; subnets can be a subset, leaving unused space for future expansion.

Azure reserves a few IP addresses from each subnet. Like standard IP networks, Azure reserves the first and last IP addresses in each subnet for network identification, broadcast. Additionally, it reserves another three addresses at the beginning for the range for internal Azure services, for a total of five unusable IP addresses.

You are required to define one subnet when creating a VNet using the Azure portal. VNets can typically have multiple subnets, and you can add new subnets to your VNet at any time.

You can't change the address range if there are resources already associated with or deployed to the subnet. If you want to make a change to a subnet's address range, you first must delete all the objects in that subnet. If the subnet is empty, you can change the range of addresses to any range that is within the address space of the VNet not assigned to any other subnets.

Subnets can be deleted from VNets only if they are empty. Once a subnet is deleted, the addresses that were part of that address range are released and available again for use within new subnets that you create.

# Additional virtual network settings

So far, this section has focused on the most important settings of each VNet and subnet: the IP address ranges. There are some additional settings and features of VNets and subnets to be aware of. Table 4-1 provides a summary of a few settings supported by virtual networks.

**TABLE 4-1** Properties of a virtual network

| Property | Description |
| --- | --- |
| Name | The VNet name must be unique within the resource group, have between 2 and 64 characters and may contain letters (case insensitive), numbers, underscores, periods, or hyphens. It must start with a letter or number and end with a letter, number, or underscore. |
| Location | Each VNet is tied to a single Azure region and can only be used by resources (such as virtual machines) in the same region. |
| Address Space | An array of IP address ranges available for use by subnets. |
| DNS settings | Contains an array of DNS servers. If specified, these DNS servers are configured on virtual machines in the virtual network in place of the Azure-provided DNS servers. |
| Subnets | The list of subnets configured for this VNet. |
| Peerings | The list of peerings configured for this VNet. Peerings are used to create network connectivity between separate VNets. |

Table 4-2 provides a summary of the settings supported by virtual network subnets.

**TABLE 4-2** Settings of a virtual network subnet

| Property | Description |
| --- | --- |
| Name | The subnet name must be unique within the VNet. It must have between 2 and 80 characters and may contain letters (case insensitive), numbers, underscores, periods, or hyphens. It must start with a letter or number and must end with a letter, number, or underscore. |
| Address Range | The IP address range for a subnet, specified in CIDR notation. All subnets must sit within the VNet address space and cannot overlap. |
| Network Security Group | Reference to the network security group (NSG) for the subnet. NSGs can be associated to a subnet and are used to control which inbound and outbound traffic flows are permitted. |
| Route Table | Route table applied to the subnet and used to override the default system routes. These are used to send traffic to destination networks that are different than the routes that Azure uses by default. |
| Service Endpoints (And Policies) | An array of service endpoints for this subnet. Service endpoints provide a direct route to various Azure PaaS services (such as Azure Storage), without requiring an internet-facing endpoint. Service endpoint policies provide further control over which instances of those services may be accessed. |
| Delegations | An array of references to delegations on the subnet. Delegations allow subnets to be used by certain Azure services, which will then deploy managed resources (such as an Azure SQL Database Managed Instance) into the subnet. Access to these resources is private and can be controlled using NSGs. Delegations also support access to and from on-premises networks when hybrid networking is used. |

## Create a virtual network and subnets using the Azure portal

To create a new VNet using the Azure portal, search for **virtual networks**. On the Virtual Networks blade, click Create.

The Create Virtual Network blade opens. Here you can provide configuration information about the virtual network. This blade requires the following inputs, as shown in Figure 4-1:

- Subscription in which the VNet is created
- The resource group where the VNet is created
- Name of the virtual network
- The location for VNet

The following values are set automatically, though you can override them as needed:

- Address space to be used for the VNet using CIDR notation
- Subnet name for the first subnet in the VNet
- The address range of the first subnet

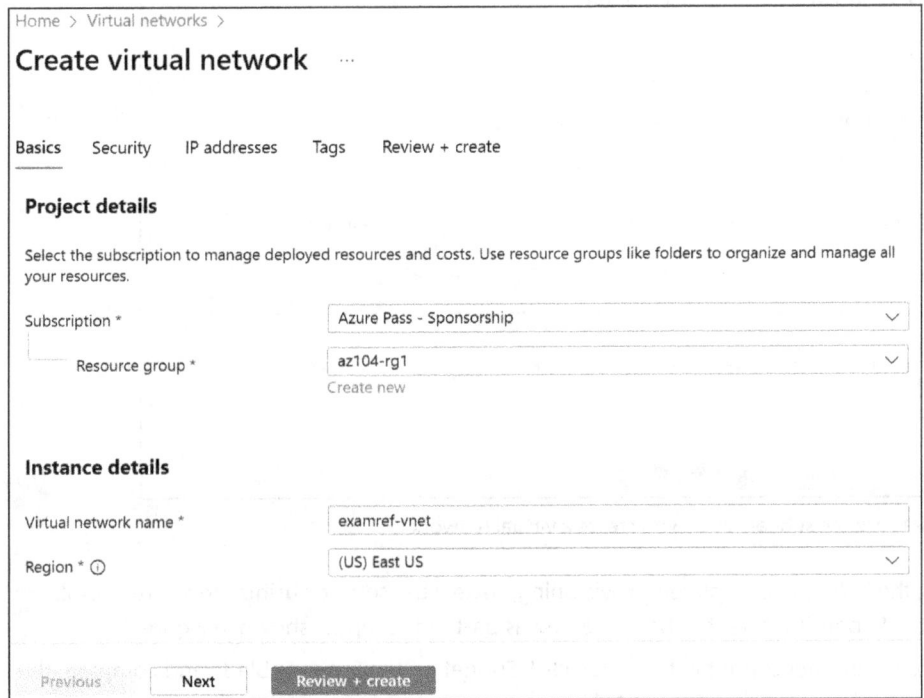

**FIGURE 4-1**  Basics tab of the Create Virtual Network blade

On the Security tab, you can enable common security services for the virtual network, including

- **Azure Bastion**  This provides secure RDP and SSH connectivity to VMs in the virtual network.
- **Azure Firewall**  This provides a layer-7 managed firewall in the virtual network.

- **Azure DDoS Network Protection** This provides enhanced DDoS attack mitigation with adaptive tuning, telemetry, and notifications for DDoS attacks.

On the IP Addresses tab, you can supply address spaces to be used for the VNet using CIDR notation. When creating a VNet using the Azure portal, you can specify multiple IP address ranges, and you can specify one or more subnets (see Figure 4-2). When you create a subnet, you can also create the service endpoints if you want to use any of the Azure Services.

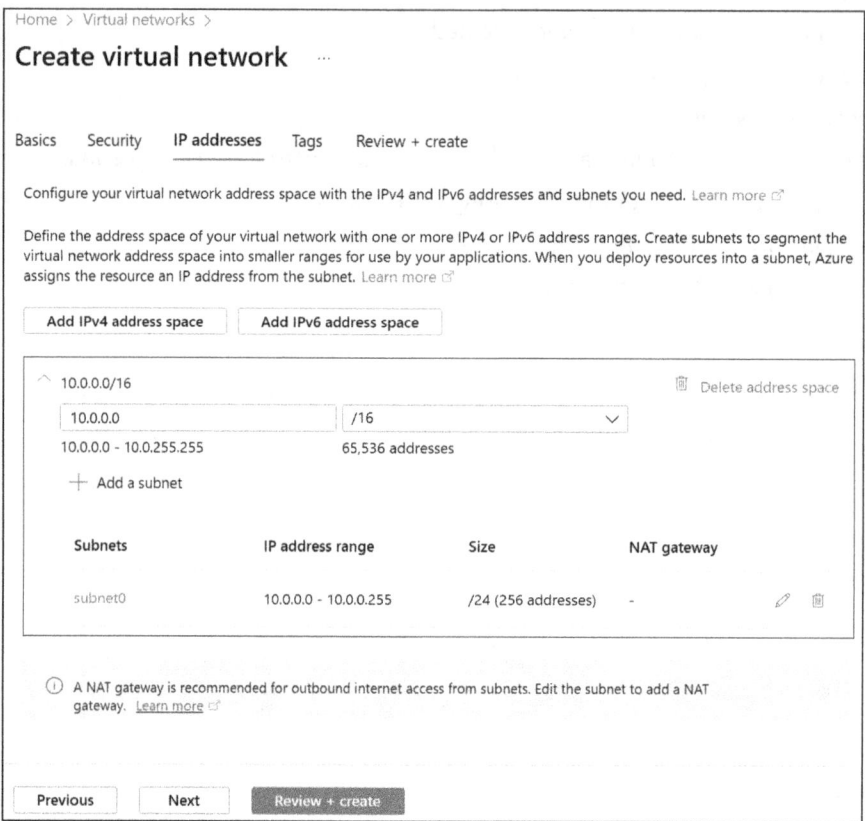

**FIGURE 4-2** Define subnets when you create a virtual network

After the VNet has completed provisioning, review the settings using the Azure portal. Notice the Subnet0 subnet has been created as part of the inputs shown in Figure 4-3.

To create another subnet in the VNet, click Subnet on the Subnets blade and complete the following fields, as shown in Figure 4-4:

- Name (of the subnet)
- Subnet Address Range (CIDR Block)
- NAT Gateway
- Network Security Group (if any)

- Route Table (if any)
- Service Endpoints
- Subnet Delegation

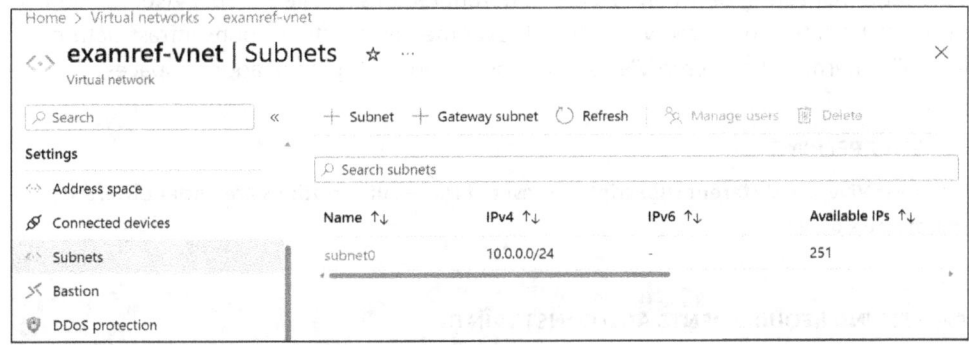

**FIGURE 4-3**  Subnets for the ExamRef-VNet virtual network

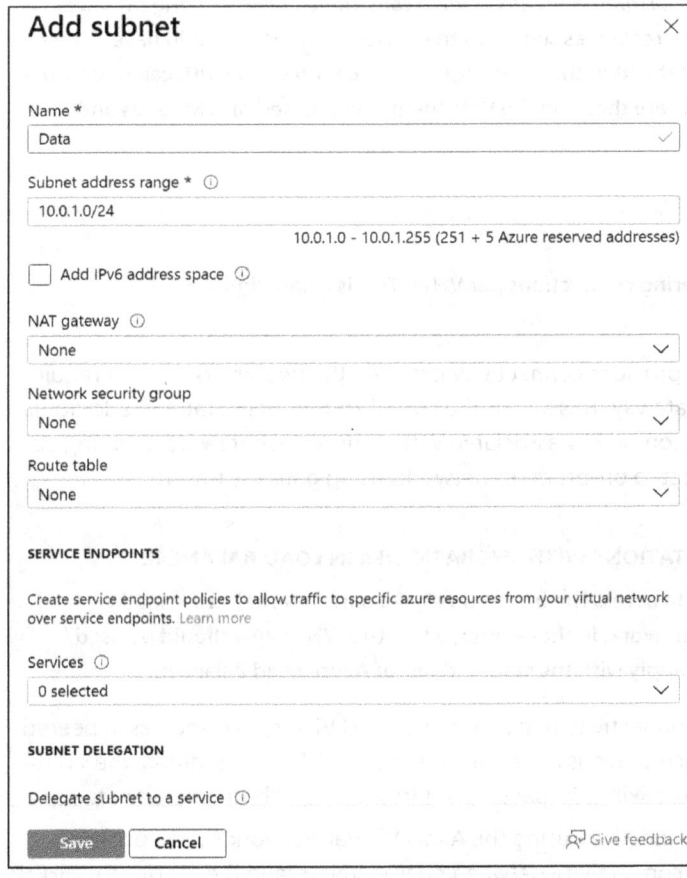

**FIGURE 4-4**  Add Subnet blade

# Create and configure virtual network peering

VNet peering allows resources in two separate virtual networks to communicate directly by using their private IP addresses. The VNets can either be in the same Azure region or in separate Azure regions. Peering between VNets in different regions is called Global VNet peering. In all cases, traffic between peered VNets travels over the Microsoft backbone infrastructure, not the public internet. The peered VNets must have nonoverlapping IP address spaces.

> **NOTE  VNET PEERING**
>
> You can peer VNets in different subscriptions, even if those subscriptions are under different Microsoft Entra tenants.

> **NOTE  PEERING REQUIREMENTS AND CONSTRAINTS**
>
> There are a few requirements and constraints to keep in mind while peering the VNets, which are found here: *https://learn.microsoft.com/en-us/azure/virtual-network/virtual-network-manage-peering?tabs=peering-portal#requirements-and-constraints*. VNet peering provides a network performance between resources similar to the performance they would have if they were placed in a single large VNet within the same region. There is no bandwidth cap imposed on peered VNets. The only limits are those on the VMs themselves, based on VM series and size.

> **NOTE  PEERING LIMITS**
>
> Be aware of the limit of 500 peering connections per VNet. This is a hard limit.

Peering two virtual networks provides connectivity between the two VNets without requiring a VPN and virtual network gateway. This avoids the cost, throughput limitations, additional latency, and additional incurred complexity associated with using VNet gateways, though you can use VNet gateways to connect to on-premises networks using gateway transit.

> **NOTE  GLOBAL PEERING LIMITATIONS WITH THE BASIC TIER IN LOAD BALANCER**
>
> Global peering cannot be used to access the front-end IP of a basic internal Azure load balancer in the remote virtual network. In these cases, a VNet-to-VNet VPN should be used instead. This limitation doesn't apply with the standard tier of Azure Load Balancer.

There are no restrictions on connectivity between the peered VNets, so resources in peered VNets can communicate with each other as if they are in the same VNet. In addition, the `VirtualNetwork` service tag (described in Skill 4.2) spans the address space of both peered networks.

Alternatively, you can limit connectivity using the Allow Virtual Network Access option; there is no automatic outbound connectivity between peered VNets, and the `VirtualNetwork` service tag does not include the address space of the peered VNet. In this case, you control the connectivity between peered virtual networks using network security groups.

A simple example of VNet peering is shown in Figure 4-5. This shows two VNets that have been connected using VNet peering. This allows (for example), the WEB1 virtual machine in VNetA to connect to the MYSQL1 database in VNetB.

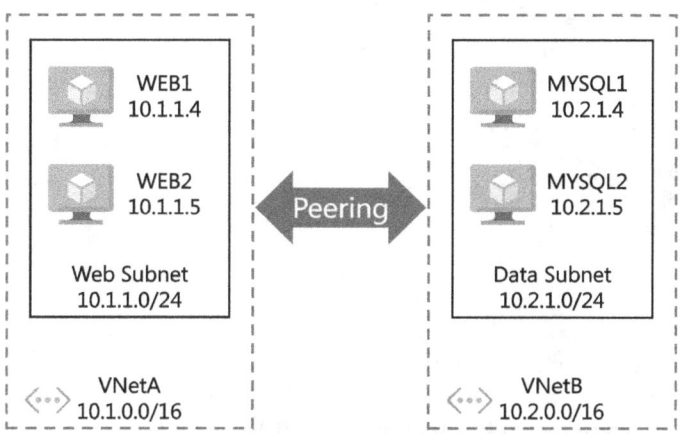

**FIGURE 4-5**  VNet peering between two virtual networks

After peering is established, traffic between VMs is routed through the Microsoft backbone infrastructure. Traffic does not pass over the public internet, even when using global VNet peering to connect VNets in different Azure regions.

While global VNet peering allows for open connectivity between virtual machines across VNets in different Azure regions, a VM can only connect to the front-end IP address of a basic internal Azure Load Balancer in the same region.

It is important to understand that VNet peering is a one-to-one relationship between two virtual networks. To create connectivity across three virtual networks (VNetA, VNetB, and VNetC), all three pairs must be peered (VNetA to VNetB; VNetB to VNetC; and VNetA to VNetC). This is illustrated in Figure 4-6.

## Service chaining and hub-and-spoke networks

A common way to reduce duplication of resources is to use a hub-and-spoke network topology. In this approach, shared resources (such as domain controllers, DNS servers, monitoring systems, and so on) are deployed into a dedicated hub VNet. These services are accessed from multiple applications, each deployed to their own separate spoke VNets.

As discussed earlier in this skill, VNet peering is not transitive. This means there is no automatic connectivity between spokes in a hub-and-spoke topology. Where such connectivity is required, one approach is to deploy additional VNet peerings between spokes. However, with a large number of spokes, this can quickly become unwieldy.

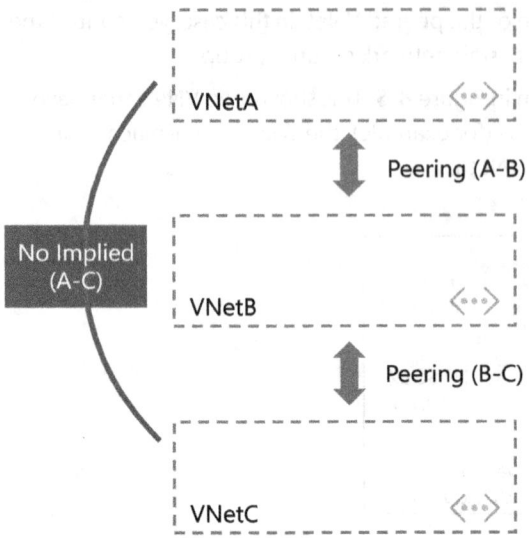

**FIGURE 4-6**  VNet peerings do not have a transitive relationship

An alternative approach is to deploy a network virtual appliance (NVA) into the hub through user-defined routes (UDRs) to route inter-spoke traffic through the NVA. This is known as *service chaining*, and it enables spoke-to-spoke communication without requiring additional VNet peerings, as illustrated in Figure 4-7.

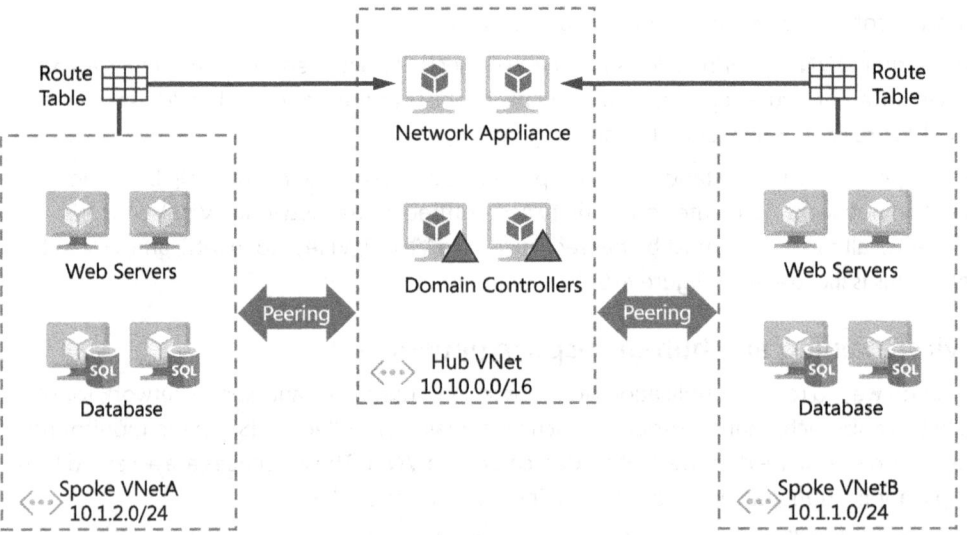

**FIGURE 4-7**  Service chaining allows for the use of common services across VNet peerings

To transit traffic from one spoke VNet to another spoke VNet via an NVA in the hub VNet, the VNet peerings must be configured correctly. By default, a peering connection will only accept traffic originating from the VNet to which it is connected. This will not be the case for traffic forwarded between spoke VNets via an NVA in a hub VNet. To permit such traffic, the Allow Forwarded Traffic setting must be enabled for those VNet peerings.

## Shared virtual network gateways

Suppose you want two peered VNets, called VNet-A and VNet-B, to send traffic to an external network via a virtual network gateway. Rather than deploy two virtual network gateways, it is much simpler and more cost-efficient for both VNets to share a single gateway. This can be achieved with local or global peering.

Suppose the virtual network gateway is deployed to VNet-A, allowing VNet-A to communicate with the external network. By default, only traffic originating in VNet-A is permitted to use this gateway, and the external network is only able to connect to VMs in VNet-A. To allow connectivity between VNet-B and the external network, the following settings must be configured:

- **Enable VNet-B To Use VNet-A's Remote Gateway**  This setting must be enabled on the peering connection from VNET-B to VNET-A. This informs VNET-B of the availability of the gateway in VNET-A. Note that to enable this setting, VNET-B cannot have its own virtual network gateway.

- **Allow Gateway In VNet-A To Forward Traffic To VNet-B**  This option must be enabled on the peering connection from VNET-A to VNET-B. This permits traffic from VNET-B to use VNET-A's gateway to send traffic to the external network. Gateway transit can be used for S2S, P2S, and VNet to VNet.

Note that in this case, the Allow VNet-B To Receive Forwarded Traffic From VNet-A peering option is not required.

## Create a VNet peering using the Azure portal

To create a peering connection between two VNets, the VNets must already have been created and must not have overlapping address spaces.

To create a new VNet peering from VNet-hub to VNet-spoke, connect to the Azure portal and locate VNet-hub. Under Settings, click Peerings, and then click Add to open the Add Peering blade. Use the following steps to set up a standard peering connection, as shown in Figure 4-8:

1. Under This Virtual Network, choose a name for the peering from VNet-hub to VNet-spoke. This example uses "hub-to-spoke."

2. Under This Virtual Network, select Allow Gateway In 'Vnet-hub' To Forward Traffic To The Peered Virtual Network.

3. Under Remote Virtual Network, you can choose Resource Manager or Classic. In this example, choose Resource Manager.

4. Select the subscription for VNet-spoke from the Subscription drop-down menu.

5. From the Virtual Network drop-down menu, choose VNet-spoke.

6. Under Remote Virtual Network, type **spoke-to-hub** for Peering Link.

7. Under Virtual Network, select Enable 'Vnet-spoke' To Use 'Vnet-hub's' Remote Gateway.

Home > Virtual networks > vnet-hub | Peerings >

# Add peering   ...

vnet-hub

This virtual network

Peering link name *

> hub-to-spoke

☑ Allow 'vnet-hub' to access 'vnet-spoke' ⓘ

☐ Allow 'vnet-hub' to receive forwarded traffic from 'vnet-spoke' ⓘ

☑ Allow gateway in 'vnet-hub' to forward traffic to 'vnet-spoke' ⓘ

☐ Enable 'vnet-hub' to use 'vnet-spoke's' remote gateway ⓘ

Remote virtual network

Peering link name *

> spoke-to-hub

Virtual network deployment model ⓘ
- ⦿ Resource manager
- ◯ Classic

☐ I know my resource ID ⓘ

Subscription * ⓘ

> Azure Pass - Sponsorship

Virtual network *

> vnet-spoke

☑ Allow 'vnet-spoke' to access 'vnet-hub' ⓘ

☐ Allow 'vnet-spoke' to receive forwarded traffic from 'vnet-hub' ⓘ

☐ Allow gateway in 'vnet-spoke' to forward traffic to 'vnet-hub' ⓘ

☑ Enable 'vnet-spoke' to use 'vnet-hub's' remote gateway ⓘ

[ Add ]

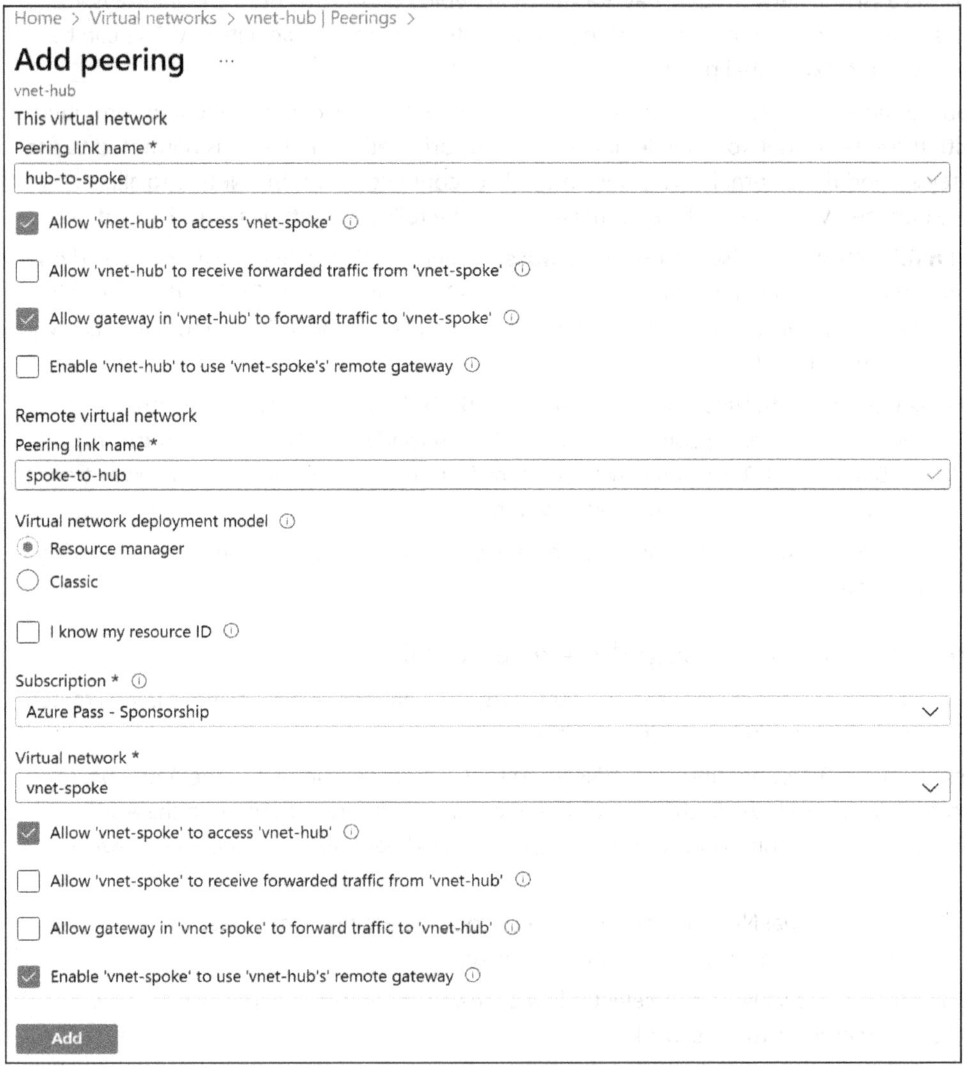

**FIGURE 4-8**   Adding peering from VNet-hub to VNet-spoke using the Azure portal

8.   Click Add to create the peering between VNet-hub and VNet-spoke. Once the peering has completed provisioning, it will appear in the Azure portal with the peering status Connected to peer network VNet-spoke, as shown in Figure 4-9.

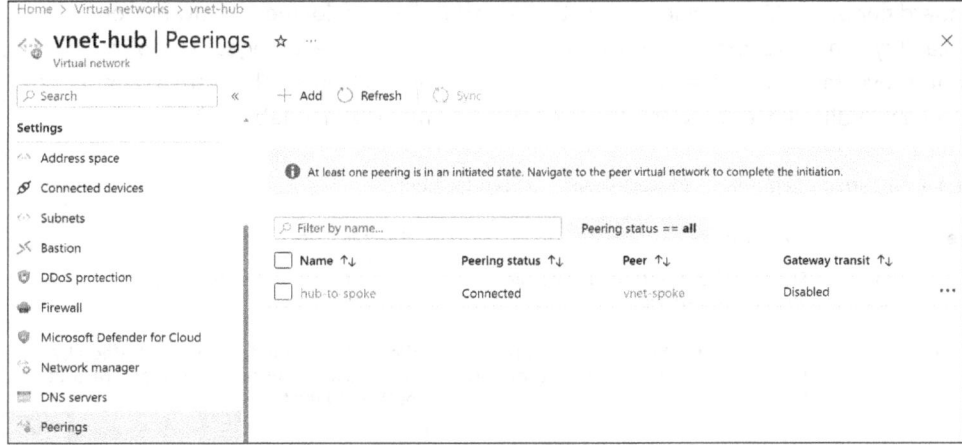

**vnet-hub | Peerings** ☆ ⋯
Virtual network

🔍 Search  «    + Add   ↻ Refresh   ↻ Sync                                                  ✕

**Settings**

⌢ Address space

🖉 Connected devices

⌢ Subnets

✂ Bastion

🛡 DDoS protection

🐘 Firewall

🛡 Microsoft Defender for Cloud

⌀ Network manager

▦ DNS servers

🔗 Peerings

ⓘ At least one peering is in an initiated state. Navigate to the peer virtual network to complete the initiation.

🔍 Filter by name...          Peering status == **all**

☐ Name ↑↓          Peering status ↑↓          Peer ↑↓          Gateway transit ↑↓

☐ hub-to-spoke          Connected          vnet-spoke          Disabled          ⋯

**FIGURE 4-9**   Hub-to-spoke peering showing as Connected in the Azure portal

9.  If you return to the Peering blade of VNet-spoke, you will see that the Peering Status of VNet2 to VNet1 is Connected.

Now, VNet-hub and VNet-spoke are peered, and VMs on these networks can communicate with each other as if this were a single virtual network.

# Configure public IP addresses

Associating a public IP address with a network interface creates an internet-facing endpoint, allowing your virtual machine to receive network traffic directly from the internet.

A public IP address is a standalone Azure resource. This contrasts with a private IP address that exists only as a collection of settings on another resource, such as a network interface or a load balancer.

To associate a public IP address with a virtual machine, the IP configuration of the network interface must be updated to contain a reference to the public IP address resource. As a standalone resource, public IP addresses can be created and deleted independently as well as moved from one virtual machine to another.

## Basic vs Standard pricing tiers

Public IP addresses are available at two pricing tiers (or SKUs): Basic or Standard. All public IP addresses created before the introduction of these tiers are mapped to the Basic tier.

> **NOTE   BASIC IP ADDRESSES**
>
> Basic SKU public IP addresses cannot be created after March 31, 2025 and will be retired in September 2025. It is possible that the AZ-104 exam might have questions related to Basic SKU public IP addresses until then. For more information, visit *https://azure.microsoft.com/en-us/updates/upgrade-to-standard-sku-public-ip-addresses-in-azure-by-30-september-2025-basic-sku-will-be-retired/.*

Standard tier public IP addresses support zone-redundant deployment, allowing you to use availability zones to protect your deployments against potential outages caused by data center-level failures (such as fire, power failure, or cooling failure). There are a number of other important differences between the two tiers, as summarized in Table 4-3.

TABLE 4-3 Comparison of public IP address Basic and Standard tiers

| Feature | Basic Tier | Standard Tier |
|---------|-----------|---------------|
| Allocation method | Supports both static and dynamic allocation methods. | Supports static allocation only. |
| Traffic restrictions | Open by default for inbound traffic. Use NSGs to restrict inbound or outbound traffic. | Closed by default for inbound traffic. Use NSGs to allow inbound traffic and restrict outbound traffic. |
| Redundancy | Not zone-redundant and doesn't support availability zone. | Zone-redundant by default, or it can instead be assigned to a specific availability zone. |
| Public IP prefixes | Does not support public IP prefixes (discussed later). | Supports public IP prefixes, allowing IP addresses to be assigned from a contiguous IP address block. |

## Public IP address allocation

As with private IP addresses, public IP addresses support both dynamic and static IP allocation. The Basic tier supports both static and dynamic allocation, with the default being dynamic. The Standard tier supports only static allocation.

Under dynamic allocation, an actual IP address is allocated to the public IP address resource only when the resource is in use—that is, when it is associated with a resource such as a running virtual machine. If the virtual machine is stopped (deallocated) or deleted, the IP address assigned to the public IP address resource is released and returned to the pool of available IP addresses managed by Azure. When you restart the virtual machine, a different IP address will most likely be assigned.

If you want to retain the IP address, the public IP address resource should be configured to use static IP allocation. An IP address will be assigned immediately (if one was not already dynamically assigned). This IP address will never change, regardless of whether the associated virtual machine is stopped or deleted.

Typically, static public IP addresses are used in scenarios where a dependency is defined by a particular IP address. For example, static IP addresses are commonly used in the following scenarios:

- Where firewall rules specify an IP address
- Where a DNS record would need to be updated when an IP address changes
- Where the source IP address is used as a (weak) form of authentication of the traffic source
- Where an SSL certificate specifies an explicit IP address rather than a domain name

With private IP addresses, static allocation allows you to specify the IP address to use from the available subnet address range. In contrast, static allocation of public IP addresses does not allow you to specify which public IP address to use. Azure assigns the IP address from a pool of IP addresses in the Azure region where the resource is located.

## Public IP address prefixes

When using multiple public IP addresses, it can be convenient to have all of the IP addresses allocated from a single IP range or prefix. For example, when configuring firewall rules, this allows you to configure a single rule for the prefix, rather than separate rules for each IP address.

To support this scenario, Azure allows you to reserve a public IP address prefix. Public IP address resources associated with that prefix will have their IP addresses assigned from that range, rather than from the general-purpose Azure pool.

When creating a prefix, specify the prefix resource name, subnet size (for example, /28 for 16 IP addresses), and the Azure region where the IP addresses will be allocated.

Once the prefix is created, individual public IP addresses can be created that are associated with this prefix. Note that only Standard-tier public IP addresses support allocation from a prefix, and thus only static allocation is supported. The IP address assigned to these resources will be taken from the prefix range—you cannot specify a specific IP address from the range.

> **NOTE** **PREFIXES BENEFITS AND CONSTRAINTS**
>
> See the following links to the benefits and constraints of public IP address prefixes:
>
> - Benefits: *https://learn.microsoft.com/en-us/azure/virtual-network/ip-services/ public-ip-address-prefix#benefits*
>
> - Constraints: *https://learn.microsoft.com/en-us/azure/virtual-network/ip-services/ public-ip-address-prefix#constraints*

### DNS LABELS

The domain name system (DNS) can be used to create a mapping from a domain name to an IP address so you can reference IP address endpoints using a domain name rather than using the assigned IP address directly.

There are four ways to configure a DNS label for an Azure public IP address:

1. By specifying the DNS name label property of the public IP address resource
2. By creating a DNS A record in Azure DNS or a third-party DNS service hosting a DNS domain
3. By creating a DNS CNAME record in Azure DNS or a third-party DNS service hosting a DNS domain
4. By creating an alias record in Azure DNS

### SPECIFY THE DNS NAME LABEL PROPERTY

With this option, you specify the left-most part of the DNS label as a property in the public IP address resource. Azure provides the DNS suffix, which will be of the form <region>.cloudapp.azure.com. The DNS label you provide is concatenated with this suffix to form the fully qualified domain name (FQDN), which can be used to look up the IP address via a DNS query.

For example, if your public IP address is deployed to the Central US region, and you specify the DNS label contoso-app, then the FQDN will be contoso-app.centralus.cloudapp.azure.com.

The major limitation of this approach is that the DNS suffix is taken from an Azure-provided DNS domain. It does not support the use of your own vanity domain, such as contoso.com. To address this, you will need to use one of the other approaches.

### CREATE A *DNS A* RECORD

In this approach, you will have already hosted your vanity domain either in Azure DNS or a third-party DNS service. Using your hosting service, you can create a DNS entry in your vanity domain mapping to your public IP address resource. If you use a DNS A record, which maps directly to an IP address, you will need to update the DNS record if the assigned IP address changes. To avoid this, use static rather than dynamic IP allocation.

### CREATE A *DNS CNAME* RECORD

In this approach, you start by creating a DNS label for your public IP address. You then create a CNAME record in your vanity domain, which maps your chosen domain name to the Azure-provided DNS name. For example, you might map www.contoso.com to contoso-app.centralus.cloudapp.azure.com. This approach has the advantage of avoiding the need for static IP allocation because the Azure-provided DNS entry updates automatically if the assigned IP address changes. However, the downside of this approach is that the domain name system does not support CNAME records at the root of a DNS domain, which means while you can create a CNAME record for *www.contoso.com*, you cannot create one for *contoso.com* (without the "www").

### CREATE AN ALIAS RECORD

In this approach, your vanity domain must be hosted in Azure DNS. You can then create an alias record, which works the same as an A record, except that rather than specifying the assigned IP address value explicitly in the DNS record, you simply reference the public IP address resource. The assigned IP address is taken from this resource and automatically configured in your DNS alias record. With alias records, the DNS record is automatically updated if the assigned IP address changes, avoiding the need for static IP allocation.

## Outbound internet connections

When a public IP address is assigned to a virtual machine's network interface, outbound traffic to the internet will be routed through that IP address. The recipient will see your public IP address as the source IP address for the connection.

However, the virtual machine itself does not see the public IP address in its network settings—it only sees the private IP address. Traffic leaves the virtual machine via the private IP address, and Source Network Address Translation (SNAT) is used to map the outbound traffic from the private IP address to the public IP address.

Note that, historically, a public IP address is not required for outbound internet traffic. Even without a public IP address assigned, virtual machines can still make outbound internet connections. In this case, SNAT is used to map the private IP address to an internet-facing IP address. However, beginning in September of 2025, default outbound access will be disabled for new virtual networks and will require a NAT Gateway or other outbound appliance.

> **NEED MORE REVIEW?** **DEFAULT OUTBOUND ACCESS**
>
> For more information regarding default outbound access, visit *https://learn.microsoft.com/ en-us/azure/virtual-network/ip-services/default-outbound-access.*

### Create a public IP address using the Azure portal

Creating a new public IP address is a simple process using the Azure portal. Search for **Public IP addresses**, and then click Create. Like all resources in Azure, some details will be required, including the name of the resource, the SKU (or pricing tier), the IP version, idle time-out, subscription, resource group, and location/region. For the Basic SKU, you also specify the IP version and static or dynamic assignment. For the Standard SKU, choose between zone-redundant deployment or a specific availability zone.

The location is critical because an IP address must be in the same location/region as the virtual machine or other resource that will use it. Figure 4-10 shows the Create Public IP Address blade.

## Configure user-defined network routes

Network routes control how traffic is routed in your network. Azure provides default routing for common scenarios, with the ability to configure your own network routes where necessary.

### System routes

Azure VMs in the same VNet can communicate automatically with each other and with the internet without any explicit configuration changes, even when they are in different subnets. This is also the case for communication from the VMs to your on-premises network when a hybrid connection from Azure to your data center has been established.

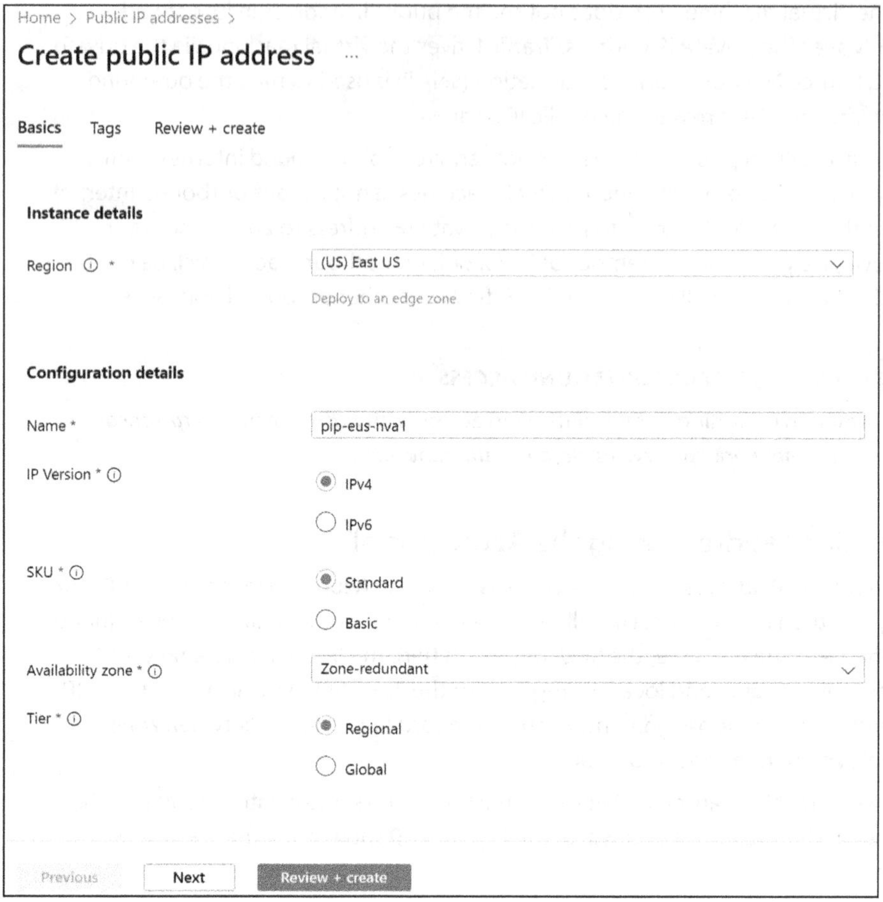

**FIGURE 4-10** Creating a public IP address in the Azure portal

This ease of setup is made possible by what is known as system routes, which define how IP traffic flows in Azure VNets. The following are the default system routes that Azure will use and provide for you:

- Within the same subnet
- From one subnet to another within a VNet
- VMs to the internet
- One VNet to another VNet through a VPN gateway (optional)
- One VNet to another VNet through VNet peering (optional)
- A VNet to your on-premises network through a VPN gateway or ExpressRoute (optional)
- VirtualNetworkServiceEndpoint (optional)

Figure 4-11 shows an example of how these system routes make it easy to get up and running. System routes provide for most typical scenarios by default, without you having to make any routing configuration.

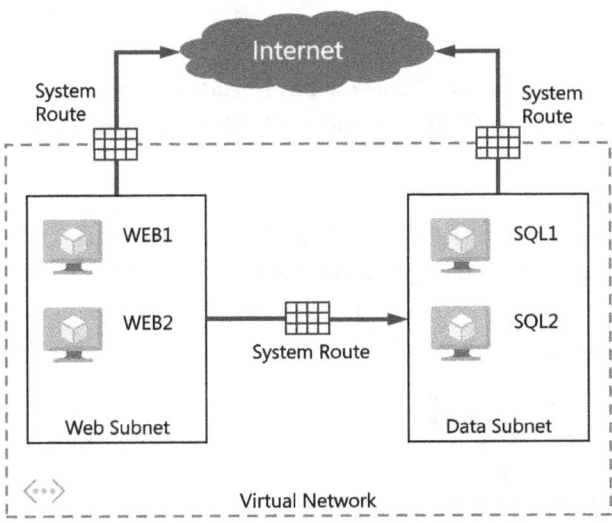

**FIGURE 4-11** N-Tier application deployed to Azure VNet using system routes

## User-defined routes

In some cases, you will want to configure the routing of packets differently from what is provided by the default system routes. One of these scenarios is when you want to send traffic through a network virtual appliance, such as a third-party load balancer, firewall, or router deployed into your VNet from the Azure Marketplace.

To make this possible, create what are known as user-defined routes (UDRs). The UDR is implemented by creating a route table resource. Within the route table, a number of routes are configured. Each route specifies the destination IP range (in CIDR notation) and the next hop IP address. A variety of different types of next hop are supported:

- **Virtual appliance** A virtual machine running a network application such as a load balancer or firewall. With this next hop type, you also specify the IP address of the appliance, which can be a virtual machine or internal load balancer for high-availability virtual appliances.

- **Virtual network gateway** Used to route traffic to a VPN gateway (but not an ExpressRoute gateway, which uses BGP for custom routes). Because there can be only one VPN gateway associated with a VNet, you are not prompted to specify the actual gateway resource.

- **Virtual network** Used to route traffic within the VNet.

- **Internet** Used to route a specific IP address or prefix to the internet.

- **None** Used to drop all traffic sent to a given IP address or prefix.

This route table is then associated with one or more subnets. Traffic originating in the subnet whose destination matches the destination IP range of a route table rule will instead be routed to the corresponding next hop IP address. The service running at this IP address is responsible for all onward routing.

Figure 4-12 shows a UDR that has been created to direct outbound traffic via a virtual appliance. In this case, the appliance is a firewall running as a VM in Azure in the DMZ subnet.

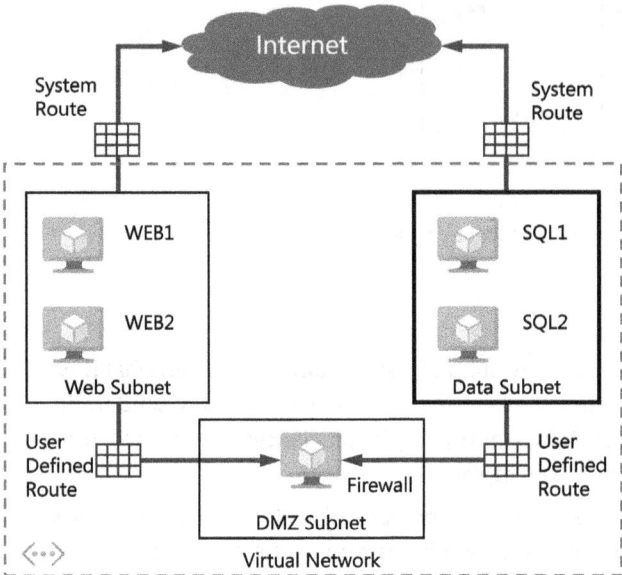

**FIGURE 4-12**   N-Tier application deployed with a firewall using user-defined routes

The same appliance can also be used to filter traffic between the Apps and Data subnets. An example route table implementing this design is shown in Figure 4-13.

Home > Route tables > ExamRef-UDR

**ExamRef-UDR | Routes** ☆ ⋯
Route table

| Search | ≪ | + Add ◯ Refresh 🗟 Give feedback | | |
|---|---|---|---|---|
| 🔆 Overview | | 🔎 Search routes | | |
| 📋 Activity log | | **Name ↑↓** | **Address prefix ↑↓** | **Next hop type ↑↓** | **Next hop IP address ↑↓** |
| 🗝 Access control (IAM) | | AppToData | 10.0.0.0/24 | VirtualAppliance | 10.10.99.4 | ⋯ |
| 🏷 Tags | | DataToApps | 10.0.1.0/24 | VirtualAppliance | 10.10.99.4 | ⋯ |
| 🛠 Diagnose and solve problems | | ToInternet | 0.0.0.0/32 | VirtualAppliance | 10.10.99.4 | ⋯ |

Settings

🗄 Configuration

🔆 Routes

**FIGURE 4-13**   Route table rules forcing network traffic through a firewall

## IP forwarding

User-defined routes (UDR) change the default system routes that Azure creates for you in an Azure VNet. In the virtual appliance scenario, UDRs forward traffic to a virtual appliance such as a firewall, which is running as an Azure virtual machine.

By default, a virtual machine in Azure will not accept a network packet addressed to a different IP address. For that traffic to be allowed to pass into that virtual appliance, you must enable IP forwarding on the network interface of the virtual machine. This configuration doesn't typically involve any changes to the Azure UDR or VNet, but depending on the scenario, you might need to make some configuration changes in the VM's operating system to enable this to work correctly.

IP forwarding can be enabled on a network interface by using the Azure portal, PowerShell, or the Azure CLI. In Figure 4-14, Enable IP Forwarding is selected for the network interface of the NGFW1 VM. This VM is now able to accept and send packets that were not originally intended for this VM.

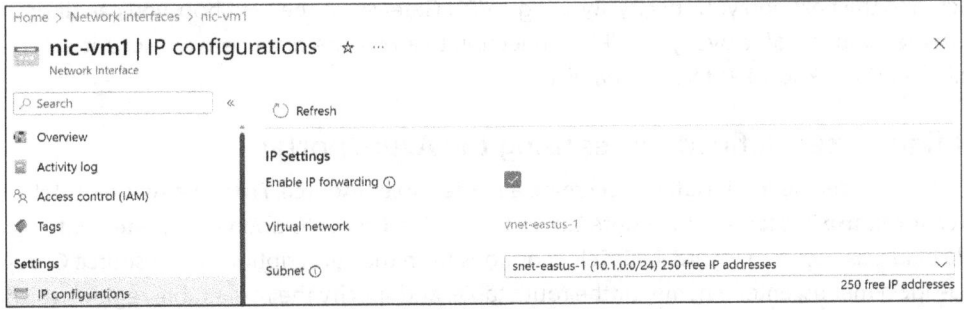

**FIGURE 4-14**   IP forwarding enabled on network interface

## How routes are applied

A given network packet may match multiple route table rules. When designing and implementing custom routes, it's important to understand the precedence rules that Azure applies.

If multiple routes contain the same address prefix, Azure selects the route type, based on the following priorities:

1. User-defined routes
2. System routes for traffic in a virtual network, across a virtual network peering, or to a virtual network service endpoint

3. BGP routes

4. Other system routes

Within a single route table, a given network packet may match multiple routing rules. There is no explicit precedence order on the rules in a route table. Instead, precedence is given to the rule with the most specific match to the destination IP address. If an IP address matches two rules, the longest prefix match algorithm is used to select the route.

For example, if a route table contains one rule for prefix 10.10.0.0/16, and another rule for 10.10.30.0/28, then any traffic to IP address 10.10.30.4 will be matched against the second rule in preference to the first.

When troubleshooting networking issues, it can be useful to have deeper insight into exactly which routes are being applied to a given network interface. Using the Effective Routes feature of each network interface, you can see the full details of every network route applied to that network interface, giving you full insight into how each outbound connection will be routed based on the destination IP address.

## Forced tunneling

A special case is when routes are configured with the destination IP prefix 0.0.0.0/0. Given the precedence rules described earlier, this route controls traffic destined for any IP address not covered by any other rules.

By default, Azure implements a system route directing all traffic matching 0.0.0.0/0 (and not matching any other route) to the internet. If you override this route, this traffic is instead directed to the next hop you specify. By using a VPN Gateway as the next hop, you can direct all internet-bound traffic over your VPN connection to an on-premises network security appliance. This is known as *forced tunneling*.

## Configure user-defined routes using the Azure portal

To configure user-defined routes, first create a Route Table resource. From the Azure portal, search for **Route Tables.** On the Route Tables blade, click Create to open the Create Route Table blade, as shown in Figure 4-15. Select options from the Subscription and Resource Group drop-down menus, enter a name for the route table, and specify the route table region, which must be the same Azure region that the subnets use with this route table.

Having created the route table, the next step is to define the routes. Open the Route Table blade, and under Settings, click Routes to open the list of routes in the route table. Then click Add to open the Add Route blade, as shown in Figure 4-16.

Home > Route tables >
Create Route table ...
Basics
Tags
Review + create
Project details
Select the subscription to manage deployed resources and costs. Use resource groups like folders to organize and manage all your resources.
Subscription *
Azure Pass - Sponsorship
Resource group *
az104-rg1
Create new
Instance details
Region *
East US
Name *
rt-examref
Propagate gateway routes *
Yes
No
Previous
Next
Review + create

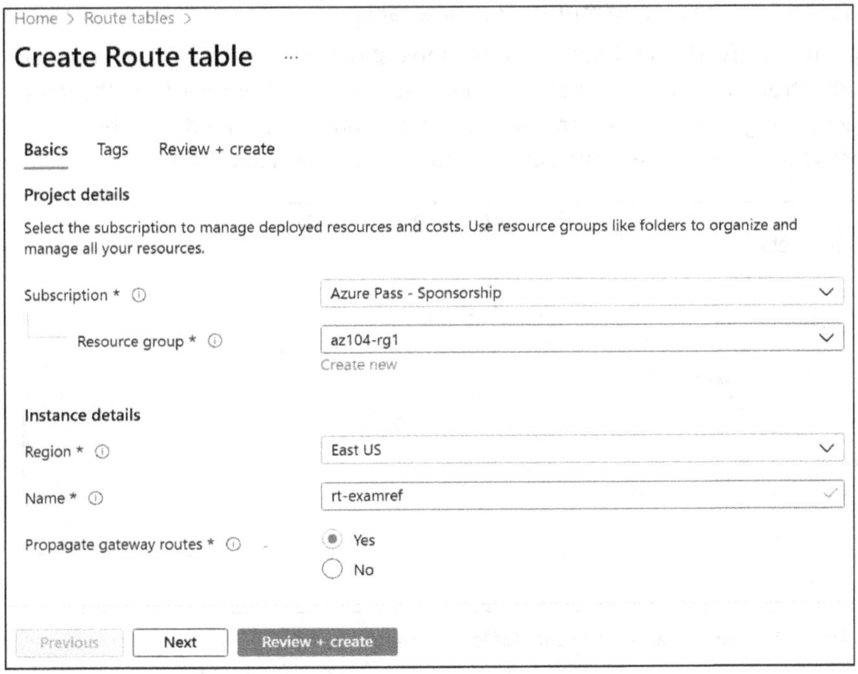

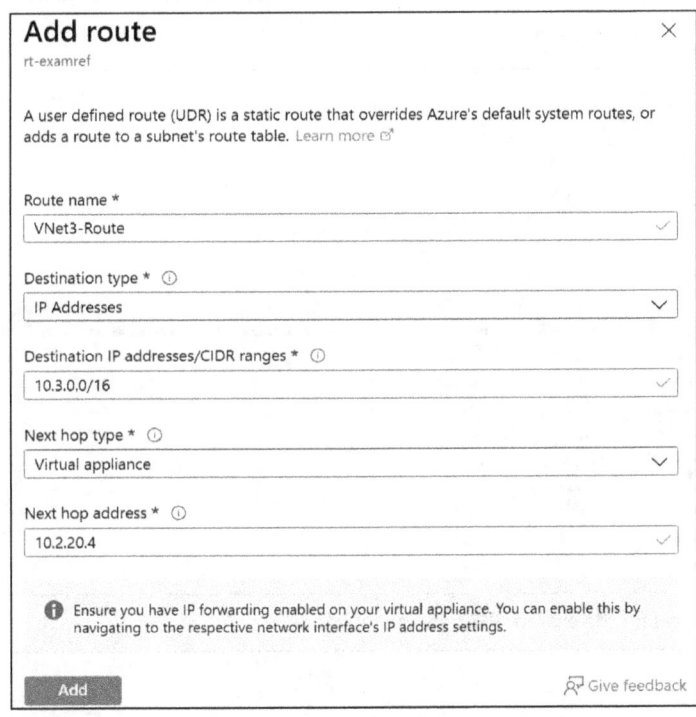

**FIGURE 4-15** The Create Route Table blade in the Azure portal

Add route
rt-examref
A user defined route (UDR) is a static route that overrides Azure's default system routes, or adds a route to a subnet's route table. Learn more
Route name *
VNet3-Route
Destination type *
IP Addresses
Destination IP addresses/CIDR ranges *
10.3.0.0/16
Next hop type *
Virtual appliance
Next hop address *
10.2.20.4
Ensure you have IP forwarding enabled on your virtual appliance. You can enable this by navigating to the respective network interface's IP address settings.
Add
Give feedback

**FIGURE 4-16** The Add Route blade in the Azure portal

Repeat this process for each custom route in the route table.

The final step is to specify which subnets this route table should be associated with. This can be configured either from the subnet, or from the route table. In the latter case, from the Route Table blade, under Settings, click Subnets to open the list of subnets associated with the route table. Click Associate to open the Associate Subnet blade, as shown in Figure 4-17.

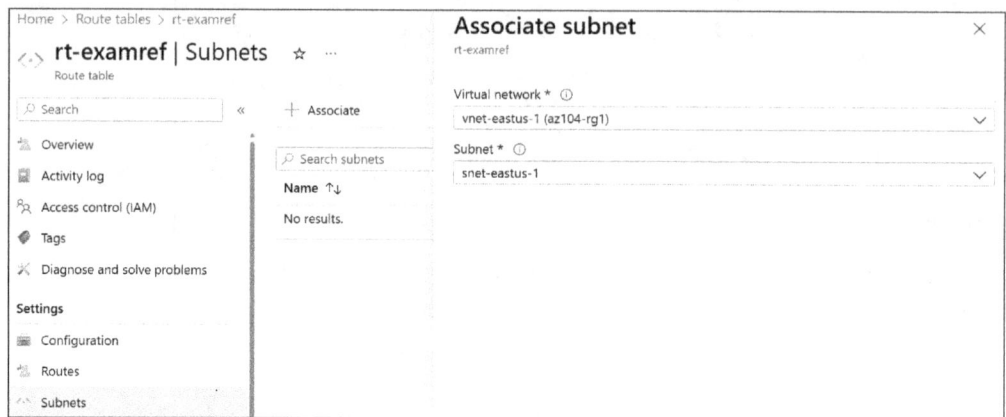

**FIGURE 4-17** The Associate Subnet blade for a route table in the Azure portal

To see the effective routes for a given network interface, navigate to the network interface blade in the Azure portal and then click Effective Routes to open the Effective Routes blade, as shown in Figure 4-18.

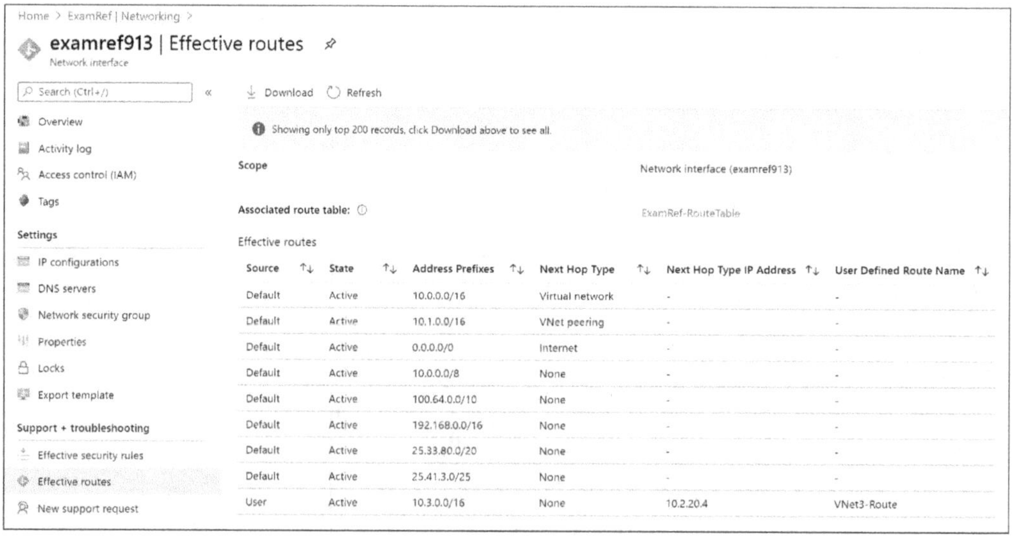

**FIGURE 4-18** The list of effective routes for the examref913 network interface

# Troubleshoot network connectivity

Azure provides several built-in tools to troubleshoot network connectivity, with most of them available through Network Watcher. This section focuses on two of the tools within Network Watcher that can help you troubleshoot network connectivity.

## Connection Troubleshoot

Connection Troubleshoot is a Network Watcher feature designed to test the connectivity between an Azure VM or an App Gateway and another endpoint—either another Azure VM, or an arbitrary internet or intranet endpoint. This diagnostic tool can identify a range of problems, including guest VM issues, such as guest firewall configuration, low memory, or high CPU; Azure configuration issues such as network security groups blocking traffic; or routing issues diverting traffic. It can also diagnose other network issues, such as DNS failures.

To use Connection Troubleshoot from the Azure portal, open Network Watcher, and then click Connection Troubleshoot. Specify the source VM, then specify the destination, either as another VM or by giving a URI, FQDN, or IPv4 address. Specify the protocol to use (either TCP or ICMP). For TCP, you can specify the destination port, and, under Advanced Settings, the source port. An example configuration is shown in Figure 4-19.

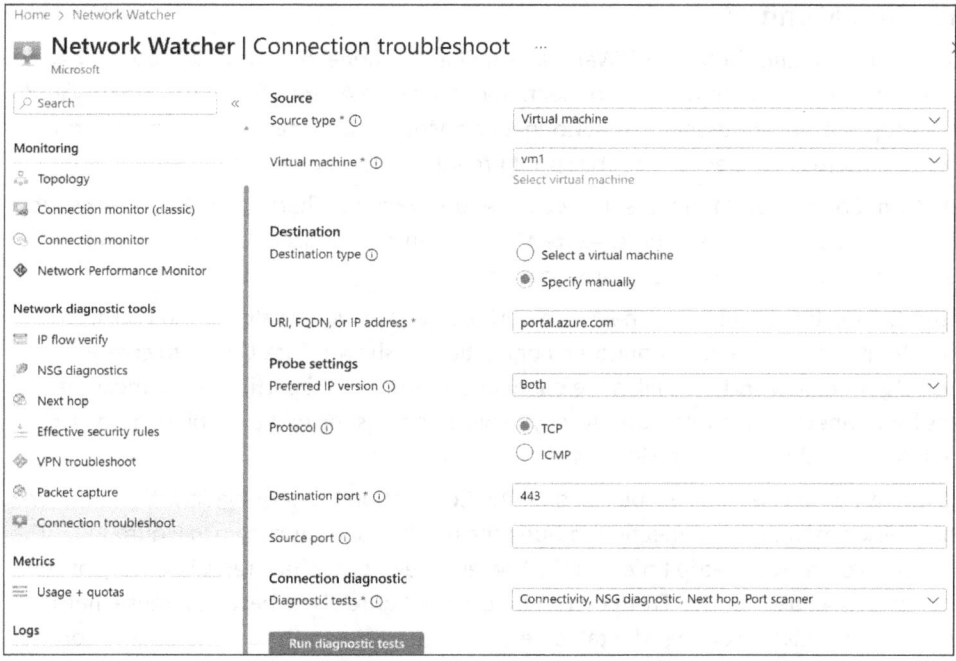

**FIGURE 4-19** Network Watcher Connection Troubleshoot configuration

The test takes a few minutes to run. Upon completion, the results will be shown at the bottom of the page. An example output is shown in Figure 4-20.

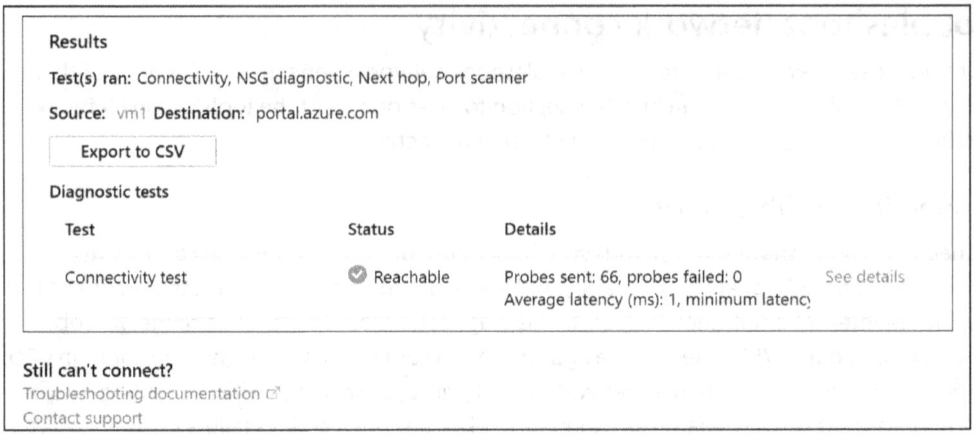

**Results**

**Test(s) ran:** Connectivity, NSG diagnostic, Next hop, Port scanner

**Source:** vm1 **Destination:** portal.azure.com

Export to CSV

**Diagnostic tests**

| Test | Status | Details | |
|------|--------|---------|---|
| Connectivity test | ✔ Reachable | Probes sent: 66, probes failed: 0<br>Average latency (ms): 1, minimum latency | See details |

**Still can't connect?**
Troubleshooting documentation ↗
Contact support

**FIGURE 4-20**   Network Watcher Connection Troubleshoot results

Connection Troubleshoot is also available via PowerShell using the `Test-AzNetwork-WatcherConnectivity` cmdlet and via the Azure CLI using the Azure Network Watcher `az network watcher test-connectivity` command.

## Connection Monitor

The Connection Monitor in Network Watcher is similar to Connection Troubleshoot in that it uses the same mechanism to test the connection between an Azure VM or App Gateway and another endpoint. The difference is that Connection Monitor provides ongoing connection monitoring, whereas Connection Troubleshoot provides only a point-in-time test.

Data from Connection Monitor is surfaced in Azure Monitor. Charts show key metrics such as round-trip time and probe failures. Azure Monitor can also be used to configure alerts, triggered by connection failures or a drop in performance.

To use Connection Monitor via the Azure portal, open Network Watcher and click Connection Monitor. A list of active monitored connections is shown. Click Create to create a new monitored connection and then fill in the connection settings. The settings are almost the same as for Connection Troubleshoot. Also, you will need to specify the probing interval in seconds. An example is shown in Figure 4-21.

The monitored connection will be listed on the Connection Monitor blade within Network Watcher. Click a monitored connection to open the results panel, as shown in Figure 4-22. The chart shows average round-trip time and the percentage of probe failures. Click the chart to view the data in Azure Monitor. From there, you can configure alerts based on these metrics exceeding thresholds you define. The table below the chart shows the current connection status; clicking each line provides further details about the status, which is similar to how Connection Troubleshoot results are shown.

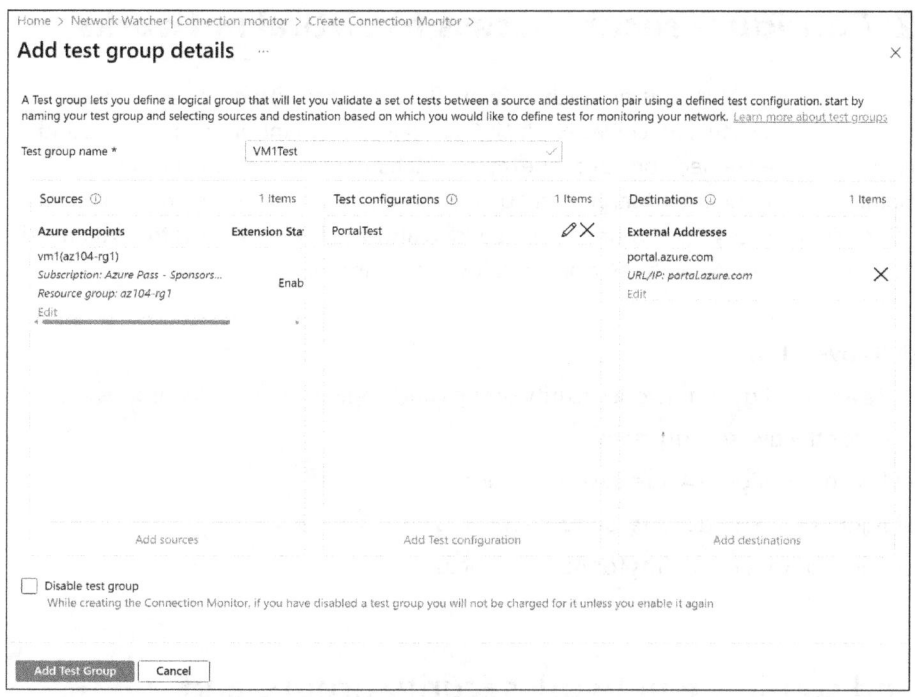

**FIGURE 4-21**  Network Watcher Connection Monitor configuration

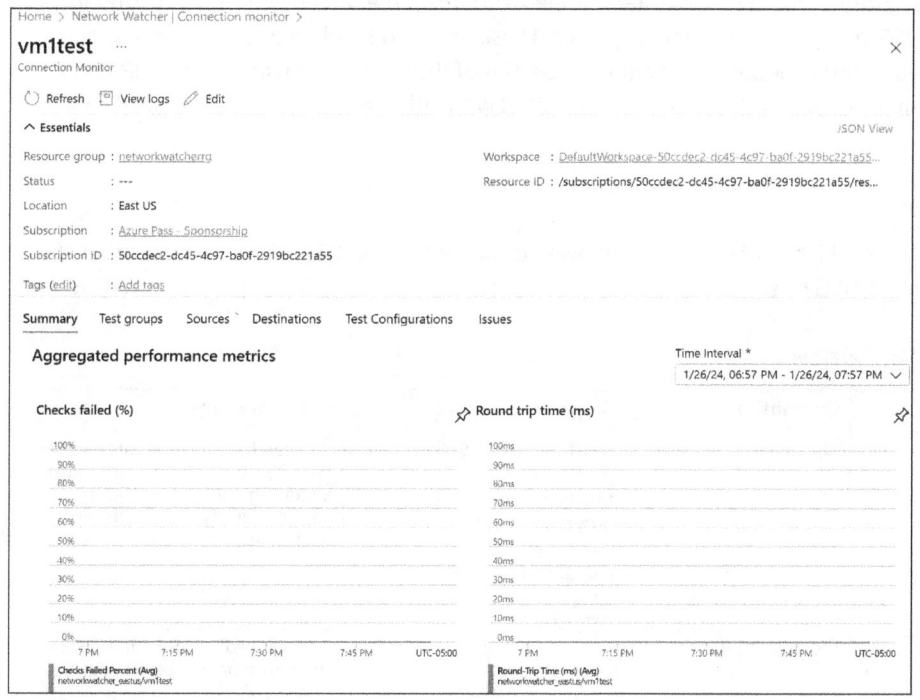

**FIGURE 4-22**  Network Watcher Connection Monitor status

# Skill 4.2: Configure secure access to virtual networks

Network security groups (NSGs) control which network flows are permitted into and out of your virtual networks and virtual machines. Each NSG contains lists of inbound and outbound rules, which give you fine-grained control over exactly which network flows are allowed or denied. Combine this with the use of service and private endpoints, which provide connectivity to Azure PaaS services directly from your virtual network, and you have the tools that you need to configure secure, private connectivity from your VMs to Azure services.

> **This skill covers how to:**
> - Create and configure network security groups and application security groups
> - Evaluate effective security rules
> - Deploy and configure Azure Bastion Service
> - Configure service endpoints for Azure services
> - Configure private endpoints for Azure services

## Create and configure network security groups and application security groups

A network security group (NSG) is a standalone Azure resource, which acts as a networking filter. Each NSG contains a list of security rules. These are used to allow or deny inbound or outbound network traffic, depending on the properties of that traffic, such as protocol, IP address, and port. When you apply the NSG, it is associated with either a subnet or with a specific VM's network interface.

### NSG rules

NSG rules define which traffic flows are allowed or denied by the NSG. Table 4-4 describes the properties of an NSG rule.

**TABLE 4-4** NSG properties

| Property | Description | Constraints | Considerations |
|---|---|---|---|
| **Name** | The name of the rule. | Must be unique within the region.<br>Must end with a letter, number, or underscore.<br>Cannot exceed 80 characters. | You can have several rules within an NSG, so make sure you follow a naming convention that allows you to identify the purpose of each rule. |
| **Protocol** | The network protocol the rule applies to. | TCP, UDP, or *. | Using * as a protocol includes ICMP as well as TCP and UDP. In the Azure portal, select 'Any' instead of '*'. |

| Property | Description | Constraints | Considerations |
|---|---|---|---|
| **Source port range(s)** | Source port range(s) to match for the rule. | Single port number from 1 to 65535; a port range (example: 1–65535); a list of port or port ranges; or * (for all ports). | The source ports could be ephemeral, so unless your client program is using a specific port, use * in most cases.<br>Try to reduce the number of rules by specifying multiple ports or port ranges in a single rule. |
| **Destination port range** | Destination port range(s) to match for the rule. | Single port number from 1 to 65535, port range (such as 1–65535), a list of port or port ranges, or * (for all ports). | Try to reduce the number of rules by specifying multiple ports or port ranges in a single rule. |
| **Source address prefix(es)** | Source address prefix(es) or service tag(s) to match for the rule. | Single IP address (such as 10.10.10.10), IP subnet (such as 192.168.1.0/24), a service tag, a list of the above, or * (for all addresses). | Consider using ranges, service tags, and lists to reduce the number of rules.<br>The IP addresses of Azure VMs can also be specified implicitly using application security groups. |
| **Destination address prefix(es)** | Destination address prefix(es) or service tag(s) to match for the rule. | Single IP address (such as 10.10.10.10); IP subnet (such as 192.168.1.0/24); a service tag; a list of the above; or * (for all addresses). | Consider using ranges, default tags, and lists to reduce the number of rules.<br>The IP addresses of Azure VMs can also be specified implicitly using application security groups. |
| **Direction** | Direction of traffic to match for the rule. | Inbound or outbound. | Inbound and outbound rules are processed separately, based on traffic direction. |
| **Priority** | Rules are checked in the order of priority. Once a matching rule is found, no more rules are tested. | Unique number between 100 and 4096. Uniqueness is only within this NSG. | Consider creating rules and jumping priorities by 100 for each rule to leave space for new rules you might create in the future. |
| **Action** | Type of action to apply if the rule matches. | Allow or deny. | Keep in mind that if an allow rule is not found for a packet, the packet is dropped. |

**NOTE    NSG RULE PRIORITY**

NSG rules are enforced based on their priority. Priority values start from 100 and go to 4096 (and from 65001 to 65003 for default rules). Rules will be read and enforced starting with 100 and are followed by 101, 102, and so on. When a rule is found that matches the traffic under consideration, the rule is applied, and all further processing stops—subsequent rules are disregarded.

For example, suppose you had an inbound rule that allowed TCP traffic on any port with a priority of 250 and another that denied TCP traffic on Port 80 with a priority of 125. An inbound TCP connection on port 80 would be denied, since the deny rule has a lower priority value and would be applied before the allow rule is considered.

## Service tags

Many Azure services are accessed via internet-facing endpoints. These endpoints can change over time, for example as new Azure regions are built. This makes it difficult to use NSG rules to control access to those services—it's hard to identify the list of IP ranges to use, and even harder to keep the list up to date.

To address this problem, Azure provides *service tags*. These are platform-defined shortcuts that map to the IP ranges of various Azure services. The IP ranges associated with each service tag are updated automatically whenever the IP addresses used by the service change.

Service tags are used in NSG rules as a quick and reliable way of creating rules that control traffic to each service. Typically, they are used in outbound rules to control which other Azure services the VMs in a VNet can or cannot access.

Note that service tags control access to the service, but not to a specific resource within that service. For example, a service tag might be used in an NSG rule allowing a VM to connect to Azure Storage. This rule cannot control which account in Azure Storage the VM will attempt to use.

Service tags are provided for more than 60 Azure services, and the list is growing. Here are some of the most commonly used service tags.

- **VirtualNetwork** Controls access to the virtual network address space where the NSG is assigned. It refers to the entire virtual network (not just the subnet), plus all connected virtual networks and any on-premises address space connected via site-to-site VPN or ExpressRoute. Note that the network address space of peered virtual networks is only included if the Allow Virtual Network Access property is set to Enabled.

- **Internet** Denotes the public internet address space. This includes the internet-facing Azure IP address ranges that are used for public IP addresses and Azure platform services.

- **AzureCloud** Denotes the Azure data center public IP space. This service tag can be scoped to a specific Azure region, such as by specifying AzureCloud.EastUs.

- **AzureLoadBalancer** Denotes the IPs where Azure Load Balancer health probes will originate. Traffic from these addresses should be allowed for any load-balanced VMs. Note that this service tag cannot be used to control traffic coming through the Load Balancer from elsewhere. This traffic can be filtered using the originating source IP, which is not modified as it passes through the Azure Load Balancer

- **AzureTrafficManager** Performs a similar role for Azure Traffic Manager. It is used to allow traffic from the source IP addresses of Traffic Manager health probes.

- **Storage** Represents the IP addresses used by the Azure Storage service. As with the Azure Cloud Service Tag, the Storage service tag can be region scoped. For example, you can specify Storage.WestUS to allow access only to Storage accounts in the West US region.

- **Sql** Represents the IP addresses used by the Azure Database for MySQL, Azure Database for PostgreSQL, and Azure Synapse Analytics. This service tag can also be scoped to a specific region.

# Default rules

All NSGs have a set of default rules. You cannot add to, edit, or delete these default rules. However, since they have the lowest possible priority, they can be overridden by other rules you create.

The default rules allow and disallow traffic as follows:

- **Virtual network**  Traffic originating and ending in a virtual network is allowed in both inbound and outbound directions.
- **Internet**  Outbound traffic is allowed, but inbound traffic is blocked.
- **Load balancer**  Allows the Azure Load Balancer to probe the health of your VMs and role instances. If you are not using a load-balanced set, you can override this rule.

> **NOTE  LOAD BALANCER TRAFFIC**
>
> The Load Balancer default rule uses the `AzureLoadBalancer` service tag. This applies only to Azure Load Balancer health probes, which originate at the load balancer. It does not apply to traffic received through the load balancer, which retain their original source IP addresses and ports.

Table 4-5 shows the default inbound rules for each NSG.

**TABLE 4-5**  Default inbound rules

| Name | Priority | Source | Source Port | Destination | Destination Port | Protocol | Access |
|------|----------|--------|-------------|-------------|------------------|----------|--------|
| AllowV-NetInBound | 65000 | VirtualNetwork | Any | Virtual-Network | Any | Any | Allow |
| AllowAzureLoadBalancerInBound | 65001 | AzureLoad-Balancer | Any | Any | Any | Any | Allow |
| DenyAllInBound | 65500 | Any | Any | Any | Any | Any | Deny |

Table 4-6 shows the default outbound rules for each NSG.

**TABLE 4-6**  Default outbound rules

| Name | Priority | Source | Source Port | Destination | Destination Port | Protocol | Access |
|------|----------|--------|-------------|-------------|------------------|----------|--------|
| AllowVNet OutBound | 65000 | Virtual Network | Any | Virtual Network | Any | Any | Allow |
| AllowInternet OutBound | 65001 | Any | Any | Internet | Any | Any | Allow |
| DenyAll OutBound | 65500 | Any | Any | Any | Any | Any | Deny |

# Application security groups

As you have seen, NSG rules are like traditional firewall rules and are defined using source and destination IP blocks. They enable you to segment your network traffic into application tiers, which are segmented into separate subnets.

This creates some management challenges:

- The IP blocks for each subnet must be carefully planned in advance. To ensure additional servers may be added in future, each subnet must be bigger than you really need, which results in inefficient use of the IP space.

- If you make a subnet too small and run out of space, it can be time-consuming to reconfigure the network to free up additional space, especially without application downtime.

- Each subnet requires a separate NSG, making it difficult to get an overall picture of the permitted and blocked traffic at an application level.

*Application security groups* (ASGs) address these challenges by offering an alternative approach to network segmentation. You can use them to segment your application into separate tiers, and they strictly control the permitted network flows between tiers. However, ASGs do not require that you associate each tier with a separate subnet, so all the challenges associated with planning and managing subnets fall away. With ASGs, you explicitly—rather than implicitly—define which application tier each VM belongs to, based on the subnet in which the VM has been placed. All VMs can be placed in a single subnet, and a single NSG is used to define all permitted network flows between application tiers. Because a single subnet is used, the IP space can be managed much more flexibly, and because there is a single NSG with rules referring to named application tiers, the network rules are easier to understand and can all be managed in one place.

Figure 4-23 shows a standard three-tier application architecture with web servers, application servers, and database servers. These servers have been grouped by associating each server with the appropriate application security group. All servers are placed in the same subnet without having to think about how the network space is subdivided. A single network security group contains rules defining the permitted traffic flows between application tiers.

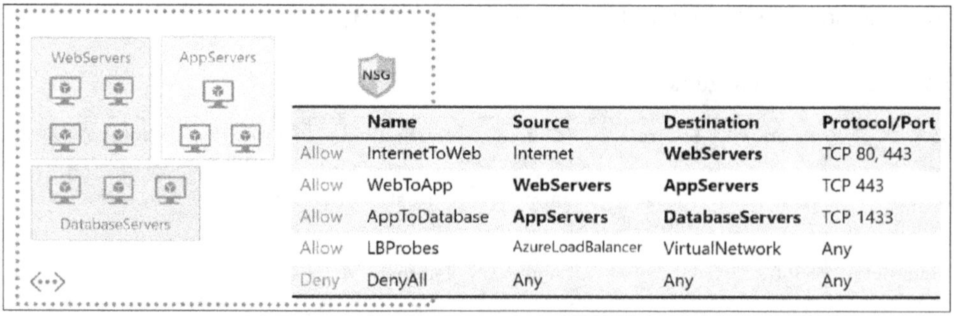

| | Name | Source | Destination | Protocol/Port |
|---|---|---|---|---|
| Allow | InternetToWeb | Internet | **WebServers** | TCP 80, 443 |
| Allow | WebToApp | **WebServers** | **AppServers** | TCP 443 |
| Allow | AppToDatabase | **AppServers** | **DatabaseServers** | TCP 1433 |
| Allow | LBProbes | AzureLoadBalancer | VirtualNetwork | Any |
| Deny | DenyAll | Any | Any | Any |

**FIGURE 4-23** Using application security groups to simplify subnet and NSG management

Application security groups enable you to configure network security as a natural extension of an application's structure, so you can group virtual machines and define network security policies based on those groups. You can reuse your security policy at scale without the manual maintenance of explicit IP addresses. The platform handles the complexity of explicit IP addresses and multiple rule sets, which allows you to focus on your business logic.

Configuring application security groups is straightforward:

1. First, create an application security group resource for each server group. This resource has no properties, other than its name, resource group, and location.

2. Next, associate the network interface from each VM with the appropriate application security group. This defines which group (or groups) each VM belongs to.

3. Finally, define your network security group rules using application security group names instead of explicit IP ranges. This is similar to how rules are configured using named service tags.

## Create an NSG using the Azure portal

To create an NSG using the Azure portal, follow these steps:

1. Search for **Network Security Groups**. On the Network Security Groups blade, click Create.

2. On the Create Network Security Group blade, provide a name, the subscription where your resources are located, the resource group for the NSG, and the location. (The location must be the same as the resources you want to apply the NSG.) In Figure 4-24, the NSG will be created to allow HTTP traffic into the Apps subnet and be named AppsNSG.

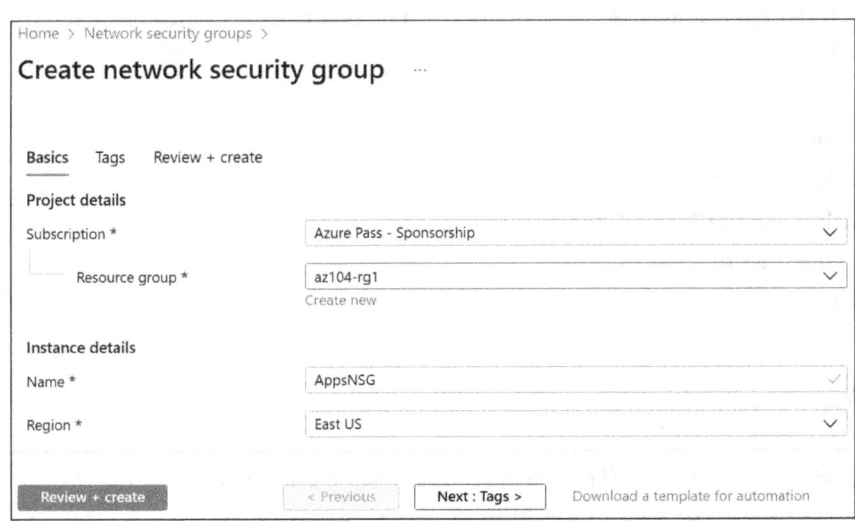

**FIGURE 4-24** Creating a network security group using the Azure portal

3. After the NSG has been created, open the NSG Overview blade, as shown in Figure 4-25. Here, you see that the NSG has been created, but there are no inbound or outbound security rules beyond the default rules.

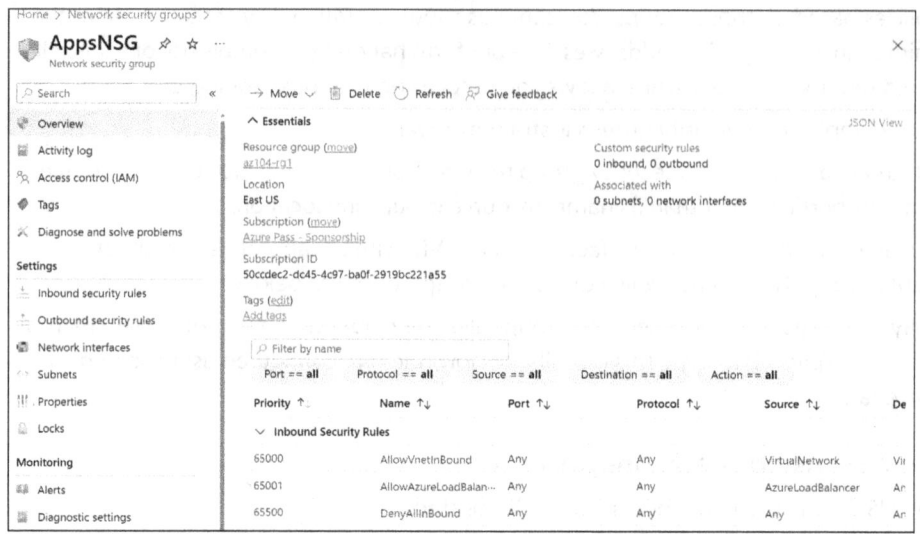

**AppsNSG** 📌 ⭐ ⋯
Network security group

🔍 Search    «    → Move ∨   🗑 Delete   ↻ Refresh   ⧉ Give feedback

| | |
|---|---|
| Overview | ∧ Essentials                                     JSON View |

| | |
|---|---|
| 📋 Activity log | Resource group (move)                         Custom security rules |
| ஃ Access control (IAM) | az104-rg1                                  0 inbound, 0 outbound |
| 🏷 Tags | Location                                 Associated with |
| ⚒ Diagnose and solve problems | East US                                0 subnets, 0 network interfaces |
| | Subscription (move) |
| **Settings** | Azure Pass - Sponsorship |
| ⊥ Inbound security rules | Subscription ID |
| ⊥ Outbound security rules | 50ccdec2-dc45-4c97-ba0f-2919bc221a55 |
| 🔌 Network interfaces | Tags (edit) |
| ⊹ Subnets | Add tags |

🔍 Filter by name

Port == **all**    Protocol == **all**    Source == **all**    Destination == **all**    Action == **all**

| Priority ↑ | Name ↑↓ | Port ↑↓ | Protocol ↑↓ | Source ↑↓ | De |
|---|---|---|---|---|---|
| ∨ **Inbound Security Rules** | | | | | |
| 65000 | AllowVnetInBound | Any | Any | VirtualNetwork | Vir |
| 65001 | AllowAzureLoadBalan⋯ | Any | Any | AzureLoadBalancer | An |
| 65500 | DenyAllInBound | Any | Any | Any | An |

**FIGURE 4-25**   The NSG Overview blade, showing the inbound and outbound security rules

4. Create the inbound rule for HTTP and HTTPS traffic: In the Settings area on the left, click Inbound Security Rules, and then click Add to open the Add Inbound Security Rule blade. Notice that you can choose Basic or Advanced mode, depending on the level of control required.

5. To allow HTTP/HTTPS traffic on Port 80 and 443, complete the fields as shown here and in Figure 4-26:

- **Source**   Any
- **Source Port Ranges**   *
- **Destination**   Service Tag
- **Destination Service Tag**   VirtualNetwork
- **Destination Port Ranges**   80,443
- **Protocol**   TCP
- **Action**   Allow
- **Priority**   100
- **Name**   Allow_HTTP_HTTPS
- **Description**   Allow HTTP and HTTPS inbound traffic on ports 80 and 443

6. When you have completed the fields, click Add to create the NSG rule.

> **NOTE**   **APPLYING NSGS TO VIRTUAL NETWORKS**
>
> The destination IP ranges refer to the VNet, which allows the NSG to be applied to any subnet in any VNet and avoids coupling the NSG to a specific IP range. Traffic will be permitted only to those subnets where the NSG is applied.

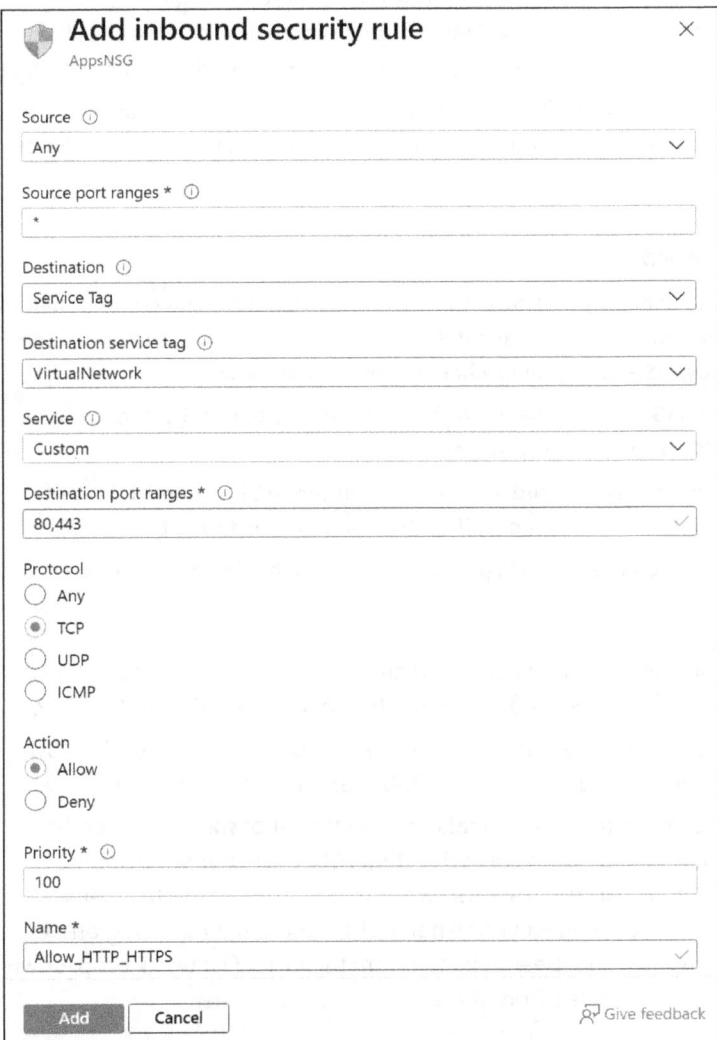

**Add inbound security rule**
AppsNSG

Source ⓘ
Any

Source port ranges * ⓘ
*

Destination ⓘ
Service Tag

Destination service tag ⓘ
VirtualNetwork

Service ⓘ
Custom

Destination port ranges * ⓘ
80,443

Protocol
○ Any
● TCP
○ UDP
○ ICMP

Action
● Allow
○ Deny

Priority * ⓘ
100

Name *
Allow_HTTP_HTTPS

[ Add ]  [ Cancel ]          ⓠ Give feedback

**FIGURE 4-26**  Adding an inbound rule to allow HTTP traffic

7.    Once the inbound rule has been saved, it will appear in the Azure portal. Review your rule to ensure it has been created correctly.

## Associate NSG to a subnet or network interface

NSGs are used to define the rules for how traffic is filtered for your IaaS deployments in Azure. You've learned how to create NSG resources and define the NSG rules. However, these NSGs, by themselves, are not effective until they are associated with a resource in Azure.

NSGs can be associated with network interface cards (NICs), which are associated to the VMs, or they can be associated with a subnet. Each NIC or subnet can only be associated with a single NSG. However, a single NSG can be associated with multiple NICs and/or subnets.

When an NSG is associated with a NIC, it applies to all IP configurations in that NIC. All inbound and outbound traffic to and from the NIC must be allowed by the NSG. It is possible to have a multi-NIC VM, and you can associate the same or different NSG to each network interface.

Alternatively, NSGs can be associated with a subnet; in that case, they apply to all traffic to and from resources in that subnet. This approach is useful when applying the same rule across multiple VMs.

---

*NOTE* **HOW NSGS ARE APPLIED**

**Microsoft does not recommend deploying NSGs to subnets and NICs within the same subnet. However, although Microsoft does not recommend it, this configuration is supported, and it's important to understand how NSGs are applied when deployed in this way.**

**For inbound traffic, first the NSG at the subnet is applied, followed by the NSG at the NIC. Traffic flows only if both NSGs allow the traffic to pass.**

**For outbound traffic, the sequence is reversed. First, the NSG at the NIC is applied, followed by the NSG at the subnet. Again, traffic flows only if both NSGs allow the traffic to pass.**

**In all cases, rules within each NSG are applied in priority order, with the first matching rule applicable first.**

---

You have seen how to create an NSG and how to add an inbound rule for HTTP and HTTPS traffic. Yet, unless the NSG has been associated with subnets or NICs, that rule is not in effect.

The next task will be to associate a rule with the Apps subnet. You can use either the NSG blade or the virtual network subnet blade for this task. This example uses the former.

In the NSG blade of the Azure portal, click Subnets to show the list of subnets currently associated with the NSG, which should be empty at this stage. Click Associate to open the Associate Subnet blade. The Azure portal will ask for two configurations: the virtual network, and the subnet. Note that you can only select virtual networks in the same Azure region as the NSG. In Figure 4-27, vnet-eastus-1 has been selected from the Virtual Network drop-down menu, and snet-eastus-1 has been selected from the Subnet drop-down menu.

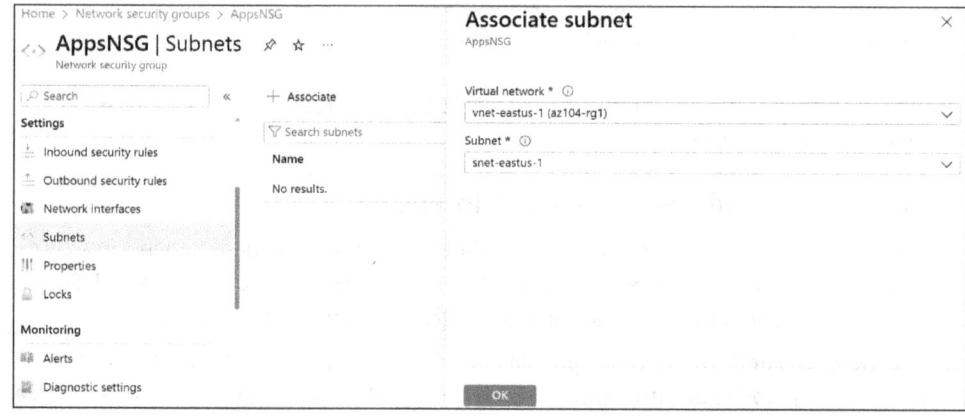

**FIGURE 4-27** The ExamRef-VNet virtual network and Apps subnet have been selected

After you save them, the rules of the NSG are enforced for all network interfaces that are associated with this subnet. This will allow inbound TCP traffic on ports 80 and 443 for all VMs that are connected to this subnet. Of course, in order for it to respond, you need to have a webserver VM configured and listening on ports 80 or 443.

## Create and configure an application security group

Application security groups (ASGs) are separate objects that you create in your Azure subscription. You can think of an ASG like a group object in an identity system—members of the group have the permissions and access that are assigned to the group. ASGs work in a similar way— you add NICs to the group, then create NSG rules that apply to the ASG. Then, any network interface that is a member of the ASG will have the NSG rules applied.

To create an ASG, do the following:

1. From the Azure portal, search for **Application security groups**.
2. On the ASG blade, click Create.
3. On the Create An Application Security Group blade, complete the required fields: Subscription, Resource Group, Name, and Region.
4. Click Review + Create, and then click Create. A sample is displayed in Figure 4-28.

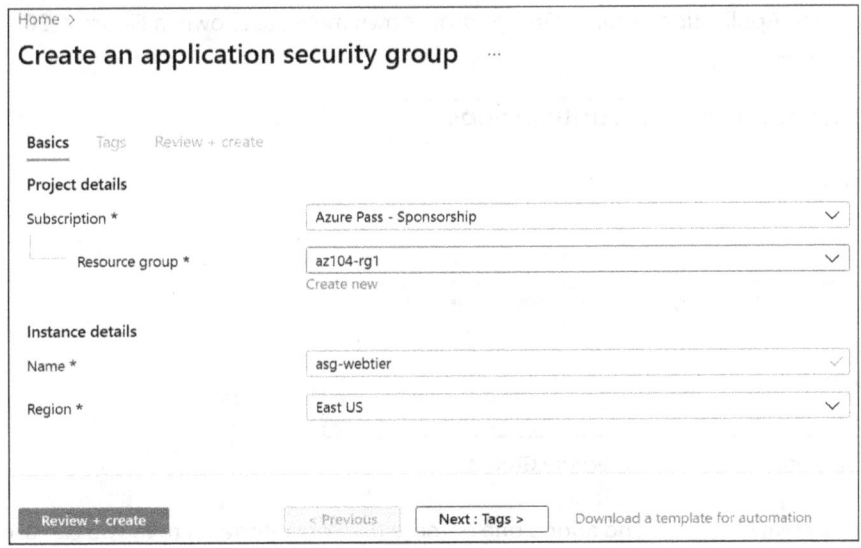

**FIGURE 4-28**  Creating a new ASG named asg-webtier

After you create an ASG, the next logical step is to associate network interfaces with the group. This action is performed from the virtual machine object that has the NIC associated with it. To add the NIC to an ASG, navigate to a virtual machine.

1. On the VM object, click Networking.
2. On the VM Networking blade, ensure that the NIC you want to configure is selected, and then click the Application Security Groups tab, as shown in Figure 4-29.

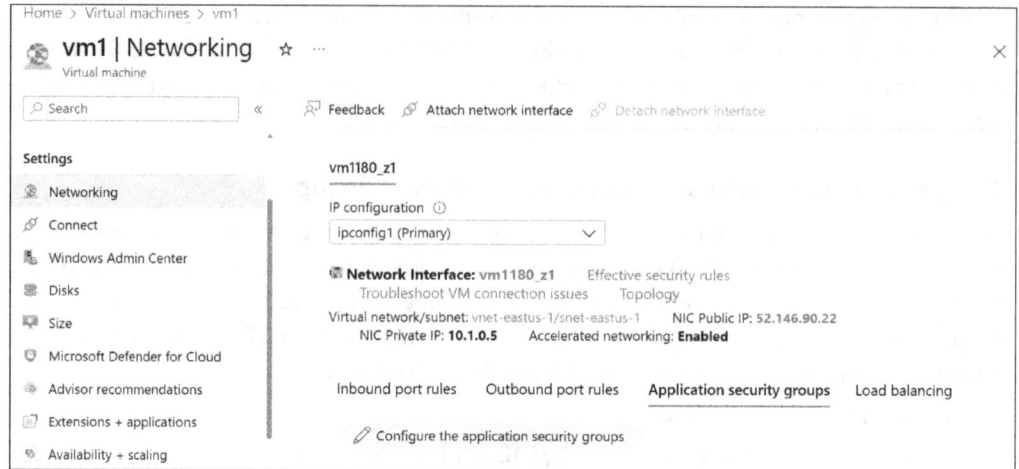

FIGURE 4-29   The Networking blade of a VM

3. On the Application Security Groups tab, click Configure The Application Security Groups.

4. On the Configure The Application Security Groups blade, select the ASG you have created from the Application Security Groups drop-down menu, as shown in Figure 4-30. Then click Save.

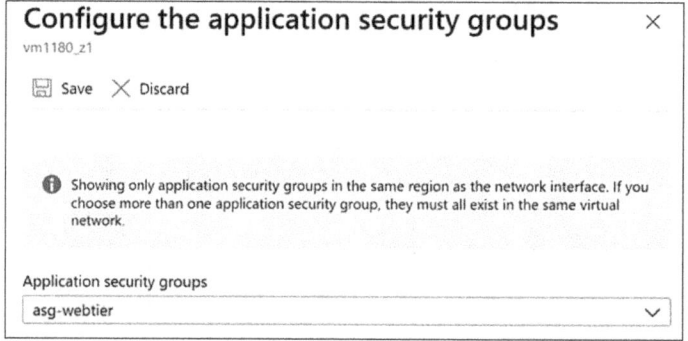

FIGURE 4-30   Configure the Application Security Groups blade

After you have created an ASG and added one or more network interfaces to it, you can use the ASG as part of an NSG rule. To do so, select Application Security Group as the Destination Type. Then, from the Destination Application Security Groups menu, select the ASG that you created. An example of modifying the rule that was created earlier in this skill is displayed in Figure 4-31.

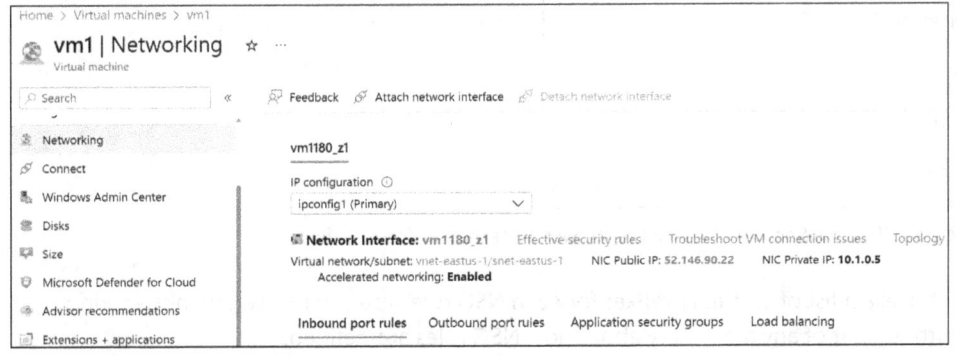

**Allow_HTTP_HTTPS**  
AppsNSG                                                          ✕

Source ⓘ

| Any | ⌄ |

Source port ranges * ⓘ

| * |

Destination ⓘ

| Application security group | ⌄ |

   Destination application security groups

| asg-webtier | ⌄ | 🗑 |

| No application security groups found | ⌄ |

**FIGURE 4-31**  Using an ASG as part of an NSG rule

## Evaluate effective security rules

When troubleshooting networking issues, it can be useful to have deeper insight into exactly how NSGs are being applied. When NSG rules are defined using service tags and application security groups, instead of explicit IP addresses or prefixes, it sometimes isn't clear whether a particular flow matches a particular rule.

The Effective Security Rules view is designed to provide this insight. You can drill into each NSG rule and see the exact list of source and destination IP prefixes that have been applied, regardless of how the NSG rule was defined.

To access the Effective Security Rules view, your virtual machine must be running because the data is taken directly from the configuration of the running VM.

Using the Azure portal, open the Virtual Machine blade and then click Networking. This will show the networking settings, including the NSG rules and a convenient Add Inbound Port Rule button. Near the top of this blade, click Effective Security Rules, as shown in Figure 4-32, to open the Effective Security Rules blade.

Home > Virtual machines > vm1

**vm1 | Networking** ☆ ⋯  
Virtual machine

| 🔍 Search | « |

- 🚊 Networking
- 🖉 Connect
- 🖳 Windows Admin Center
- 🖴 Disks
- 🖥 Size
- 🛡 Microsoft Defender for Cloud
- ☁ Advisor recommendations
- 🗐 Extensions + applications

🗨 Feedback  🖉 Attach network interface  🖉 Detach network interface

vm1180_z1

IP configuration ⓘ

| ipconfig1 (Primary) | ⌄ |

🖧 **Network Interface:** vm1180_z1    Effective security rules    Troubleshoot VM connection issues    Topology  
Virtual network/subnet: vnet-eastus-1/snet-eastus-1    NIC Public IP: 52.146.90.22    NIC Private IP: **10.1.0.5**  
   Accelerated networking: **Enabled**

**Inbound port rules**    Outbound port rules    Application security groups    Load balancing

**FIGURE 4-32**  Azure virtual machine Networking blade

The Effective Security Rules blade (see Figure 4-33) looks very similar to the Networking blade shown in Figure 4-32. It shows the name of the network interface and associated NSGs, along with a list of NSG rules.

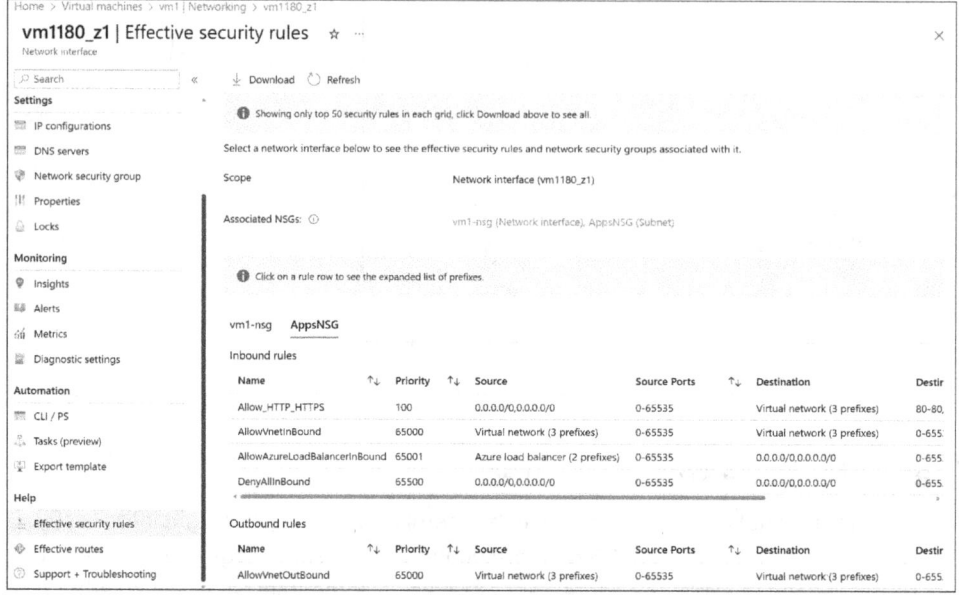

**FIGURE 4-33**   Azure virtual machine Effective Security Rules blade

The difference becomes clear when you click one of the NSG rules, which displays the exact source and destination IP address prefixes used by that rule. For example, in Figure 4-34, you can see the exact list of IP address prefixes used for the Allow_HTTP_HTTPS rule.

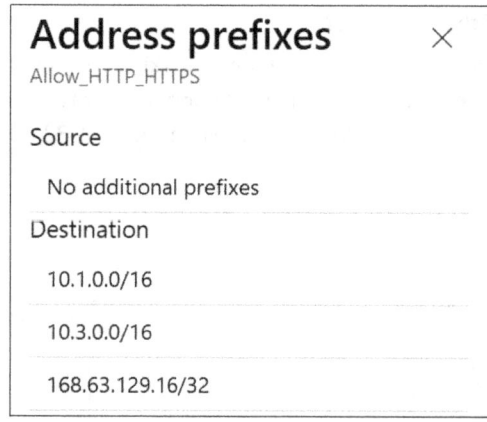

**FIGURE 4-34**   Effective Security Rules blade showing internet address prefixes

With the exact list of address prefixes for each NSG rule, you can investigate networking issues without fear of any ambiguity about how NSG rules are defined.

# Deploy and configure Azure Bastion Service

Generally, you connect to remote virtual machines with either RDP or SSH. To do so, you either need to assign a public IP address (with the RDP/SSH port exposed) to the VM to which you are trying to connect, or you need to provision an additional jump server, assign a public IP address to that jump server, and then connect to the other virtual machines using private IP addresses internally.

You can also try implementing network security groups (NSGs) to restrict the source IP addresses and ports allowed for your network traffic. Still, you are exposing RDP/SSH ports to the source servers over the internet, which could be a potential security threat.

To overcome this issue, Microsoft has created a managed PaaS service called Azure Bastion to provide secure connections to Azure VMs using the SSL channel through a browser directly without using any external client. This service helps you limit threats like port scanning and other malware.

> **NOTE** **AZURE BASTION REGIONS**
>
> The Azure Bastion service is only available in selected regions across the globe. You can find the supported regions at *https://learn.microsoft.com/en-us/azure/bastion/bastion-faq*.

The Azure Bastion service is provisioned within a VNet within a separate subnet called AzureBastionSubnet. If you have multiple VNets in your environment, you can deploy the service once in a hub virtual network, and access VMs in other VNets that are peered with the hub. If the VNets are not peered, then each VNet would need its own Bastion subnet and service.

In the following example, it is assumed that you have already created the Exam-Ref-VNet VNet with a subnet named AzureBastionSubnet and with a prefix of at least /27. Refer to Skill 4.1 for detailed instructions on how to create a virtual network and subnet.

To create a Bastion service using the Azure portal, follow these steps:

1. Search for **Bastion**. On the Bastion blade, click Create.

2. On the Create A Bastion blade, provide a name, the subscription where your resources are located, the resource group for the Bastion, and the region (select the supported region).

3. You also need to select the virtual network and subnet and create a public IP address, as shown in Figure 4-35.

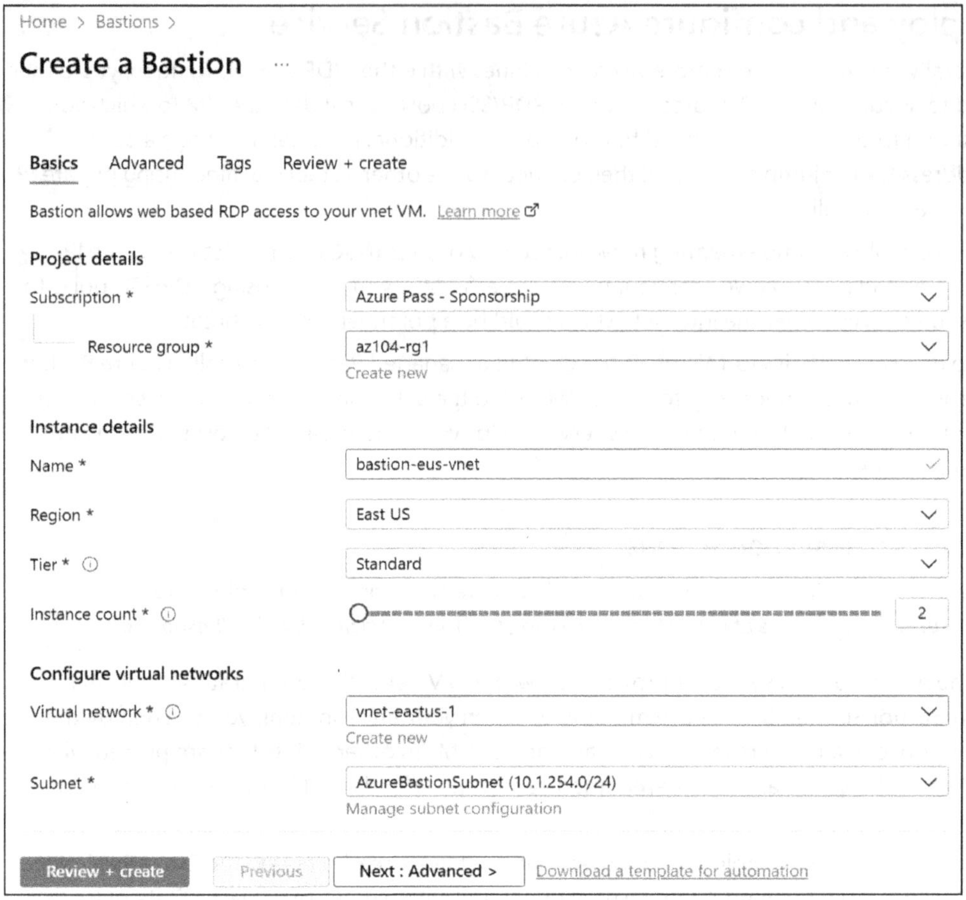

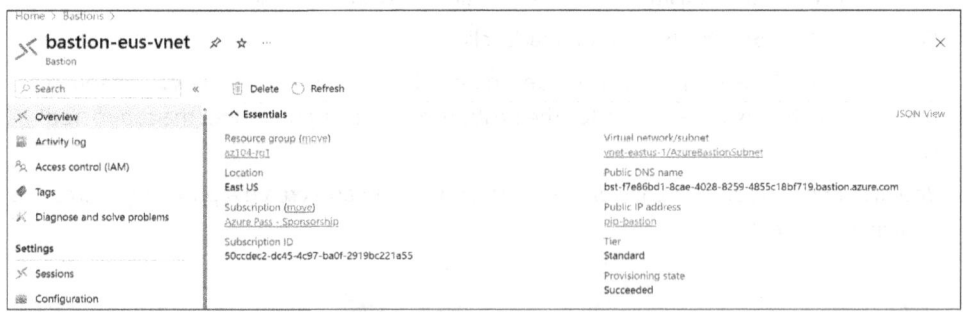

Home > Bastions >
Create a Bastion
Basics
Advanced
Tags
Review + create
Bastion allows web based RDP access to your vnet VM. Learn more
Project details
Subscription *
Azure Pass - Sponsorship
Resource group *
az104-rg1
Create new
Instance details
Name *
bastion-eus-vnet
Region *
East US
Tier *
Standard
Instance count *
2
Configure virtual networks
Virtual network *
vnet-eastus-1
Create new
Subnet *
AzureBastionSubnet (10.1.254.0/24)
Manage subnet configuration
Review + create
Previous
Next : Advanced >
Download a template for automation

**FIGURE 4-35** Creating a Bastion

bastion-eus-vnet
Bastion
Search
Delete
Refresh
Overview
Activity log
Access control (IAM)
Tags
Diagnose and solve problems
Settings
Sessions
Configuration
Essentials
JSON View
Resource group (move)
az104-rg1
Location
East US
Subscription (move)
Azure Pass - Sponsorship
Subscription ID
50ccdec2-dc45-4c97-ba0f-2919bc221a55
Virtual network/subnet
vnet-eastus-1/AzureBastionSubnet
Public DNS name
bst-f7e86bd1-8cae-4028-8259-4855c18bf719.bastion.azure.com
Public IP address
pip-bastion
Tier
Standard
Provisioning state
Succeeded

**FIGURE 4-36** Overview blade of the Bastion resource

4. Once created, the Bastion-eus-vnet overview blade will appear, as shown in Figure 4-36.
5. To test this Bastion, browse to the overview blade of your VM, click Connect, and click the Bastion tab, as shown in Figure 4-37.

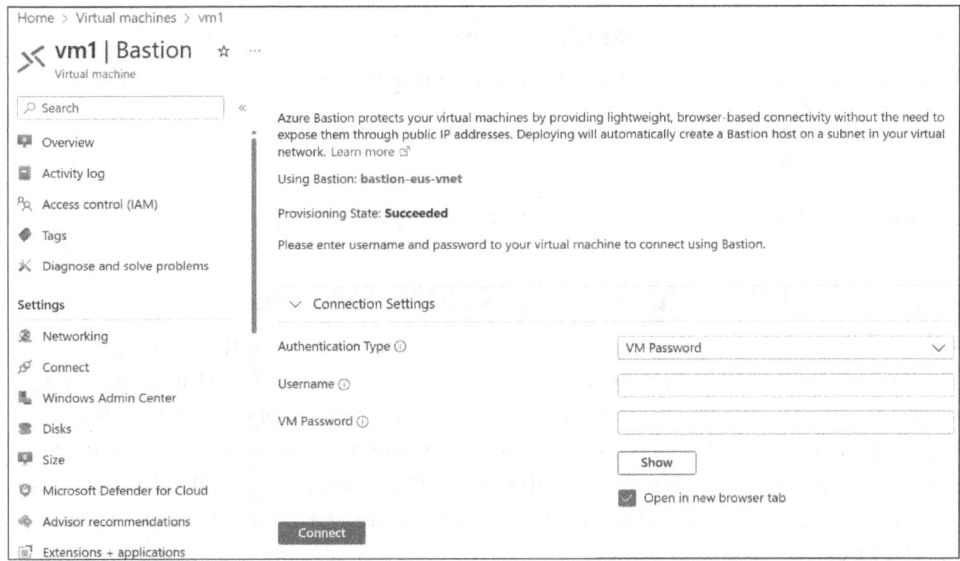

**FIGURE 4-37**   Connecting to a VM using Azure Bastion

6.   Provide the credentials and click Connect. You will be redirected to the interactive browser session to the VM through Bastion, as shown in Figure 4-38.

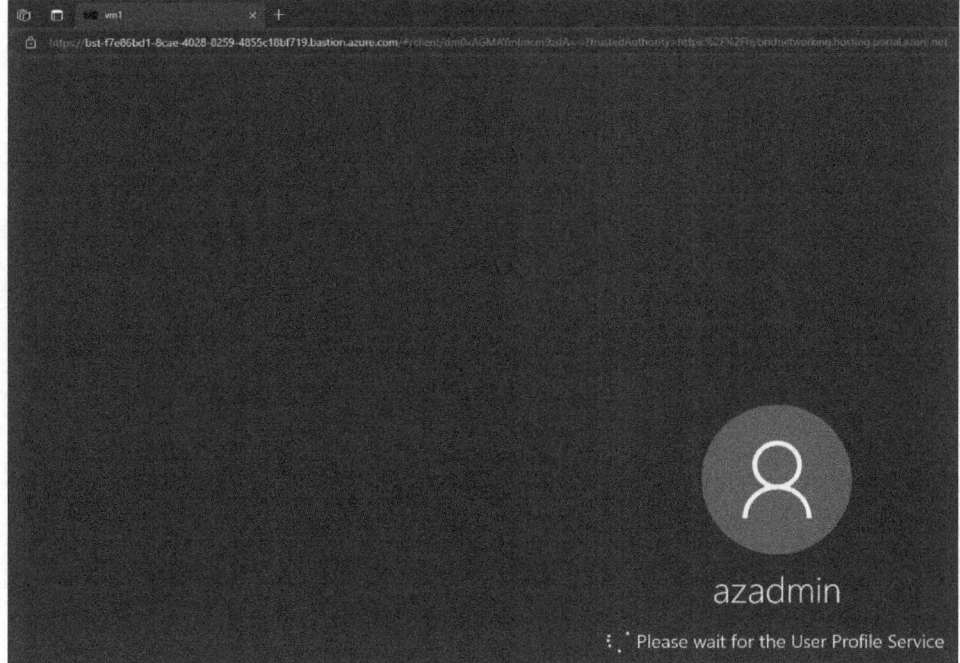

**FIGURE 4-38**   Managing a VM using Bastion

## Configure service endpoints for Azure services

By default in Azure, PaaS services are accessible using a public endpoint that resolves to a public IP address. When a VM in a subnet accesses this service, for example a storage account, the network and routing from the VM translates the source IP to the IP of the VNet or NAT Gateway on the virtual network as the traffic egresses the network. This means that if you were to capture the packet before the PaaS service, the source IP would be a public IP address of the VNet. If a second VM on the same VNet were to access the storage account, it too would have the public IP address as the source IP.

A service endpoint changes two things about how a VM might access a PaaS service, such as a storage account. First, the routing is optimized to ensure that the Microsoft backbone is used to communicate from the VNet to the service. Second, the VNet does not translate the IP address of the packet from the VM. This means that the source IP of the request shows the private IP address of the VM that is trying to access the service. However, the service is still using the public endpoint and public IP address that was assigned, in this case to the storage account.

Service endpoints are created at the subnet level of a virtual network. Suppose you have two VMs: VM1 and VM2, that exist in two subnets: Subnet1 and Subnet2. Subnet1 has a service endpoint for storage. Subnet2 does not have any service endpoints defined. If VM2 tries to access a storage account, the source IP address will be a public IP address. If VM1 tries to access the same storage account, the source IP address will be the private IP address of VM1.

1. A service endpoint can be configured from the subnet of a virtual network. To configure a service endpoint, navigate to your VNet and then click Subnets.

2. Select the name of the subnet to modify its properties.

3. In the Services drop-down menu, select the services that you want to enable a service endpoint on. Figure 4-39 displays a new service endpoint for storage being created for subnet0 of the vnet-hub VNet.

4. Click Save. The process might take a few minutes to reflect in routing and packet traces.

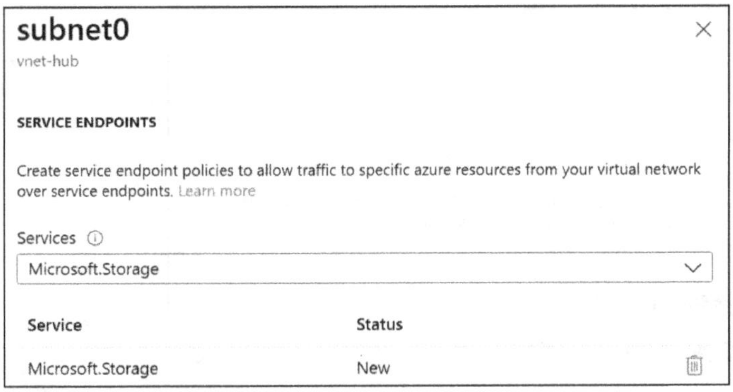

**subnet0** ✕
vnet-hub

**SERVICE ENDPOINTS**

Create service endpoint policies to allow traffic to specific azure resources from your virtual network over service endpoints. Learn more

Services ⓘ

| Microsoft.Storage | ∨ |

| Service | Status | |
|---------|--------|---|
| Microsoft.Storage | New | 🗑 |

**FIGURE 4-39** Creating a service endpoint

## Configure private endpoints for Azure services

Private endpoints take the concept of service endpoints one step further. In the same scenario of a VM in subnet0 trying to communicate with a storage account, in addition to using a private IP address as the source IP address, the destination IP address will also be private. Private endpoints create a dedicated network interface for the specific PaaS service resource that you create it for and make that interface accessible from the subnet that you configure.

In this example, suppose you create a private endpoint for blob storage in the storage account. You could create that private endpoint in the same subnet, subnet0, that the VM is associated with. The VM NIC might have an IP address of 10.0.0.4/24, and the private endpoint for blob storage could be 10.0.0.5/24. When the VM connects to the endpoint of the blob storage, the DNS name resolves from the VNet to the private IP address of the private endpoint. Therefore, all communication uses private IP addresses as both the source and the destination.

You can typically create a private endpoint directly from the resource you want to configure it on, or from the Private Link Center. To create a private endpoint for blob storage, follow these steps:

1. From the Azure portal, search for **Private endpoints**.
2. On the Private Endpoints blade, click Create.
3. On the Basics tab of the Create A Private Endpoint blade, select your subscription, resource group, and region. Then provide a name for the endpoint, such as "pe-blobstorage1." The network interface should autopopulate with a name based on the resource name. Figure 4-40 displays the Basics tab.
4. Click Next: Resource.
5. On the Resource tab, select the Azure PaaS resource that you want to create the private endpoint for. In this example, select Microsoft.Storage/storageAccounts as the resource type, a storage account as the resource, and blob as the target subresource, as shown in Figure 4-41.

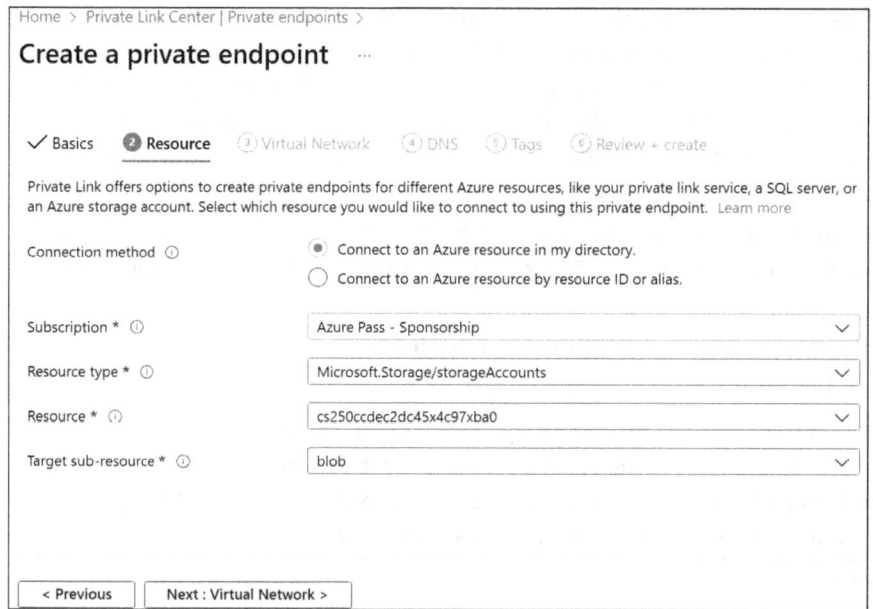

**FIGURE 4-40** The Basics tab of the Create A Private Endpoint blade

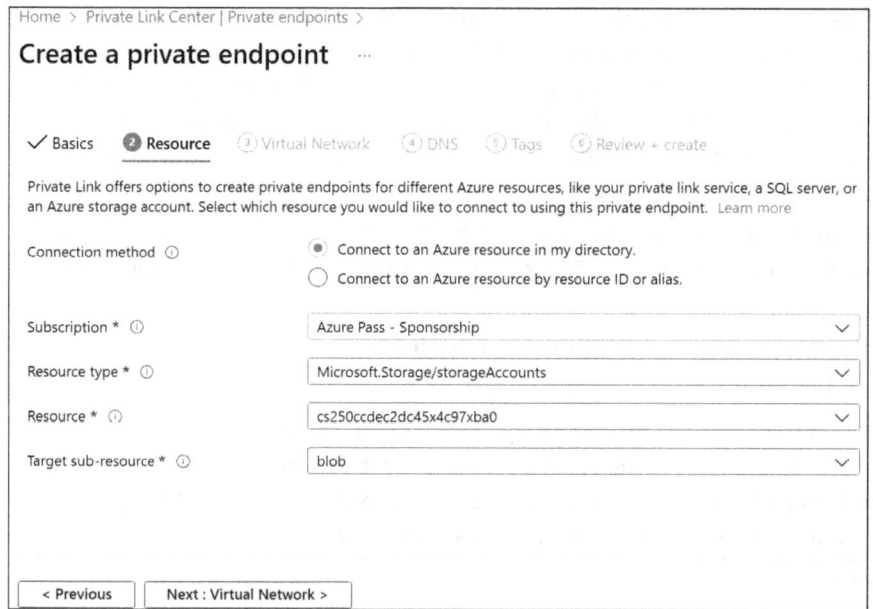

**FIGURE 4-41** The Resource tab of the Create A Private Endpoint blade

6. Click Next: Virtual network.

7. On the Virtual Network tab, select the VNet and subnet that you want the private endpoint to be associated with. In the example shown here, vnet-hub is selected for the

VNet and subnet0 as the subnet. By default, an IP address will be allocated to the private endpoint dynamically, as shown in Figure 4-42.

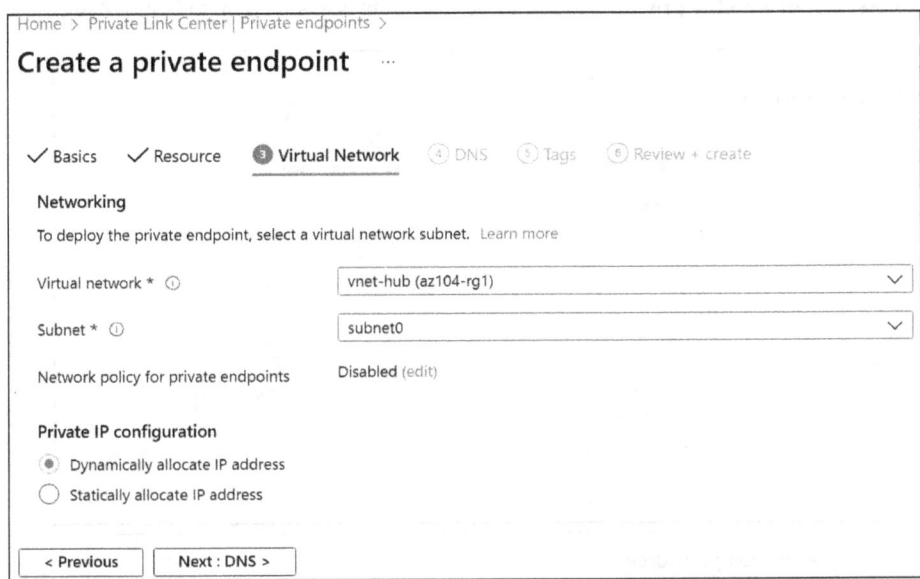

**FIGURE 4-42** The Virtual Network tab of the Create A Private Endpoint blade

8. Click Next: DNS.

9. On the DNS tab, choose whether to integrate the private endpoint network interface with a DNS zone. This is recommended so that the private IP address is resolved when accessing the storage account, as shown in Figure 4-43.

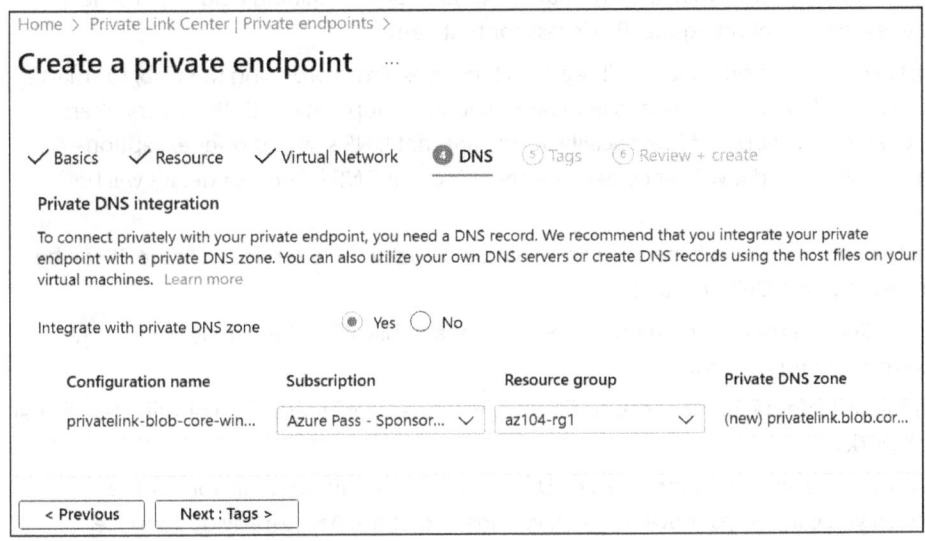

**FIGURE 4-43** The DNS tab of the Create A Private Endpoint blade

10. Accept the remaining defaults on the Tags and Review + Create tabs to create the private endpoint. After the private endpoint is created, you can view the private IP address that was associated with the private endpoint network interface, as shown in Figure 4-44.

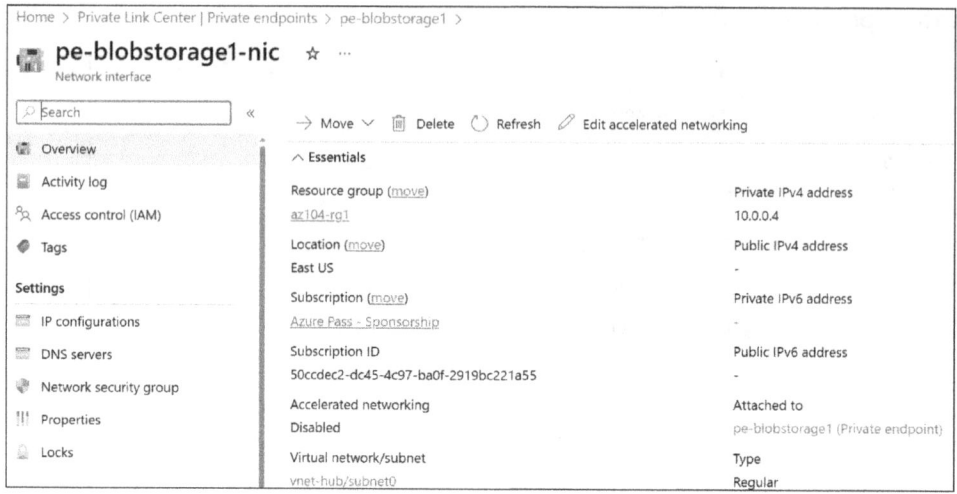

**FIGURE 4-44**  Private endpoint resource

# Skill 4.3: Configure name resolution and load balancing

Humans work with names, but computers prefer IP addresses. Fundamentally, DNS is about mapping names to IP addresses, making name-based rather than IP-based networking possible. Simplifying somewhat, a client makes a DNS query containing a domain name and receives a response containing the IP address for that name.

Almost everywhere you look, you'll see DNS scenarios. From browsing the web, to smart-phone apps, to IoT devices, to database lookups within an application, DNS is everywhere. Because DNS is so universal, it is especially important that DNS services offer exceptionally high availability and low latency because the effects of DNS failures or delays will be widespread.

Azure DNS provides a high-performance, highly available DNS service in Azure. It can be used for two separate DNS scenarios:

- Providing internet-facing name resolution for a public DNS domain by hosting the corresponding DNS zone
- Providing internal name resolution between virtual machines within or between virtual networks

Also, with Azure, you can control which DNS servers are configured on your virtual machines, so you can use your own DNS servers instead of the Azure-provided service.

This skill will also discuss load balancing. Load balancing is one of the crucial requirements of a network design. Azure offers various options to design load balancing solutions. In this section, you will learn how to configure load balancers in Azure.

Azure Load Balancer is a fully managed load-balancing service, which is used to distribute inbound traffic across a pool of back-end servers running in an Azure virtual network. It can receive traffic on either internet-facing or intranet-facing endpoints and supports both UDP and TCP traffic.

Azure Load Balancer operates at the transport layer (OSI layer 4) to route inbound and outbound connections at the packet level. It does not terminate TCP connections, and thus, it does not have visibility into application-level constructs. For example, it cannot support SSL offloading, URL path-based routing, or cookie-based session affinity.

Azure Load Balancer provides low latency and high throughput, scaling to millions of network flows. It also supports automatic failover between back-end servers based on health probes and enables high-availability applications.

---

**This skill covers how to:**
- Configure Azure DNS
- Configure load balancing
- Troubleshoot load balancing

---

# Configure Azure DNS

This section describes how Azure DNS is configured to host internet-facing domains. We start with a summary of how the domain name system works because understanding DNS is a prerequisite to understanding Azure DNS.

## How DNS Works

To properly understand the various DNS services and features available in Azure, it is first necessary to understand how the domain name system works. In particular, it is important to understand the different roles played by recursive and authoritative DNS servers, and how a DNS query is routed to the correct DNS name servers using DNS delegation.

First, it's important to understand the distinction between a domain name, and a DNS zone. The internet-facing domain name system is a single global name hierarchy. A domain name is just a name within that hierarchy. Owning a domain name gives you the legal right to control the DNS records within that name and any subdomains of that name.

You purchase a domain name from a domain name registrar. The registrar then lets you control which name servers (NS) receive the DNS queries for that domain, by letting you configure the NS records for the domain.

A *DNS zone* is the representation of a domain name in an authoritative DNS server. It contains the collection of DNS records for a given domain name. The service hosting the

DNS zone lets you manage the DNS records within the zone and hosts the data on authoritative name servers, which answer DNS queries with DNS responses based on the configured DNS records.

In Azure, you can purchase domain names using the App Service Domains service. DNS zone hosting is provided by Azure DNS.

The DNS settings on the user's device point to a recursive DNS server, also sometimes known as a local DNS service (or LDNS) or simply as a DNS resolver. The recursive DNS service is typically hosted by your company (if you're at work) or by your ISP (if you're at home). There are also public recursive DNS services available, such as the Google public DNS (8.8.8.8) service. The recursive DNS service doesn't host any DNS records, but it allows your device to off-load most of the work associated with resolving DNS queries.

To understand the role of recursive and authoritative DNS servers, consider Figure 4-45, which describes the DNS resolution process for a single DNS query, *www.contoso.com*.

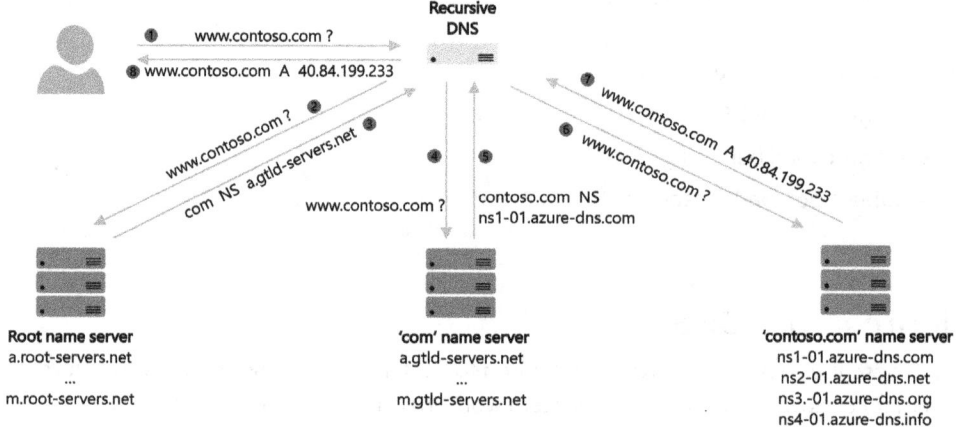

**FIGURE 4-45** The DNS resolution process

This resolution process is described here:

1. Your PC makes a DNS query to its locally configured recursive DNS server. This query is simply a packet sent over UDP port 53, although TCP can also be used (typically when responses are too big to fit in a UDP packet).

2. Assume the recursive DNS server has just been switched on, so there is nothing in its cache. It passes the query to one of the root name servers (the addresses of the root name servers are preconfigured). The root name servers are authoritative name servers—they host the actual DNS records for the root zone. A zone is simply the data representing a node in the DNS hierarchy.

3. The root name servers don't know anything about the contoso.com DNS zone. They do, however, know where you can find the com zone. So, they return a DNS record of type NS, which tells the recursive DNS server where to find the com zone.

4. The recursive server tries again, this time calling the com name servers. Again, these are authoritative name servers, this time for the com zone.

5. These name servers don't recognize "www.contoso.com," but they do have NS records that define where the contoso.com DNS zone can be found.

6. The recursive server tries again, this time calling the authoritative contoso.com name servers.

7. These servers *are* authoritative for the contoso.com DNS zone. And there is a record on these servers matching the www record name. The server does recognize the www.contoso.com query name and returns the A record response that maps this name to an IP address.

8. The recursive server then returns this result back to the client.

The recursive DNS server can also follow a chain of CNAME records (which map one DNS name to another name). And the recursive DNS server also caches the responses it receives, so that it can respond more quickly next time. The duration of the cache is determined by the TTL (time-to-live) property of each DNS record.

The domain name system is a distributed system, where one set of servers can refer queries to another set using NS records. The process you've just seen to map a query name to a result—perhaps via a long chain of authoritative DNS servers—is called *DNS name resolution*.

The NS records tell clients on the internet where to find the name servers for a given DNS zone. The NS records for a DNS zone are configured in the parent zone, and a copy of the records is also present in the child zone. Setting up these NS records is called delegating a DNS domain.

A fully qualified domain name (FQDN) is a domain name containing all components all the way up to the root zone. Strictly speaking, a fully qualified name ends with a "." (for example, www-dot-contoso-dot-com-DOT), which represents the root zone, although by convention, the trailing period is often omitted.

Reverse DNS is the ability to map an IP address to a name (as opposed to map a name to an IP address, which is what normal DNS provides). Some applications use reverse DNS as a weak form of authentication. For example, it's commonly used in email spam-scoring algorithms.

Reverse DNS lookups use a DNS hierarchy that is completely independent of the forward lookups. The reverse lookup for "www.contoso.com" does not sit in the contoso.com zone. Instead, it sits in a separate DNS zone hierarchy based on reversed IP addresses. For example, suppose "www.contoso.com" resolves to IP address 1.2.3.4. Typically, the reverse lookup for the IP address 1.2.3.4 will be a record named 4 in the DNS zone 3.2.1.in-addr.arpa, giving a FQDN of 4.3.2.1.in-addr.arpa (notice the reversed IP address.)

Reverse DNS lookup zones are controlled by whomever owns the IP subnet. The reverse DNS lookup zone for an IP block you own can be hosted in Azure DNS. Public IP addresses in Azure reside in Microsoft-owned IP blocks, which means the reverse DNS lookups use Microsoft-managed reverse DNS lookup zones.

There's nothing in the domain name system to ensure the reverse lookup maps to the same name as was used in the forward lookup. That's achieved simply by the correct configuration in both forward and reverse lookup zones.

## DNS services in Azure

There are several DNS-related services and features in Azure. An overview of each is given in the following list. The first three items are Azure services, which you consume by creating service-specific resources that you will be billed for. The remaining three items are Azure features, which you configure using settings on other resource types, such as a virtual network, public IP address, or network interface.

- **Azure DNS**   Allows you to host your DNS domains in Azure. It provides the ability to create and manage the DNS records for your domain and provides name servers, which answer DNS queries for your domain from other users on the internet. Azure DNS also supports private DNS zones, which are used for intranet-based name resolution for VM to VM lookups, including support for some scenarios not supported by the Azure-provided DNS service, which will be covered shortly.

- **Azure Traffic Manager**   An intelligent DNS service that uses DNS to implement global traffic management. Where Azure DNS always provides the same DNS response to any given DNS query, in Azure Traffic Manager the same query may result in one of several possible responses, depending on a number of factors you control, such as where the end-user is located or which of your service endpoints is currently available. This enables you to route traffic intelligently between Azure regions or between Azure deployments and on-premises deployments. Understanding Traffic Manager is beyond the scope of the AZ-104 exam.

- **App Service Domains**   This service is used to purchase domain names, which can then be hosted in Azure DNS. This service is integrated with Azure App Service, but can be used for any domain registration, even if App Service is not being used.

- **Azure-provided DNS**   Sometimes called Internal DNS, it allows the VMs in your virtual network to find each other, using DNS queries based on the hostname of each VM. The DNS queries are internal (private) to the virtual network.

- **Recursive DNS**   A service provided by Azure for DNS name resolution from your Azure VMs or other Azure services. You can also configure your VMs to use your own DNS server instead. This is sometimes informally called "bring your own DNS." This is common when joining your VMs to a domain controller.

- **Reverse DNS**   Provides the ability to configure the reverse DNS lookup for an Azure-assigned public IP address. (Reverse DNS lookup zones for IP blocks you own can be hosted in Azure DNS.)

## Create and delegate a DNS Zone to Azure DNS

A DNS zone is a resource in Azure DNS. Creating a DNS zone resource allocates authoritative DNS name servers to host the DNS records for that zone. Azure DNS can then be used to manage those DNS records. DNS queries directed to those DNS name servers receive a DNS response based on the DNS records configured at that time.

You do not have to own the corresponding domain name before creating a DNS zone in Azure DNS. You can create a DNS zone with any name, except for names on the public suffix list (see *https://publicsuffix.org/*). You can also create more than one DNS zone resource with the same DNS zone name, as long as they are in different resource groups. In this case, the DNS zones will be allocated to separate DNS name servers, so no conflict arises.

You can test your DNS records by directing DNS queries directly to the assigned DNS name servers for your zone. For general use, however, your DNS zone should be delegated from the parent zone. This requires you to own the corresponding domain name.

Before you can delegate your DNS zone to Azure DNS, you first need to know the names of the name servers assigned to your zone. These can be obtained using the Azure portal, PowerShell, or CLI after the DNS zone resource has been created. You can't predict in advance which name server pool will be assigned to your DNS zone. You need to create the DNS zone, and then check.

The assigned name servers vary between zones, so if you're setting up multiple zones in Azure DNS, you need to check the name servers on each one. Don't assume that the name servers will be the same across all your zones.

Each domain name registrar has their own DNS management tool allowing you to set the name server (NS) records for a domain. In the registrar's DNS management page, edit the NS records and replace the NS records with the ones Azure DNS assigned.

When delegating a domain to Azure DNS, you must use the name server names provided by Azure DNS. You should always use all four name server names, regardless of the name of your domain. Domain delegation does not require the name server name to match your domain name.

> **NOTE** **DELEGATING DNS ZONES TO AZURE DNS**
>
> When delegating a domain to Azure DNS, do not use DNS glue records to point to the Azure DNS name server IP addresses directly. A glue record is a DNS server record that is not author-itative for the zone and is used to avoid a condition of impossible dependencies for a DNS zone. These IP addresses might change in the future. Delegations using name server names in your own zone—sometimes called vanity name servers—are not currently supported in Azure DNS.

Azure DNS treats child zones as entirely separate zones. Therefore, delegating a child zone follows the same process as delegating the parent zone:

1. Create the child zone resource.

2. Identify the name servers for the child zone. These will be different from the name serv-ers assigned to the parent zone.

3. Create NS records in the parent zone to delegate the child zone. The name of the NS records should be the child zone name (excluding the parent zone name suffix), and the RDATA in the NS records should be the child zone name servers.

## Manage DNS records in Azure DNS

Each record in the domain name system includes the following properties:

- **Name**  The name of the DNS record is combined with the name of the DNS zone to form the fully qualified domain name (FQDN). For example, the record "www" in zone "contoso.com" corresponds to the FQDN "www.contoso.com."

- **Type**  The type of DNS record determines what data is associated with the record and what purpose it is used for. A list of record types supported by Azure DNS is provided in Table 4-7.

- **TTL**  The Time-to-Live (TTL) property tells recursive DNS servers how long a DNS record should be cached.

- **RDATA**  The data returned for each DNS record. The type of data returned depends on the DNS record type. For example, an A record will return an IPv4 address, whereas a CNAME record returns another domain name.

The collection of records in a DNS zone with the same name and the same type is called a *resource record set*. (These collections are also referred to as "RRSets" and as "record sets" in Azure DNS). Records in Azure DNS are managed using record sets. Record sets are a child resource of the DNS zone and can contain up to 20 individual DNS records. The name, type, and TTL are configured on the record set, and the RDATA is configured on each DNS record within the record set.

To create a DNS record set at the root (or *apex*) of a DNS zone, use the record set name @. For example, the record set named @ in the contoso.com zone will resolve against queries for "contoso.com." You can also use an asterisk (*) in the record set name to create wildcard records (subject to DNS wildcard matching rules).

Azure DNS supports all commonly used DNS record types. The full list of supported record types—together with a description of each—is provided in Table 4-7.

**TABLE 4-7**  DNS record types in Azure DNS

| DNS Record Type | Remarks |
| --- | --- |
| A | Used to map a name to an IPv4 address. |
| AAAA | Used to map a name to an IPv6 address. |
| CAA | Used to specify which certificate authorities can issue certificates for a domain. Note that CAA records are not currently available in the Azure portal, so they must be configured using the Azure CLI or Azure PowerShell. |

| DNS Record Type | Remarks |
|---|---|
| CNAME | Provides a mapping from one DNS name to another. The DNS standards do not allow CNAME records at the zone apex. In addition, you cannot create a CNAME record with the same name as a record of any other record type, and CNAME record sets only support a single DNS record rather than a list of records. These are DNS RFC constraints, not Azure DNS limitations. |
| MX | Used for mail server configuration. |
| NS | An NS record set at the zone apex containing the name servers for the DNS zone is required by the DNS standards. This is created for you when the DNS zone is created. It can be edited, for example to add additional records when co-hosting a DNS zone with more than one provider, but not deleted.<br>You can create additional NS record sets to delegate child zones. |
| PTR | Used for reverse DNS lookups in reverse lookup zones. |
| SOA | An SOA record is required at the apex of every zone. This is created and deleted with the DNS zone resource. |
| SRV | SRV records are used for service discovery for a wide range of services, from Kerberos to Minecraft to the Session Initiation Protocol used for internet telephony.<br>Note that the Service and Protocol parameters are specified as part of the record set name, such as _service._protocol.media.contoso.com. Some DNS services prompt you to enter these values separately and then merge them to form the record set name. With Azure DNS, you need to specify them as part of the record set name, but they are not entered separately. |
| TXT | Used for a wide range of applications, including email Sender Policy Framework (SPF). |

*NOTE*  **SPF RECORDS**

**Sender Policy Framework (SPF) records are used to identify legitimate mail servers for a domain and help prevent spam. The SPF record type was deprecated by RFC7208, which states that the TXT record type should be used for SPF records.**

## Alias records

Azure DNS offers integration with other services hosted in Azure via alias records.

With conventional DNS records, you explicitly specify the target, such as the IP address of an A record. If the IP address changes, you need to update the DNS record accordingly.

Using *alias records*, you can define the target of the DNS record implicitly by referencing another Azure resource. The value of the DNS record is populated automatically based on the resource it references and is updated automatically if that resource changes.

Alias records can reference three different resource types:

- **An A or AAAA**  These records can reference a public IP address, of type IPv4 or IPv6, respectively.
- **A, AAAA, or CNAME**  These records can reference a Traffic Manager profile. This exposes the dynamic, traffic-managed name resolution of the Traffic Manager directly within a record in your DNS domain. Prior to this feature, you had to create a CNAME

record from your domain to a record in the trafficmanager.net domain provided by Azure Traffic Manager.

- **An A, AAAA or CNAME**   These records can also reference another record in the same DNS zone. This lets you create synchronized records with ease.

Alias records are a very useful way to address a number of scenarios.

- First, alias records prevent orphaned DNS records. A common problem with DNS systems is that records are not cleaned up when the services they reference are deleted. The DNS record is left dangling. With alias records, the DNS record no longer resolves once the underlying service is deleted.
- Second, as has already been discussed, by updating automatically when underlying resources change, alias records reduce your management overhead and help you avoid accidental application downtime.
- Third, because alias records enable you to avoid using a CNAME record when using a vanity domain name with Azure Traffic Manager, you can implement a traffic-managed record at the apex of your domain.

## Create DNS zones and DNS records using the Azure portal

To create a DNS zone, from the Azure portal, search for **DNS Zone**. On the DNS Zones blade, click Create to open the Create DNS Zone blade. Specify the DNS domain name as the DNS zone resource name, and select your resource group, as shown in Figure 4-46.

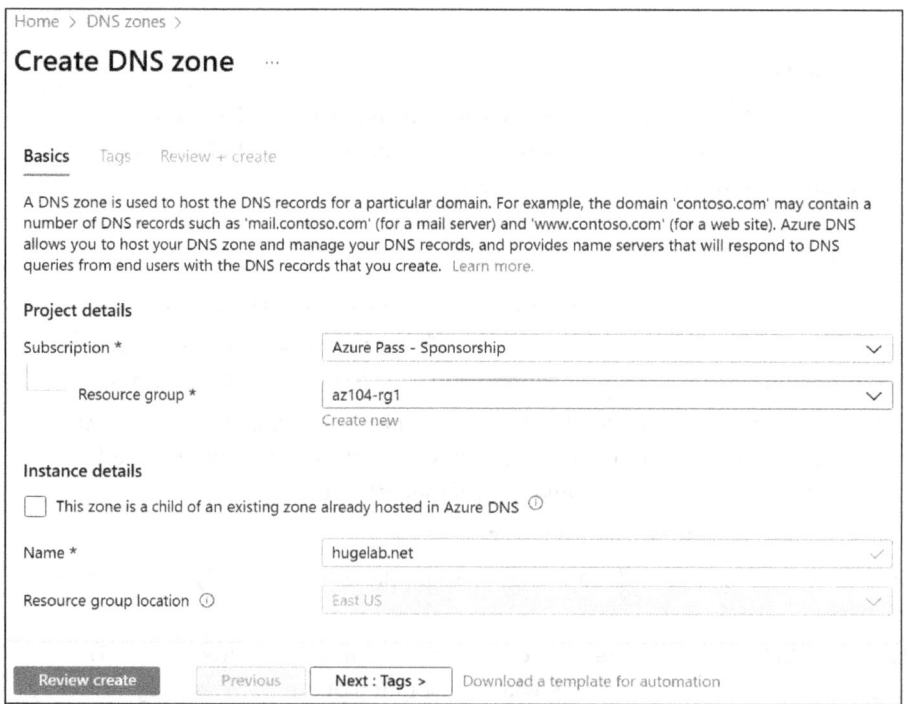

**FIGURE 4-46**   Creating a DNS zone using the Azure portal

Once the DNS zone has been created, open the DNS zone blade. The Azure DNS name
servers assigned to the zone are listed under Essentials, as highlighted in Figure 4-47.

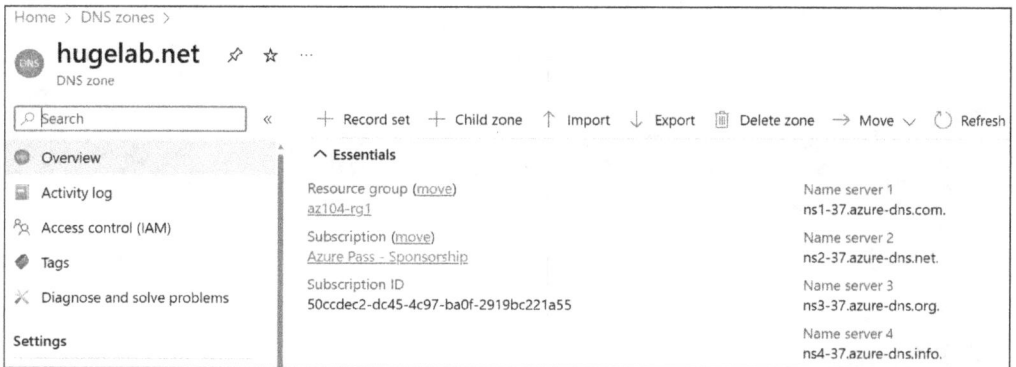

**FIGURE 4-47**   The DNS zone blade, displaying the Azure DNS name servers assigned to this zone

To set up DNS delegation for the DNS zone, these name servers must be listed in the
corresponding NS records in the parent zone. If the domain name was purchased using the
Azure App Service Domains service, this will be done automatically. Otherwise, this must be
configured at the DNS registrar where the domain name was purchased.

To create a DNS record in a new record set, click +Record Set to open the Add Record Set
blade. If there is an existing record with the same name and type as the record you want to
create, you should instead click the existing record set and add the new record there. To create
a pair of A records with name "www" (giving the fully qualified domain name "www.hugelab.
net"), use the following values, as shown in Figure 4-48.

- **Name**   www
- **Type**   A
- **Alias Record Set**   No
- **TTL**   1 hour (or choose your own value)
- **IP Addresses**   Enter A record IP addresses, one for each DNS record in the record set.

**FIGURE 4-48**  The Add Record Set blade

Suppose now you want to create a DNS record at the zone apex (so the fully qualified domain name is simply the DNS zone name "hugelab.net"), pointing to a dynamically allocated public IP address. Click +Record Set again and complete the Add Record Set blade with the following settings, as shown in Figure 4-49:

- **Name**   @ (This is a DNS convention for records at the zone apex.)
- **Type**   A
- **Alias Record Set**   Yes
- **Choose Subscription**   Choose the subscription containing the public IP address
- **Azure Resource**   Choose the public IP address resource
- **TT**   1 hour (or choose your own value)

## Configure custom DNS settings

When a virtual machine connects to a virtual network, it receives its IP address via DHCP. As part of that DHCP exchange, DNS settings are also configured in the VM. By default, VMs are configured to use recursive DNS servers in Azure. These provide name resolution for internet-hosted domains, plus private VM-to-VM name resolution within a virtual network.

The hostname of the VM is used to create a DNS record mapping to the private IP address of the VM. You specify the hostname—which is simply the VM name—when you create the virtual machine. Azure specifies the DNS suffix, using a value that is unique to the virtual network. These suffixes end with internal.cloudapp.net. The hostname and DNS suffix together form the unique fully qualified domain name.

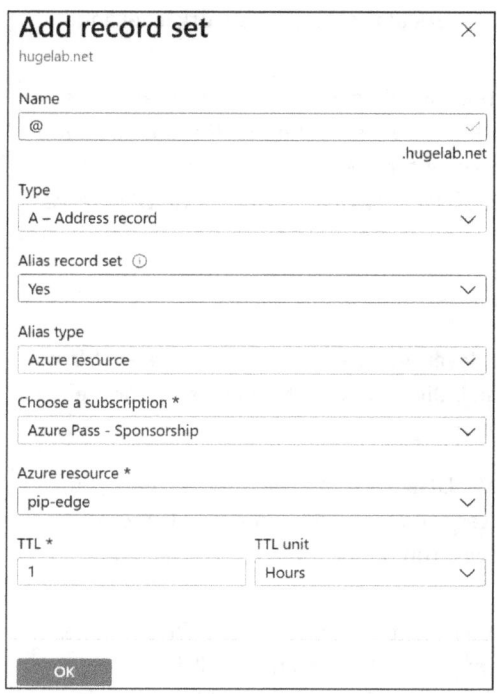

**FIGURE 4-49** The Add Record Set blade for an alias record set

Name resolution for these DNS records is private—they can only be resolved from within the virtual network. The DNS suffix is configured as a lookup suffix within each VM, so names can be resolved between VMs within the virtual network using the hostname only.

This built-in DNS service uses the IP address 168.63.129.16. This is a special static IP address that is reserved by the platform for this purpose. This IP provides both the authoritative DNS service for Azure-provided DNS as well as the Azure recursive DNS service, which is used to resolve internet DNS names from Azure VMs. This IP is used for other things as well, such as health problems from Azure Load Balancer, heartbeat messages for PaaS roles, and so on.

**BRING YOUR OWN DNS**

Alternatively, you can configure your own DNS settings, which will be configured during the DHCP exchange on the VMs instead. This enables you to specify your own DNS servers, either in Azure or running on-premises. With your own DNS servers, you can support any DNS scenario, including scenarios not supported by the Azure-provided service. Example scenarios requiring you to use your own DNS servers include name resolution between VMs in different virtual networks, name resolution between on-premises resources and Azure virtual machines, reverse DNS lookup of internal IP addresses, and name resolution for non-internet-facing domains, such as domains associated with Active Directory.

You should not specify your own DNS settings within the VM itself because the platform is unaware of the settings you have chosen. Instead, Azure provides configuration options within the virtual network settings. These DNS server settings are at the virtual network level and apply to all VMs in the virtual network.

You can also specify VM-specific DNS server settings within each network interface. This takes precedence over settings at the virtual network level. Where multiple VMs are deployed in an availability set, setting DNS servers at the network interface, all VMs in the availability set are updated. The DNS servers applied are the union of the network interface-level DNS servers from across the availability set.

> *NOTE* **DNS NAME SERVER SETTINGS**
>
> Custom DNS settings can be configured at the VNet level, and the network interface level, but not at the subnet level. To use specific settings for an individual subnet, you must configure those settings on each network interface in the subnet.

You can use these DNS settings to direct your VMs' DNS queries to any DNS servers you choose. They can point to IP addresses of on-premises servers, such as an Active Directory Domain Controller or network appliance, a DNS service running in an Azure virtual machine, or anywhere else on the internet.

If you use your own DNS servers, those servers will need to offer a recursive DNS service, otherwise name resolution for internet domains from your virtual machines will break. If you point the DNS settings directly at an internet-based recursive DNS service, such as Google 8.8.8.8, then you will not be able to perform VM-to-VM lookups.

> *NOTE* **RESTART VIRTUAL MACHINES WHEN CHANGING DNS SETTINGS**
>
> If you make changes to the DNS settings at the virtual network level, any affected virtual machines must restart to pick up the new settings. If you make changes to DNS settings at the network interface level, the affected VM (or VMs across the availability set, if used) will restart automatically to pick up the new settings.

One challenge when using your own DNS servers is that you need to register each VM in your DNS service. To do this, you can configure the DNS service to accept Dynamic DNS queries, which the VM will send when it boots. This allows the VMs to register with the DNS server automatically. A problem with this approach is that the DNS suffix in the Dynamic DNS query must match the DNS zone name configured on the DNS server, and Azure does not support configuring the DNS suffix via the Azure platform settings. As a workaround, you can configure the correct DNS suffix within each VM yourself, using a start-up script.

### CONFIGURE CUSTOM DNS SETTINGS USING THE AZURE PORTAL

To configure the DNS servers on a VNet, open the virtual network blade, and then click DNS Servers under Settings on the left, as seen in Figure 4-50. You can then enter the DNS servers you want this VM to use. After saving your changes, you need to restart the VMs in the VNet to pick up the changes.

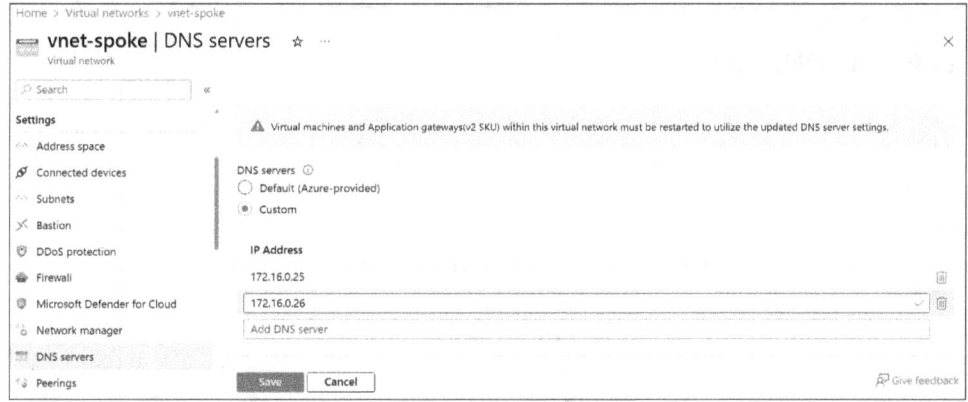

**FIGURE 4-50** Custom DNS servers for a virtual network configured using the Azure portal

The steps to configure the DNS servers on an individual VM are similar to what is displayed in Figure 4-50. Open the blade for the VM's network interface, and then click DNS Servers under Settings. You can then enter the DNS servers you want this VM to use. Note that VMs in an availability set will adopt the union of DNS servers from network interfaces across the availability set. After saving your changes, your VM (or VMs in the availability set) will automatically restart to pick up the changes.

## Configure private DNS zones

In addition to supporting internet-facing DNS domains, Azure DNS also supports private DNS domains. This provides an alternative approach to name resolution within and between virtual networks.

By using private DNS zones, you can use your own custom domain names—including the DNS suffix, rather than the Azure-provided DNS suffix—without the overhead or complexity of running your own DNS servers.

The service supports automatic registration of VMs into the private zone, but only from a single virtual network, called the registration VNet. This must be registered with the DNS zone before any VMs are created.

If you want to resolve VM names from multiple virtual networks, the VMs in any other networks must be registered with the service manually (or via a custom automation). Name resolution between VNets is independent of connectivity between VNets, so peering your virtual networks or setting up a VNet-to-VNet connection is not required.

Virtual networks that support name resolution are called *resolution VNets*. The zone name is not registered with the VMs as a DNS search suffix, so you will need to register it yourself or use fully qualified domain names in your DNS queries.

To create a private DNS zone from the Azure portal, search for Private DNS Zones. On the Private DNS Zones page, click Create to open the Create Private DNS Zone blade. Specify the DNS domain name as the DNS zone resource name and select your resource group, as shown in Figure 4-51.

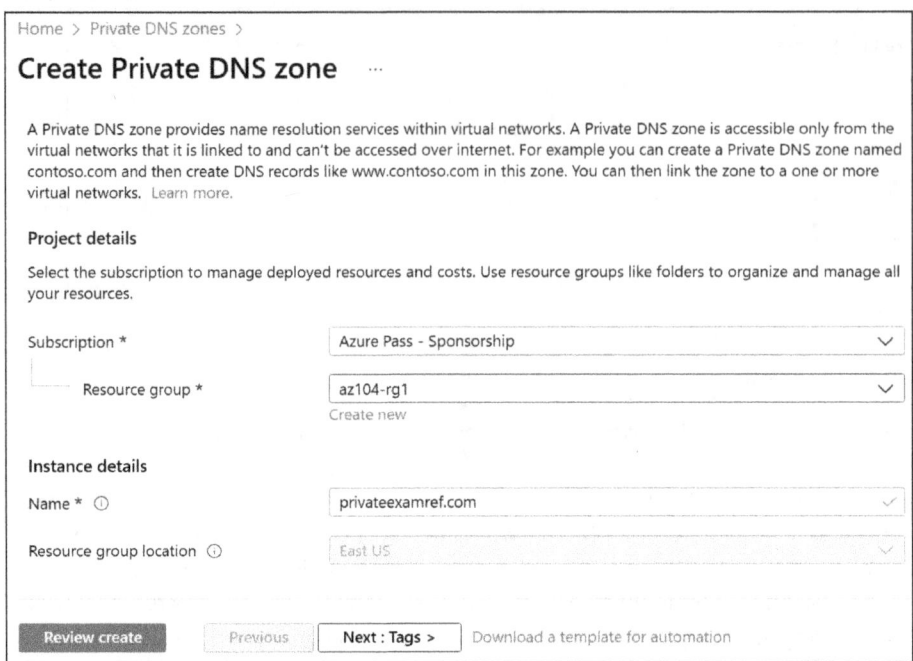

**FIGURE 4-51** Creating a private DNS zone using the Azure portal

With a private DNS zone, you can create virtual network links by clicking Virtual Network Links, and then clicking Add, as shown in Figure 4-52.

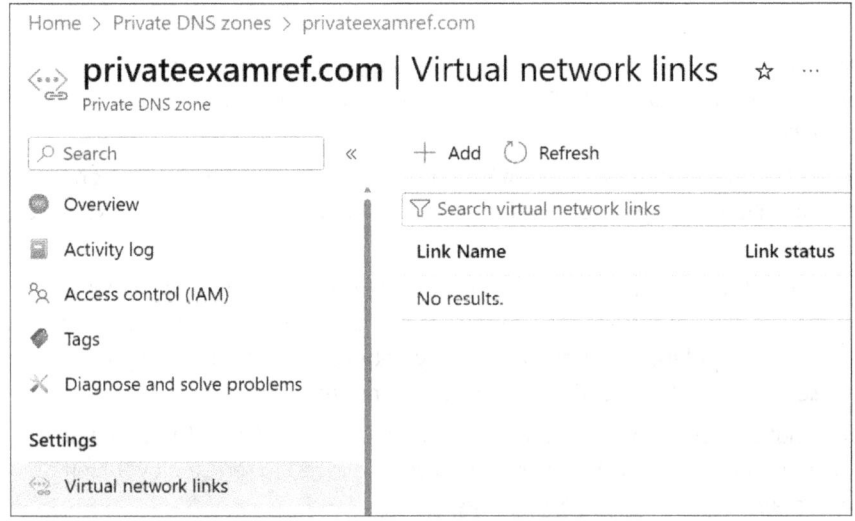

**FIGURE 4-52** Virtual network links for a private DNS zone

You only need to complete the Link Name, Subscription, and Virtual Network Name fields, as shown in Figure 4-53. You can also select the Enable Auto Registration checkbox, which will automate the creation of DNS records in the Private DNS zone for the virtual machines that are connected to the virtual network.

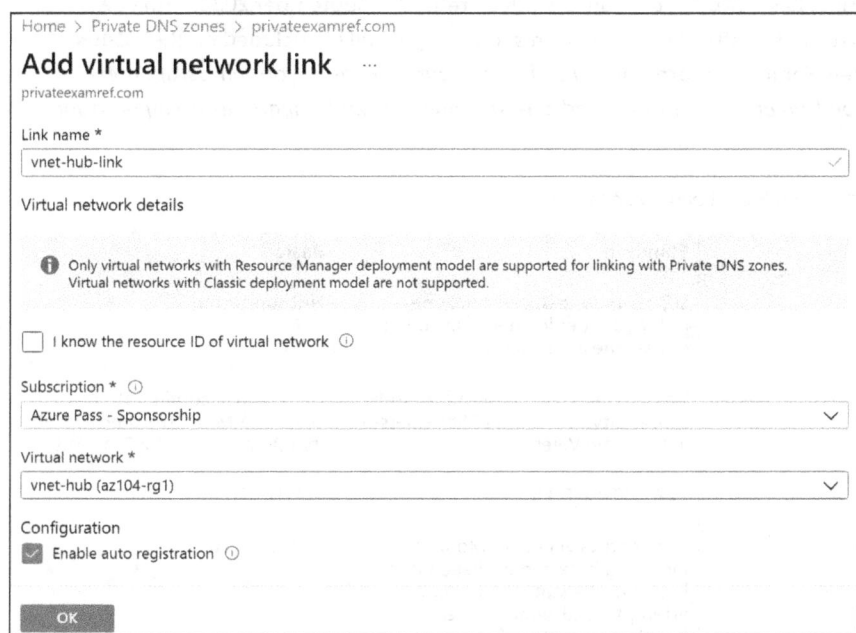

**Add virtual network link** ...
privateexamref.com

Link name *
vnet-hub-link

Virtual network details

ⓘ Only virtual networks with Resource Manager deployment model are supported for linking with Private DNS zones. Virtual networks with Classic deployment model are not supported.

☐ I know the resource ID of virtual network ⓘ

Subscription * ⓘ
Azure Pass - Sponsorship ⌄

Virtual network *
vnet-hub (az104-rg1) ⌄

Configuration
☑ Enable auto registration ⓘ

OK

**FIGURE 4-53** Add a virtual network link for a private DNS zone

Once created, a virtual network link appears on the right.

# Configure load balancing

Load balancing is one of the crucial requirements of network design. Azure offers various options to design load-balancing solutions. In this section, you will learn how to configure Azure Application Gateway and different load balancers in Azure.

## Azure Load Balancer

The deployment of Azure Load Balancer involves the coordinated configuration of several groups of settings. These settings work together to define the overall Load Balancer behavior.

### BASIC AND STANDARD LOAD BALANCER TIERS

Azure Load Balancer is available in two pricing tiers (SKUs): Basic and Standard. These tiers offer different levels of scale, features, and pricing. Table 4-8 provides a comparison of the main feature differences between the Basic and Standard tiers.

**TABLE 4-8**   Standard and Basic Load Balancer tiers

| Feature | Standard | Basic |
| --- | --- | --- |
| Availability Zones | Supports zone-specific or zone-redundant deployments, including cross-zone load balancing | Not supported |
| Backend Pools | Up to 5,000 servers, any mix of VMs, availability sets, and VM Scale Sets—in the same VNet | Up to 300 IP configurations Must be VMs in the same availability set or a single VM Scale Set |
| Health Probes | TCP, HTTP, HTTPS | TCP, HTTP |
| Diagnostics | Rich metrics via Azure Monitor, including byte and packet counters, health probe status, connection attempts, outbound connection health, and more | Not supported |
| Security | Inbound flows closed by default Whitelist-permitted inbound flows using Network Security Groups | Open by default Can optionally restrict flows using Network Security Groups |
| Outbound Connectivity | Supports multiple outbound IP addresses that are configurable via outbound rules | Single outbound IP Not configurable |
| Other Features | Supports HA Ports, TCP Reset on idle timeout, and faster management operations | N/A |
| Pricing | Based on the number of rules and data processed | Free |
| SLA | 99.99 percent availability for a data path with two healthy VMs | None |

### FRONTEND IP CONFIGURATION

Azure Load Balancer supports two modes: Internal Load Balancer and Public Load Balancer. In each case, the frontend IP configuration defines the endpoint upon which the load balancer receives incoming traffic:

- **Internal Load Balancer**   Used to load-balance traffic for intranet-facing applications, or between application tiers. The frontend IP configuration references a subnet, and an IP address from that subnet is allocated using either dynamic or static assignment to the load balancer.

- **Public Load Balancer** Used to load-balance traffic for internet-facing applications. The frontend IP configuration references a separate public IP address resource, which is used to receive inbound traffic.

When used with IaaS VMs, each load balancer can support multiple frontend IP configurations. Therefore, it can receive traffic on multiple IP addresses to load-balance traffic for multiple applications. All frontend configurations, however, must be of the same type: internal or public.

A public load balancer must be associated with a public IP address resource. If the load balancer uses the Standard pricing tier, then the public IP address must also use the Standard pricing tier. Standard-tier load balancers support both zone-specific and zone-redundant deployment options. The choice of deployment option is taken from the associated public IP address, rather than being explicitly in the load balancer properties.

### BACK-END CONFIGURATION

The back-end pool defines the back-end servers over which the load balancer will distribute incoming traffic.

When using a Basic-tier load balancer, this back-end pool must comprise either a single virtual machine, virtual machines in the same availability set, or a VM scale set. If using a VM scale set, traffic will be distributed to all virtual machines in the VM scale set. You cannot distribute traffic to multiple virtual machines unless they are members of the same availability set or VM scale set.

With a Standard-tier load balancer, these restrictions are lifted. Back-end pools can comprise a combination of virtual machines across availability sets and VM scale sets.

### HEALTH PROBES

Azure Load Balancer supports continual health probing of back-end pool instances to determine which instances are healthy and able to receive traffic. The load balancer will stop sending traffic flows to any back-end pool instance that is determined to be unhealthy. Unhealthy instances continue to receive health probes, so the load balancer can resume sending traffic to that instance once it returns to a healthy state.

Azure Load Balancer supports three types of health probes:

- **TCP** Probes attempt to initiate a connection by completing a three-way TCP handshake (SYN, SYN-ACK, ACK). If successful, the connection is then closed with a four-way handshake (FIN, ACK, FIN, and ACK).
- **HTTP** Probes issue an HTTP GET with a specified path.
- **HTTPS** Probes are similar to HTTP probes, except that a TLS/SSL wrapper is used. HTTPS probes are only supported on the Standard-tier load balancer.

All three probe types must also specify the probe port or the interval. The minimum probe interval is five seconds in length, and the minimum consecutive probe failure threshold is two seconds. For HTTP and HTTPs probes, the probe path must also be given.

An endpoint is marked as unhealthy if

- For HTTP or HTTPS probes only, the endpoint returns an HTTP status code other than 200 OK.
- The probe endpoint closes the connection using a TCP reset.
- The probe endpoint fails to respond to a consecutive number of requests during the timeout period. The number of failed requests required to mark the endpoint unhealthy is configurable.

Configuring a dedicated health check page, such as /healthcheck.php, enables each back-end server to implement custom application logic to determine whether it is healthy. Checking the availability of a back-end database is an example of this.

When configuring network security groups (NSGs) for back-end servers, it is important to allow both inbound traffic and probe traffic. Azure Load Balancer does not modify the source IP address of inbound traffic, so inbound traffic rules should be configured as if the load balancer were not in use. To whitelist inbound probe traffic, allow traffic originating from the `AzureLoadBalancer` service tag.

## Configure load balancing rules

Load-balancing rules are used to connect the frontend IP configuration to the back-end server pool, and to a health probe. With Azure Load Balancer, there are no separate back-end HTTP settings configuration; any additional HTTP settings are defined directly within the load-balancing rule itself. These include frontend and back-end ports, idle timeout, protocol (TCP or UDP), and IP version (IPv4 or IPv6).

With the load-balancing rule, you can also configure how inbound connections are distributed between back-end instances. There are three options:

- **None**   Traffic is distributed based on a 5-tuple hash of source IP, destination IP, source port, destination port, and protocol. This is the default option.
- **Source IP**   Traffic is distributed based on a 2-tuple hash of source and destination IP only.
- **Source IP And Protocol**   Traffic is distributed based on a 3-tuple hash of source IP, destination IP, and protocol.

Under the default option, new TCP sessions from a given client might be routed to a different back-end endpoint because the source port will have changed. By excluding the source port from the load-balancing algorithm, the Source IP and Source IP Protocol options provide consistent mappings between client and individual back-end servers across separate connections. This is useful in applications where traffic between the client and server uses more than one connection or protocol. Examples are media uploads that use both a TCP session to control and monitor the upload, as well as UDP packets to upload the media data.

### INBOUND NAT RULES
Azure Load Balancer can be configured to distribute inbound traffic across a pool of back-end servers. Another common scenario is where a connection must be made to a specific back-end

server via the load balancer frontend. This is useful for gaining access to a specific server, such as when diagnosing a problem without exposing a new endpoint on that server.

Direct connectivity to individual servers is achieved by creating a *port mapping* from the frontend to a specific back-end server. This mapping is also known as an *inbound NAT rule*. Each inbound NAT rule specifies a frontend IP address, frontend port, protocol (TCP or UDP), back-end server, and back-end port. Once enabled, traffic received by the front-end IP on the designated frontend port is directed to the specified back-end server and port.

### NETWORK SECURITY GROUP CONFIGURATION

The final step in configuring the Azure Load Balancer is to ensure that network security groups (NSGs) are correctly configured. These NSGs can be associated with the subnet containing the back-end virtual machines or with their network interfaces. Two inbound security rules are required.

First, an inbound rule must permit traffic from the end users to the back-end servers. Even though traffic passes through the load balancer, this does not change the source IP of the inbound traffic, hence the rule must reference the end user source IP address and port range.

A second inbound rule must permit traffic originating from the load balancer health probe. The IP addresses from which the health probes originate are defined in the `AzureLoadBalancer` service tag, which should be used to define the source IP address range for this rule.

> **NOTE**  **LOAD BALANCER AND NETWORK SECURITY GROUPS**
> Standard-tier load balancers use Standard-tier public IP addresses, which are closed to inbound traffic by default. When using a Standard-tier load balancer, traffic *must* be approved using NSGs. In contrast, when using Basic-tier load balancers, traffic *should* be approved using NSGs but will also flow if NSGs are not used.

## Configure an Azure load balancer

As discussed earlier, both internal and public load balancers involve the coordinated configuration of several groups of settings. These settings work together to define the overall load balancer behavior.

To use Azure Load Balancer, the administrator must first provision the resource, which includes the frontend IP configuration. After this step has been completed, you can create the back-end pool, the heath probes, and finally the load-balancing rule.

To create the load balancer in the Azure portal, search for **Load Balancer**. On the Load Balancers blade, click Create. This will open the Create Load Balancer blade, as shown in Figure 4-54. Complete the fields:

- **Name**   Provide a name for the load balancer resource.
- **SKU**   Select the pricing tier: Basic, Standard, or Gateway.
- **Type**   Choose Public or Internal.

- **Tier**   Specify whether the load balancer is Regional or Global.
- **Subscription, Resource Group, And Location**   Specify as required.

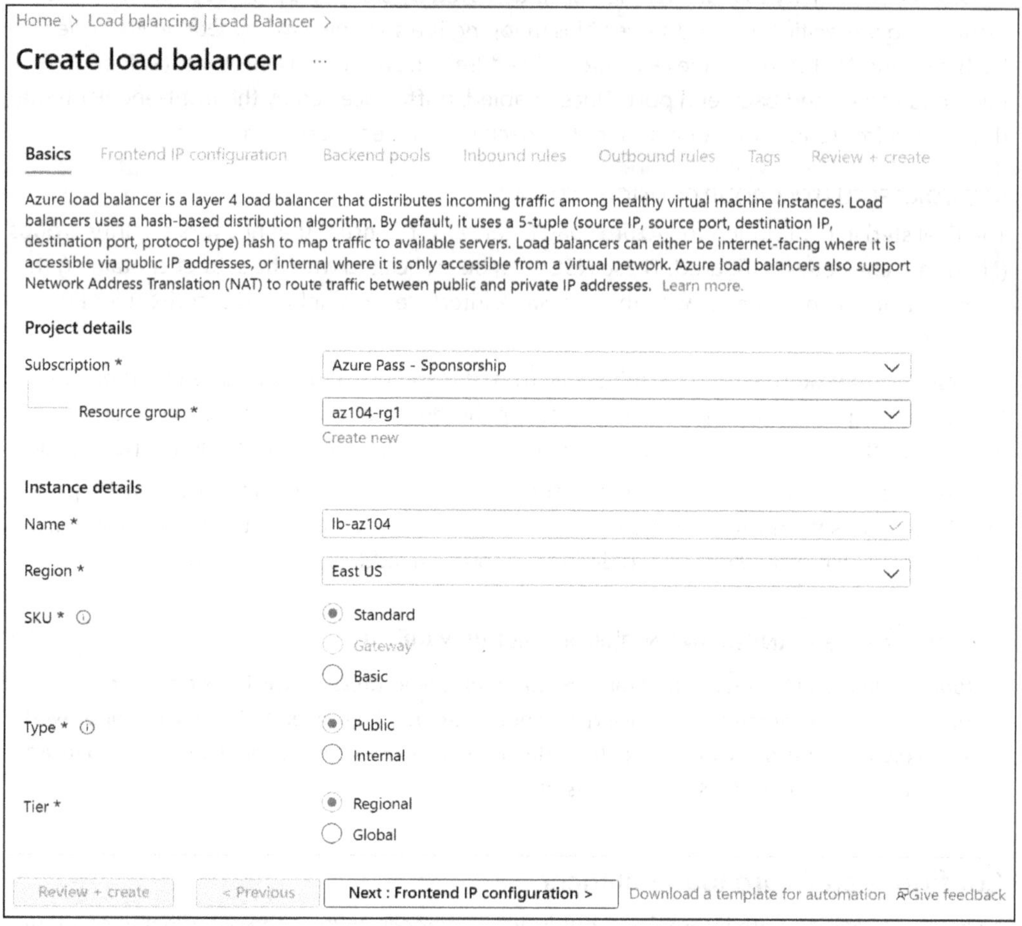

Create load balancer   ...

Basics   Frontend IP configuration   Backend pools   Inbound rules   Outbound rules   Tags   Review + create

Azure load balancer is a layer 4 load balancer that distributes incoming traffic among healthy virtual machine instances. Load balancers uses a hash-based distribution algorithm. By default, it uses a 5-tuple (source IP, source port, destination IP, destination port, protocol type) hash to map traffic to available servers. Load balancers can either be internet-facing where it is accessible via public IP addresses, or internal where it is only accessible from a virtual network. Azure load balancers also support Network Address Translation (NAT) to route traffic between public and private IP addresses.   Learn more.

**Project details**

Subscription *          Azure Pass - Sponsorship

Resource group *        az104-rg1
                        Create new

**Instance details**

Name *                  lb-az104

Region *                East US

SKU * ⓘ                 ⦿ Standard
                        ◯ Gateway
                        ◯ Basic

Type * ⓘ                ⦿ Public
                        ◯ Internal

Tier *                  ⦿ Regional
                        ◯ Global

Review + create         < Previous         Next : Frontend IP configuration >     Download a template for automation   ⏚Give feedback

**FIGURE 4-54**   Creating a public load balancer with the Azure portal

Click Next: Frontend IP configuration. On this tab, click Add A Frontend IP Configuration to configure the frontend IP address, depending on whether the load balancer is set to be a public or internal appliance. For public load balancers, create and associate the public IP address here. For internal load balancers, select the virtual network and subnet to place the load balancer in. Figure 4-55 displays the public frontend IP configuration, and Figure 4-56 displays the internal frontend IP configuration.

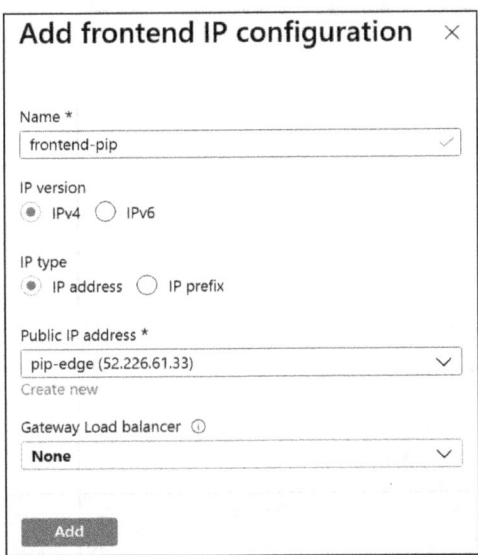

**FIGURE 4-55**  Public frontend IP configuration

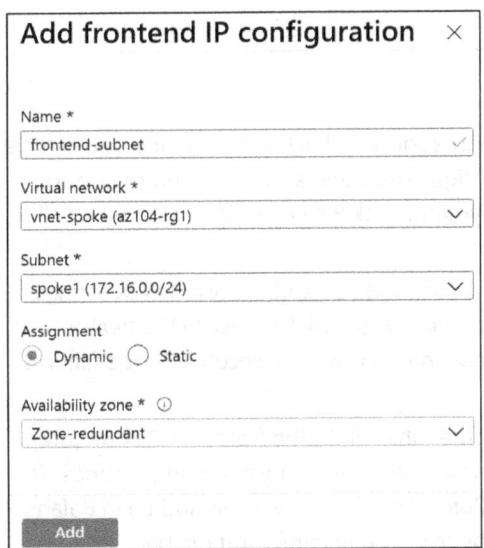

**FIGURE 4-56**  Internal frontend IP configuration

Click Next: Backend Pools. On the Backend Pools tab, click Add A Backend Pool. A back-end pool is the target of the incoming traffic. Figure 4-57 displays the Add Backend Pool blade using an IP address for the back-end resource.

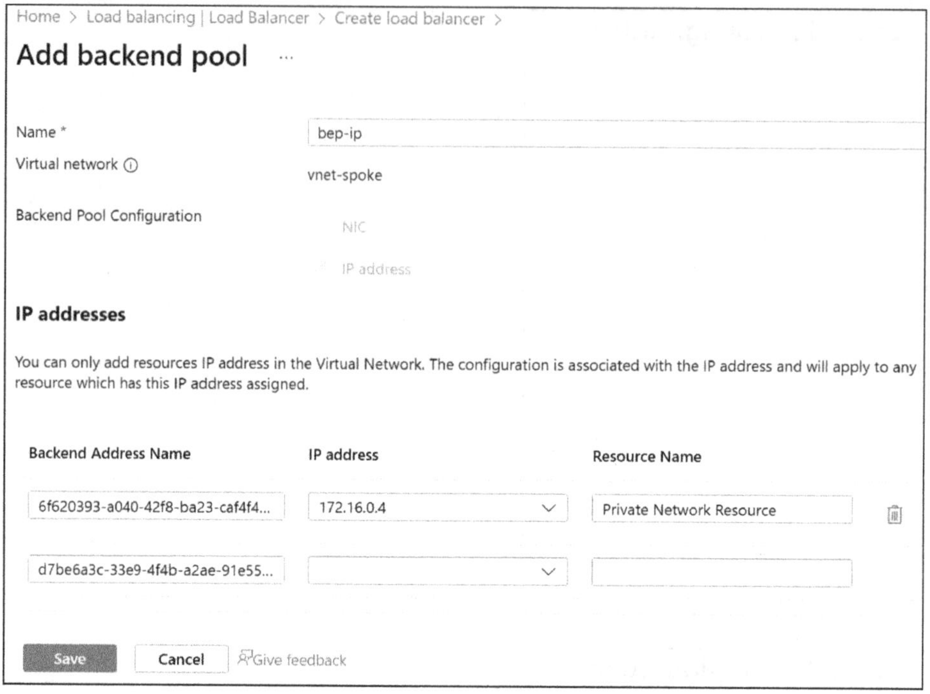

# Add backend pool ···

| | |
|---|---|
| Name * | bep-ip |
| Virtual network ⓘ | vnet-spoke |
| Backend Pool Configuration | NIC |
| | IP address |

## IP addresses

You can only add resources IP address in the Virtual Network. The configuration is associated with the IP address and will apply to any resource which has this IP address assigned.

| Backend Address Name | IP address | Resource Name | |
|---|---|---|---|
| 6f620393-a040-42f8-ba23-caf4f4... | 172.16.0.4 ∨ | Private Network Resource | 🗑 |
| d7be6a3c-33e9-4f4b-a2ae-91e55... | ∨ | | |

**Save** Cancel 🗨Give feedback

**FIGURE 4-57** Back-end pool IP addresses

After configuring the Basics, Frontend IP Configuration, and Backend Pool tabs, you can create the load balancer. Optionally, you can also configure inbound and outbound rules during creation, but those can also be added after deployment. Click Review + Create, and then click Create to deploy the load balancer.

To create a health probe, navigate to the Load Balancer blade and choose Health Probes, Add. This opens the Add Health Probe blade, as shown in Figure 4-58. Specify the health probe name, together with the protocol, port, probe interval, and consecutive probe failures threshold.

The final step is to configure a load-balancing rule, which links the frontend IP configuration to the back-end pool, specifying the health probe and other load-balancing settings. On the Load Balancer blade, choose Load Balancing Rules, Add. This opens the Add Load Balancing Rule blade, as shown in Figure 4-59. Choose the frontend IP configuration, back-end pool, and health probe selected earlier. For HTTP traffic, select TCP, and specify port 80 for both the frontend and back-end ports. Select None for Session Persistence and leave the Idle Timeout value at the default of 4 minutes.

> **NOTE** **FLOATING IP**
>
> The last setting, Floating IP (direct server return), is recommended only when load-balancing traffic for a SQL Server Always On Availability Group listener. For other scenarios, the Floating IP setting should be left disabled.

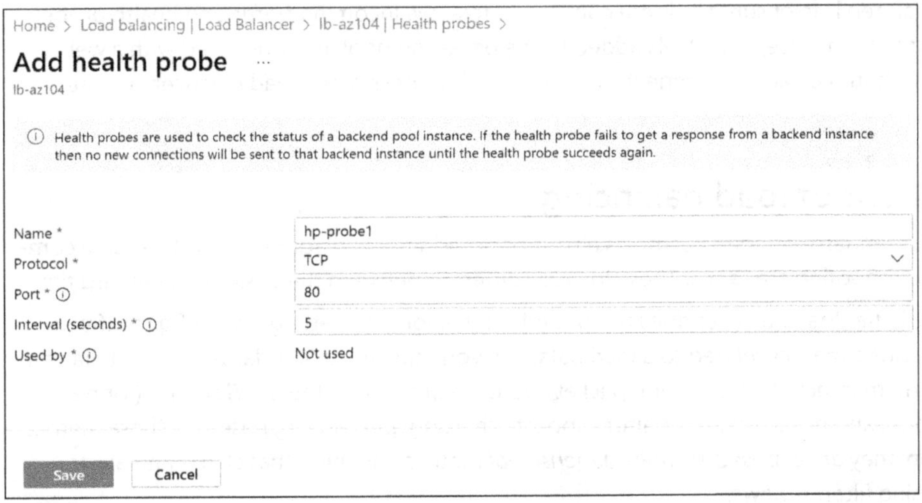

# Add health probe ...
lb-az104

ⓘ Health probes are used to check the status of a backend pool instance. If the health probe fails to get a response from a backend instance then no new connections will be sent to that backend instance until the health probe succeeds again.

| | |
|---|---|
| Name * | hp-probe1 |
| Protocol * | TCP ⌄ |
| Port * ⓘ | 80 |
| Interval (seconds) * ⓘ | 5 |
| Used by * ⓘ | Not used |

Save    Cancel

**FIGURE 4-58**  Creating a health probe in Azure Load Balancer

# Add load balancing rule ...
lb-az104

A load balancing rule distributes incoming traffic that is sent to a selected IP address and port combination across a group of backend pool instances. Only backend instances that the health probe considers healthy receive new traffic.

| | |
|---|---|
| Name * | lbrule-http |
| IP Version * | ⦿ IPv4 |
| | ◯ IPv6 |
| Frontend IP address * ⓘ | frontend-subnet (172.16.0.4) ⌄ |
| Backend pool * ⓘ | bep-ip ⌄ |
| High availability ports ⓘ | ☐ |
| Protocol | ⦿ TCP |
| | ◯ UDP |
| Port * | 80 |
| Backend port * ⓘ | 80 |
| Health probe * ⓘ | hp-probe1 (TCP:80) ⌄ |
| | Create new |
| Session persistence ⓘ | None ⌄ |
| Idle timeout (minutes) * ⓘ | 4 |
| Enable TCP Reset | ☐ |
| Enable Floating IP ⓘ | ☐ |

Save    Cancel

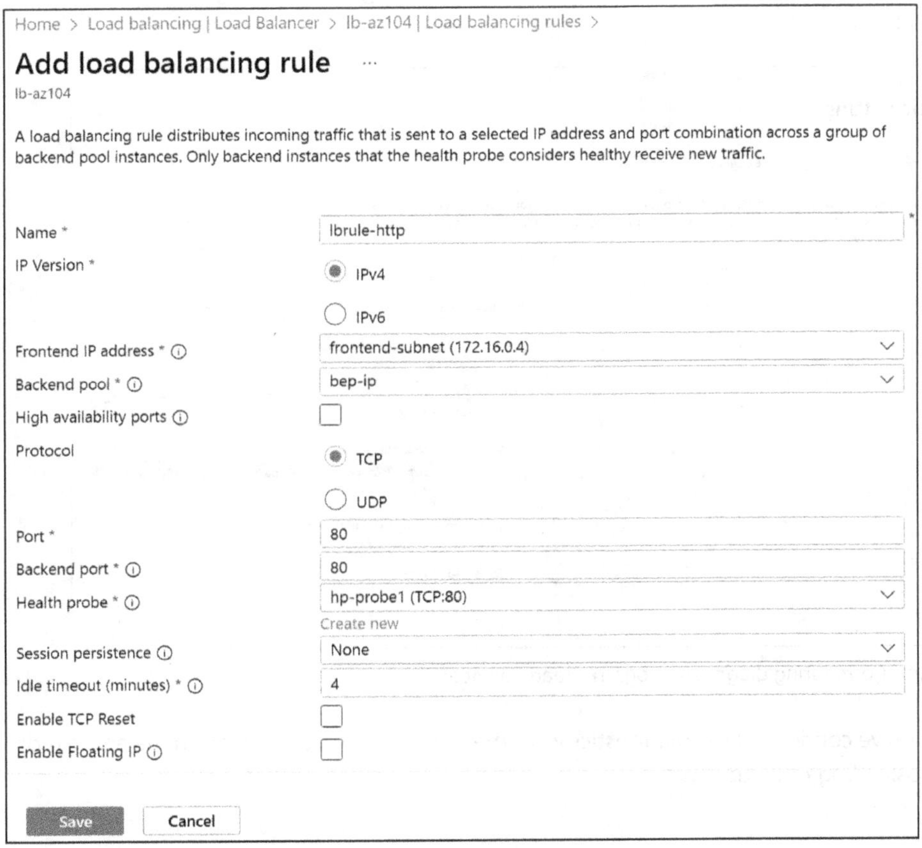

**FIGURE 4-59**  Creating a load-balancing rule in Azure Load Balancer

The final step is to ensure NSGs are configured to allow incoming traffic and health probe traffic. With this in place, if the VMs added to the back-end pool are configured with a web server, you should be able to connect to the public IP address of the load balancer and see the webpage.

## Troubleshoot load balancing

Basic- and Standard-tier load balancers also support additional diagnostic logs to enable common troubleshooting scenarios. These logs are different between the Basic and Standard tiers.

The Basic-tier load balancer does not provide any diagnostic settings to configure. If you need to capture metrics related to a load balancer, you must use a Standard-tier load balancer. Available metrics include byte count, packet count, health probe status, SYN count (for new connections), and more. Azure Monitor supports charting and alerting based on these metrics. In addition, they are exposed as *multi-dimensional* metrics, meaning that charts and alerts can be built using filtered views.

To enable load-balancer logs, open the Load Balancer blade in the Azure portal, click Diagnostic Logs, and click Add Diagnostic Setting to open the Diagnostic Setting blade, as shown in Figure 4-60.

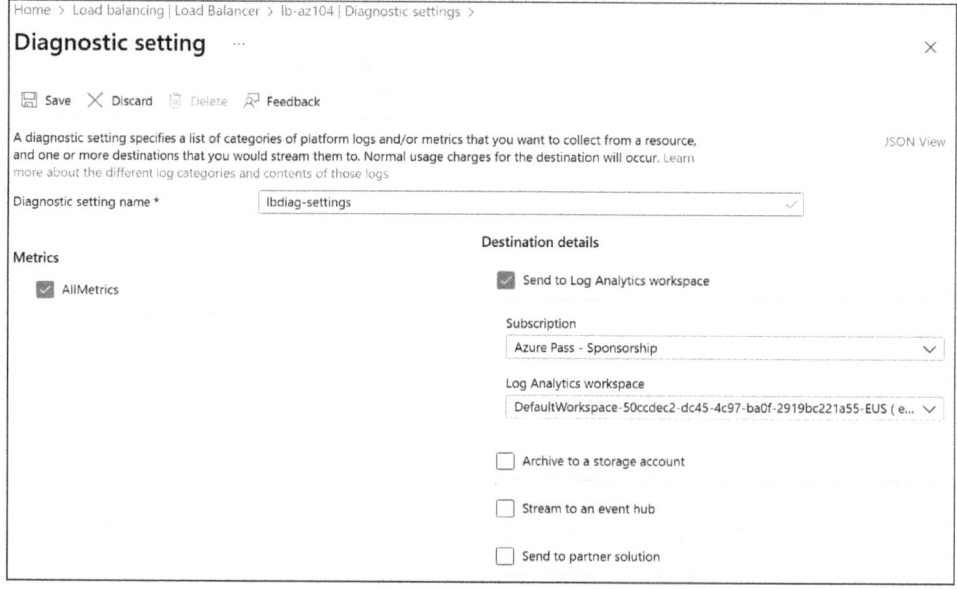

**FIGURE 4-60**   Configuring diagnostics logs in a load balancer

Once you've configured the diagnostics logs, they can be downloaded for offline analysis or analyzed using Log Analytics.

# Chapter summary

This chapter covered many of the advanced networking features available in Azure. Here are some of the key takeaways from this chapter:

- Azure virtual networks (VNets) are isolated networks using a private IP address space.
- Virtual networks are divided into subnets, so you can isolate workloads.
- Azure reserves the first four and last IP address in each subnet. The first IP address allocated to VMs is therefore typically the .4 IP address.
- Private IP addresses for a VM are assigned from a subnet and configured as settings on the IP configuration of a network interface resource.
- A VM can be associated with one or more network interfaces, and each network interface can contain multiple IP configurations.
- Private IP addresses support two allocation methods: dynamic or static. Dynamic IP addresses are released when the VM is stopped (deallocated).
- Public IP addresses are managed as a standalone resource, which can be associated with a network interface IP configuration.
- Public IP addresses support two pricing tiers (SKUs). The Basic tier supports dynamic and static assignment and provides open connectivity (which can be restricted using NSGs). The Standard tier supports zone-redundant deployments, uses static allocation only, and is closed by default (access is enabled using NSGs).
- User-defined routes (UDRs) change the default behavior of subnets so you can direct outbound traffic to other locations. Typically, traffic is sent through a virtual appliance such as a firewall.
- If a UDR is used to send traffic to a virtual appliance, IP forwarding must be enabled on the NIC of the virtual appliance VM.
- Routing outbound internet traffic via a VPN connection to a network security device is known as forced tunneling.
- The effective routes for each network interface can be reviewed to help diagnose routing issues.
- VNets can be connected using either VNet peering or VNet-to-VNet VPN connections.
- To connect two VNets, they must have nonoverlapping IP address spaces.
- Virtual networks can be connected using VNet peering. This is supported both within a region or across regions.
- By default, peered VNets appear and perform as a single network. There is an option to limit connectivity, in which case NSG rules must be used to define the permitted connections.

- VNet peering allows VMs to see each other as one network, but their relationships are nontransitive. If VNetA and VNetB are peered and VNetB and VNetC are peered, VNetA and VNetC are not peered.

- A common approach is to use a hub-and-spoke network architecture, in which separate spoke VNets are used by each application, peered to a hub VNet containing a network virtual appliance (NVA). The peering connections must enable Allow Forwarded Traffic.

- Using VNet peering to provide access to a central VNet containing shared services, such as Active Directory domain controllers, is known as service chaining.

- Azure DNS provides an authoritative DNS service for hosting internet-facing domains.

- DNS zones in Azure DNS must be delegated from the parent domain. To do this, set up appropriate NS records in the parent domain, pointing to the name servers assigned by Azure DNS.

- DNS records in Azure DNS are managed using record sets, which are the collection of records with the same name and the same type.

- DNS records at the zone apex use the record name @. You cannot create records with the CNAME record type at the zone apex.

- Azure DNS alias records allow DNS records to reference other Azure resources, such as a public IP address.

- DNS zone files are a standard format used to transfer DNS records between DNS systems. DNS zone files can only be imported into or exported from Azure DNS by using the Azure CLI.

- Azure-provided DNS, also known as Internal DNS, provides VM-to-VM DNS lookups within a virtual network.

- Alternatively, a customer can implement their own DNS servers, which can be configured either at the VNet or the network interface level.

- Azure DNS also supports private DNS zones, which can also be used to enable VM-to-VM DNS lookups.

- Network security groups are used to create firewall rules to control network flows.

- NSGs can be applied at the subnet level, or on individual VM network interfaces.

- Each NSG includes a list of default rules, which can be overridden using user-defined rules. Rules are applied in priority order (processing stops at the first rule matching the traffic in question).

- Source and destination IP address ranges in NSG rules can be specified explicitly using CIDR ranges.

- IP address ranges can also be specified using service tags that are platform shortcuts for the IP ranges for key Azure services. Commonly used service tags include VirtualNetwork, Internet, Azure Cloud, Storage, and SQL.

- IP address ranges can also be specified using application security groups (ASGs). Using ASGs, NSG rules to be defined for groups of VMs without needing to allocate the VMs into separate subnets.

- Tools to help identify the required NSG rules include service map and NSG flow logs.

- Effective security rules can be reviewed for each network interface, so you can see the exact IP ranges used by each service tag and ASG.

- The Azure Bastion service is provisioned within a virtual network within a separate subnet called AzureBastionSubnet. If you have multiple VNets in your environment, you will need to deploy Azure Bastion for each VNet separately.

- Azure Load Balancer (ALB) is a fully managed, high-performance load-balancing service for TCP and UDP traffic. It operates at the transport layer (OSI Layer 4). Unlike App Gateway, it does not have visibility into application-level traffic.

- ALB can be deployed with either a public (internet) or private (intranet) frontend IP address.

- ALB comes in two pricing tiers (SKUs): Basic or Standard. The Standard tier supports availability zones, larger and more flexible back-end pools, and a number of other features. The Basic tier is free of charge.

- An ALB load-balancing configuration comprises frontend IP configuration, back-end pool, health probes, and load-balancing rule.

- ALB also supports port forwarding, using inbound NAT rules. This maps a specific frontend port to a specific back-end port on a specific back-end server.

- Use Connection Troubleshoot to test the connectivity between two Azure VMs, or between a VM and an arbitrary external endpoint.

- Connection Monitor enables long-term connection monitoring, using similar diagnostics as used by Connection Troubleshoot.

# Thought experiment

In this thought experiment, apply what you have learned about this objective. You can find answers to these questions in the next section.

Your company, Contoso, wants to lift and shift an existing HR application to Azure. The application architecture comprises two web servers, and a database tier implemented using three servers in a SQL Server Always-On Availability Group. The web application uses an in-memory session state that requires each user to be consistently routed to the same web server instance. The application should be accessible only to the company intranet, and not exposed to the internet.

In addition, Contoso has already migrated several other applications to Azure. A recent finance review, however, has highlighted the increasing of Azure spend, and your manager has identified the duplication of infrastructure components (such as domain controller virtual machines) across each migrated application as a potential area where savings can be made.

Each of these applications is managed by a separate team, and the team should have administrative access only to their application.

1. How should the database tier be load-balanced?

2. How can you restrict network traffic between application tiers, and prevent on-premises uses from having direct access to the database tier?

3. How should the application be integrated into the company intranet, avoiding exposing an internet endpoint?

4. How can you reduce costs by consolidating duplicated components?

5. How does your design maintain administrative separation between applications?

## Thought experiment answers

This section contains the solution to the thought experiment for the chapter.

1. The database tier should be load-balanced using Azure Load Balancer. The load balancer will be configured with an internal (intranet) IP address only. Because the load balancer is being used as a SQL Server Always-On Availability Group Listener, the Floating IP (Direct Server Return) option should be enabled.

2. Network security groups should be used to restrict inbound and outbound traffic for the subnets used by each application tier. Optionally, application security groups can be used to simplify the IP address management and reduce the number of subnets and NSGs required.

3. Connectivity between the application and the on-premises network can be achieved in two ways. The simplest option is to establish a site-to-site VPN between the on-premises network and the Azure virtual network. This creates an encrypted tunnel (over the internet) linking the two networks together. A compatible on-premises VPN device with a static internet-facing IPv4 address is required, together with a VPN gateway in Azure (hosted in a dedicated gateway subnet). Alternatively, an ExpressRoute connection can be used. This provides a more reliable and consistent connection over a dedicated connection from a connectivity provider. In this case, an ExpressRoute gateway is used to connect the ExpressRoute circuit to the Azure virtual network.

4. A dedicated VNet should be created to contain common services (such as Active Directory servers), which are consumed by multiple applications. Each application should remain in its own VNet, which should only contain application-specific components. The application VNets should be peered with the shared services VNet, in a hub-and-spoke configuration (with the shared services VNet as the hub). This peering will give the applications network access to the shared components.

5. Because each application retains its own VNet containing all application-specific components, there is no loss of isolation or control for the application owners. These application components can even be deployed in separate subscriptions, making separate role-based access and billing straightforward. Peering of Resource Manager VNets is supported across subscription boundaries.

# CHAPTER 5

# Monitor and back up Azure resources

As you begin to deploy services into your Azure subscriptions, you need to decide how the environment will be monitored. To do so, you must think about all the services in your deployment. You will most likely have several services deployed, including Infrastructure-as-a-Service services (for example, virtual machines), which include compute, storage, and networking. And even without services deployed today, over time, you might have Platform-as-a-Service services for hosting applications. You will also be using the services that drive your virtual machines in more meaningful ways, such as implementing advanced configurations in Azure Storage and Azure Identity.

You will need to account for all these services—along with the Azure platform itself—in your monitoring strategy. This includes all your infrastructure, applications, and networking.

By developing a proactive monitoring strategy, you will be able to understand the operation of your environment at a component level, including resource health and resource spend. Implementing a robust strategy will help you increase your uptime through proactive notifications, so you can resolve issues before they become problems and optimize your resources for optimal performance, increasing your ROI with the services you deploy.

As you develop your strategy, there are three areas you should consider:

- **Visibility into services and the Azure Platform**   This is all about understanding how an application or set of services is performing across the board. You will need to understand what metrics you need to monitor and how those can be acted on in Azure through both alerts and visualizations in dashboards.

- **Deeper insights into applications**   This is particularly true with service or dependency maps and advanced tracing. You may even use these insights to drive automation and remediations within your environments.

- **Resource optimization**   You need to understand which metrics are important not just for the health of your applications, but also the effects on users or systems that consume those applications. By using the visibility and insights you extract from the Azure platform, you can directly correlate the effects of remediations in your environment.

Azure includes multiple services that perform specific roles for monitoring and optimization. It is critical that you understand both the out-of-the-box monitoring capabilities of Azure and the scenario-specific monitoring capabilities within the platform. This section will focus on out-of-the-box monitoring and optimization through Azure Monitor, as well as scenario-specific monitoring with Azure Monitor logs and log data that is stored in Log Analytics.

Azure Backup is another critical service that enables simplified disaster recovery for virtual machines by ensuring that data is securely backed up and easily restorable. In this chapter, you'll also review how to implement and manage Azure backup and recovery solutions with an emphasis on the Azure Backup Service and Azure Site Recovery.

### Skills covered in this chapter:

- Skill 5.1: Monitor resources in Azure
- Skill 5.2: Implement backup and recovery

## Skill 5.1: Monitor resources in Azure

Azure Monitor maximizes the availability and performance of your applications by delivering a comprehensive solution for collecting, analyzing, and acting on telemetry from your cloud and on-premises environments. It helps you understand how your applications are performing and proactively identifies issues affecting them and the resources on which they depend. The Azure Monitor landing page provides a jumping off point to configure other more-specific monitoring services, such as Application Insights, Network Watcher, Log Analytics, Management Solutions, and so on. Figure 5-1 shows some of the various data sources and how they are collected, either as metric or log data. The data is consumed, visualized, or acted on by various services in Azure.

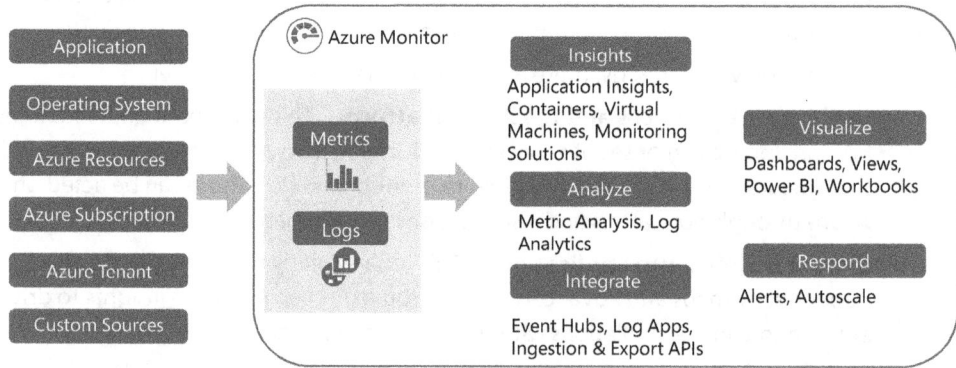

**FIGURE 5-1** Azure Monitor data sources for metric and log data and the ways you can act on the data

> **NEED MORE REVIEW?** **AZURE MONITOR**
>
> To learn more about the capabilities of Azure Monitor see *https://learn.microsoft.com/en-us/azure/azure-monitor/overview*.

Azure Monitor helps you track performance, maintain security, and identify trends by ingesting metrics and telemetry from multiple areas, including applications and the operating systems of virtual machines. You can also query your Azure resources (which emit performance counters), your Azure subscriptions, Entra tenant, and event custom sources.

The data from your Azure resources is ingested into either metrics stored within the Azure platform and accessible by the monitor service, or as logs into a Log Analytics workspace in your Azure subscription.

Comparing metrics and logs surfaces some key differentiators:

- **Retention**  Most of the metrics are retained for 93 days within the Azure service, while logs stored in Log Analytics can be retained for up to two years. However, metric queries can only span up to 30 days. There are opportunities to retain long-term metrics by storing metrics in Log Analytics as well.

- **Properties**  Metrics have a fixed set of properties (or attributes). These are time, type, resource, value, and dimensions (optional). Logs have different properties for each log type and even support rich data types, such as date and time.

- **Data availability**  Metrics are gathered over time (like once a minute) and available for immediate query. Logs are often gathered after being triggered by an event (such as when an event is written to an application log) and can take time to process before they are available for query. While both offer near real-time query capabilities, metrics will typically be used for fast alerts, and logs are used for more complex analysis.

Once the data is collected, Azure Monitor provides "a single pane of glass," or entry point, to interact with your metrics and logs. Interactions can include querying and alerting, building visualizations and dashboards, or even automated responses based on telemetry for functionality, such as autoscaling in virtual machines.

Data stored in Log Analytics can also be queried directly through a Log Analytics workspace, where you will have access to the same query interfaces as you have through Azure Monitor, but you also can make customizations to the configuration of the workspace and access workspace-specific solutions, including visualizations and queries.

All the data that you can access through Azure Monitor can be used to create alerts within Azure Monitor with alert rules. Alert rules are built based on target resources or resource types, such as virtual machines, storage account, and even PaaS services and your custom conditions. Alerts proactively notify you of the health of the resources you deploy in Azure. You are not limited to notifications; alert rules leverage actions groups so you can even implement automation based on an alert condition.

**This skill covers how to:**
- Interpret metrics in Azure Monitor
- Configure log settings in Azure Monitor
- Query and analyze logs in Azure Monitor
- Set up alert rules, action groups, and alert processing rules in Azure Monitor
- Configure Application Insights
- Configure and interpret monitoring of VMs, storage accounts, and networks by using Azure Monitor Insights
- Use Azure Network Watcher and Connection Monitor

## Interpret metrics in Azure Monitor

Recall that metrics are the numerical values that are output by resources and services within Azure. Metrics are available for a number of Azure resources, but not all resources support metrics at this time.

Metrics includes platform metrics, which are created by Azure resources and made available in Azure Monitor for querying and alerting. You can also query application metrics from Application Insights if the service is enabled and you have instrumented your applications—regardless of whether that application is hosted on a virtual machine or even in a PaaS service, such as Azure App Service. When using a PaaS service such as App Service, you can retrieve some data without additional instrumentation. Virtual machines in Azure can also push custom metrics to the monitor service using the Windows Diagnostic extension on Windows servers and with the InfluxData Telegraf Agent on Linux VMs. There is also an opportunity to push custom metrics from other sources through a REST API.

Figure 5-2 shows an example of a metrics chart displaying the end-to-end latency for a storage account.

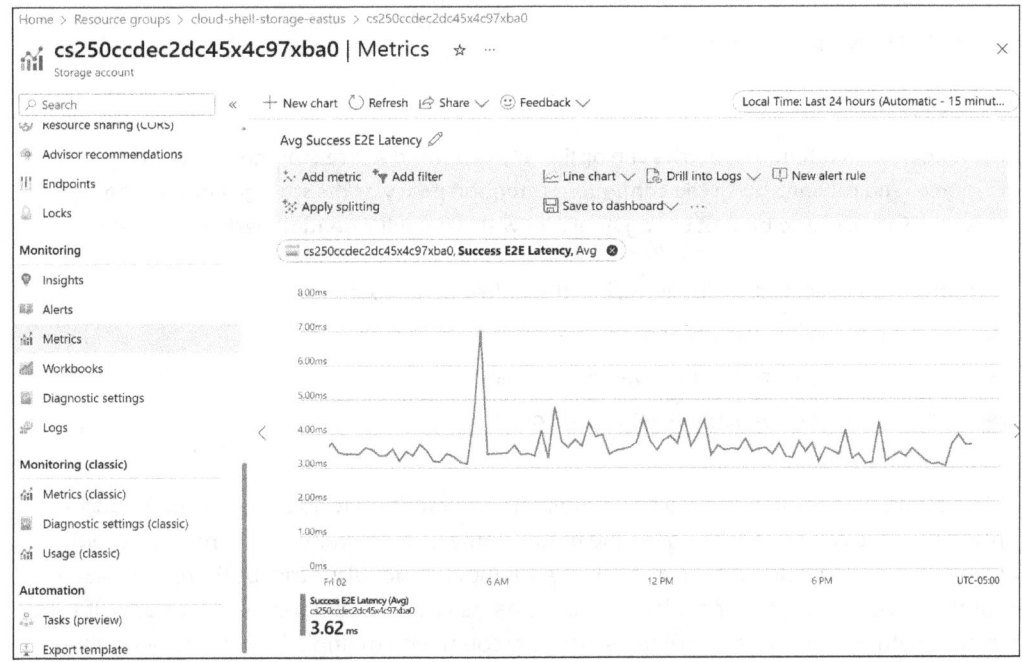

**FIGURE 5-2** Azure Metrics

Azure metrics are collected at one-minute intervals (unless otherwise specified) and are identified by a metric name and a namespace (or category). Most Azure metrics are retained for 93 days within Azure Monitor, but there are notable exceptions listed below:

Guest OS metrics

- Classic guest OS metrics

  - Collected through diagnostic extensions and sent to an Azure storage account.

  - Retention period of 14 days.

- Guest OS metrics sent to Azure Monitor metrics

  - Monitored by Windows diagnostic extensions or the InfluxData Telegraf agent and routed to an Azure Monitor data sink.

  - Retention period of 93 days.

- Guest OS metrics collected by Log Analytics agent
    - Collected by the Log Analytics agent and sent to a Log Analytics workspace.
    - Retention period of 31 days. This retention period can be extended for up to two years.
- Application insights log-based metrics
    - Log-based metrics are translated into log queries.
    - Retention period of 90 days.

---

***EXAM TIP***

**For longer-term retention, metrics can optionally be sent to Azure Storage for select resources and retained up to the configured retention policy or the storage limits of the account. They can also be sent to Log Analytics with a default retention period of 31 days.**

---

As metrics are collected, each metric has the following properties:

- The time the value was collected
- The type of measurement the value represents
- The resource with which the value is associated
- The value itself

Metrics can be one-dimensional or multidimensional with up to 10 dimensions. A nondi-mensional metric can be thought of as the metric name, and the value of the metric output is collected by the Monitor service over time. A multidimensional metric (both from an Azure resource or a custom metric) includes the metric name and an additional name-value pair with additional data. For example, imagine a storage account with multiple blob containers where you need to track the consumption of storage by container. A nondimensional metric would provide only the total consumed storage for the blob endpoint in the storage account where a multidimensional metric would provide the consumption by container, as it has the additional data stored in the metric record.

To interact with metrics from the Azure portal, search for **Monitor**. Then on the Monitor blade, click Metrics to open the Metrics blade. You will be presented with a blank chart. You can select the scope and required metrics to customize the metrics chart as needed, as shown in Figure 5-3.

To begin populating the chart, you need to select a metric. To select a metric, you must select a subscription and a resource group. Optionally, you can filter by resource type as well. When you select a resource, you can select a metric namespace (or category), a metric, and an aggregation if applicable. For example, to view the Ingress metric for a storage account, select the storage account from the Scope drop-down menu, choose Account from the Metric Namespace drop-down menu, choose Ingress from the Metric drop-down menu, and choose Sum from the Aggregation drop-down menu, as shown in Figure 5-4.

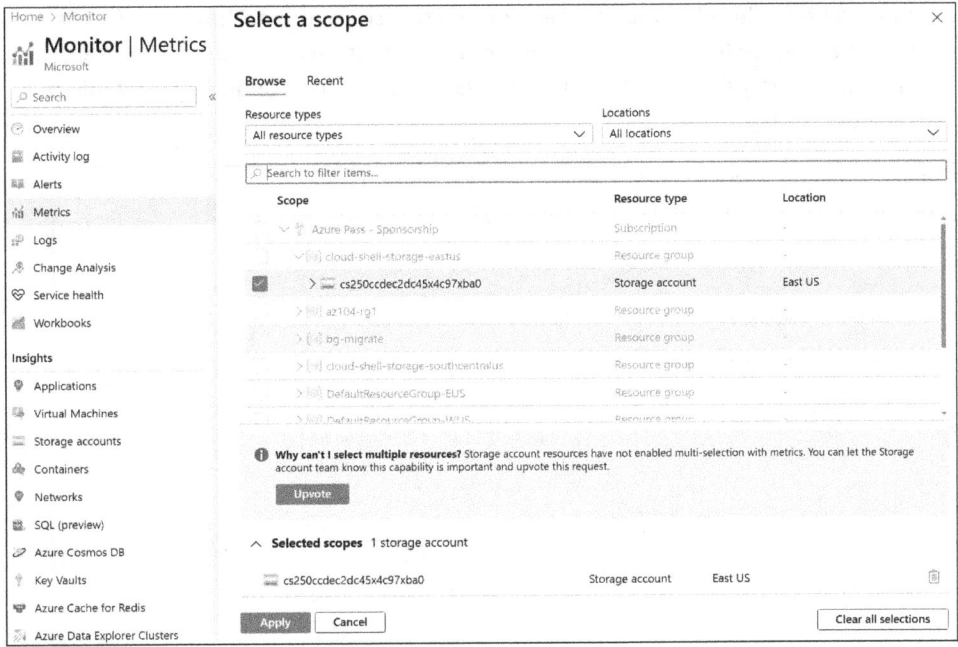

**FIGURE 5-3**  Azure Metrics Select A Scope blade

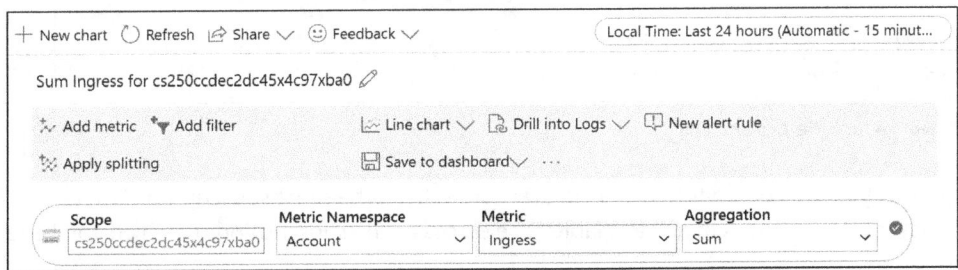

**FIGURE 5-4**  Azure metrics selection

You can add multiple metrics to the chart, and you can even mix resources, namespaces, metrics, and aggregations as shown in Figure 5-5.

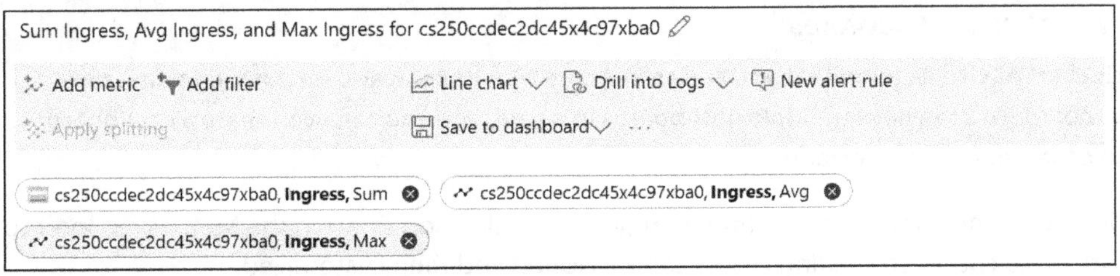

Sum Ingress, Avg Ingress, and Max Ingress for cs250ccdec2dc45x4c97xba0 ✎

🗠 Add metric   ⬥ Add filter            ⬡ Line chart ⌄  🔲 Drill into Logs ⌄   🔲 New alert rule
⋮⋮ Apply splitting                       🔲 Save to dashboard⌄  …

cs250ccdec2dc45x4c97xba0, **Ingress**, Sum ⊗    ⌁ cs250ccdec2dc45x4c97xba0, **Ingress**, Avg ⊗
⌁ cs250ccdec2dc45x4c97xba0, **Ingress**, Max ⊗

**FIGURE 5-5**  Azure multiple metrics selection for a resource

The chart will be rendered as you complete each resource selection. The period for the query can be changed up to the retention limits of the metrics service, and the chart can be rendered as a line chart (default), area chart, bar chart, scatter chart, or grid. An example of a line chart is shown in Figure 5-6.

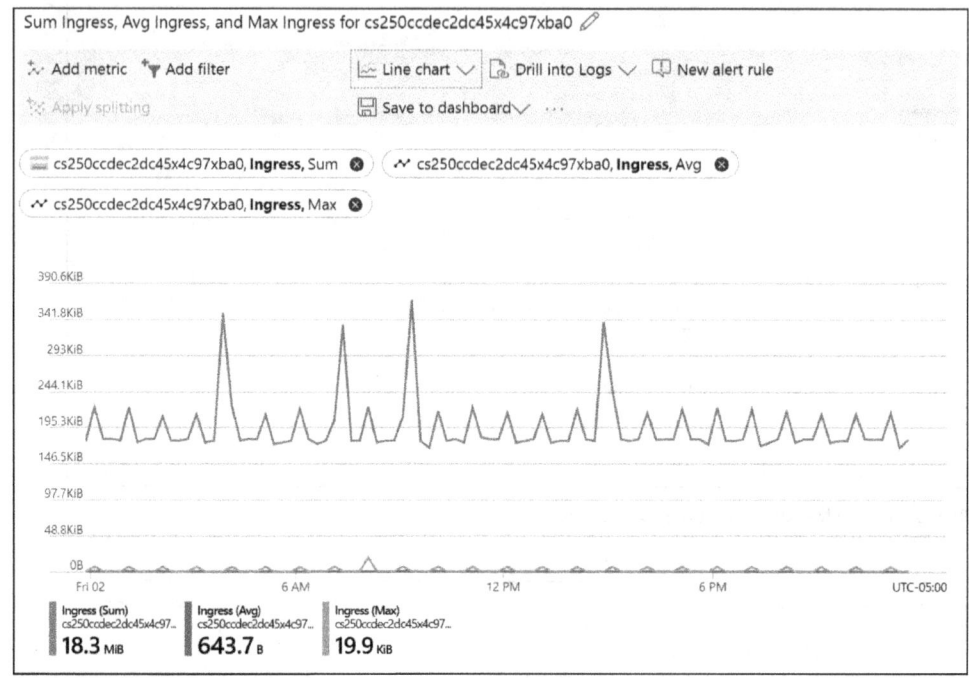

**FIGURE 5-6**   Azure Metrics line chart

Note that you are not limited to charting resources from the same subscription. You can select metrics for resources of any available type across all the subscriptions to which you have access.

From the Metrics blade, you can also create a new alert rule based on the metric query that is visualized. If you need to perform a deeper analysis, you can export the raw metric data to Excel.

> **NOTE   AZURE DASHBOARDS**
>
> Each chart or visualization that you create in Azure Monitor can also be pinned to an Azure dashboard. You can have multiple dashboards in Azure, and you can even share a dashboard with others in your organization.

You also are not limited to creating a single chart. Click Add Chart in the Metrics Explorer to stack multiple charts, so existing charts can be cloned and then customized.

# Configure log settings in Azure Monitor

Log Analytics helps you collect, correlate, search, and act on log and performance data generated by operating systems, applications, and Azure services. It gives you operational insights using rich search and visualizations. Log Analytics provides a single pane of glass for interacting with the data from the entire platform and the workloads you host on it including both Linux and Windows servers. Also, Log Analytics can be used with other cloud providers.

Logs are collected and aggregated in a Log Analytics workspace. The logs can also be queried and visualized through Log Analytics or through Azure Monitor. A workspace is an Azure resource, meaning that RBAC can be applied for granular access to the service and the data stored within it. This also means that workspaces can be in regions that meet your organization's regulatory requirements, data isolation, and scope. You can create multiple workspaces in a single subscription.

## Implement Log Analytics workspace

You can create a workspace through the Azure portal, Azure PowerShell, the Azure CLI, and ARM templates. To create a workspace through the Azure portal, search for **Log Analytics workspace**. Click Create to open the Create Log Analytics Workspace blade.

To configure a workspace, you will need to provide the following (see Figure 5-7):

- A name for the workspace
- The subscription the workspace will be associated with
- A resource group
- A location

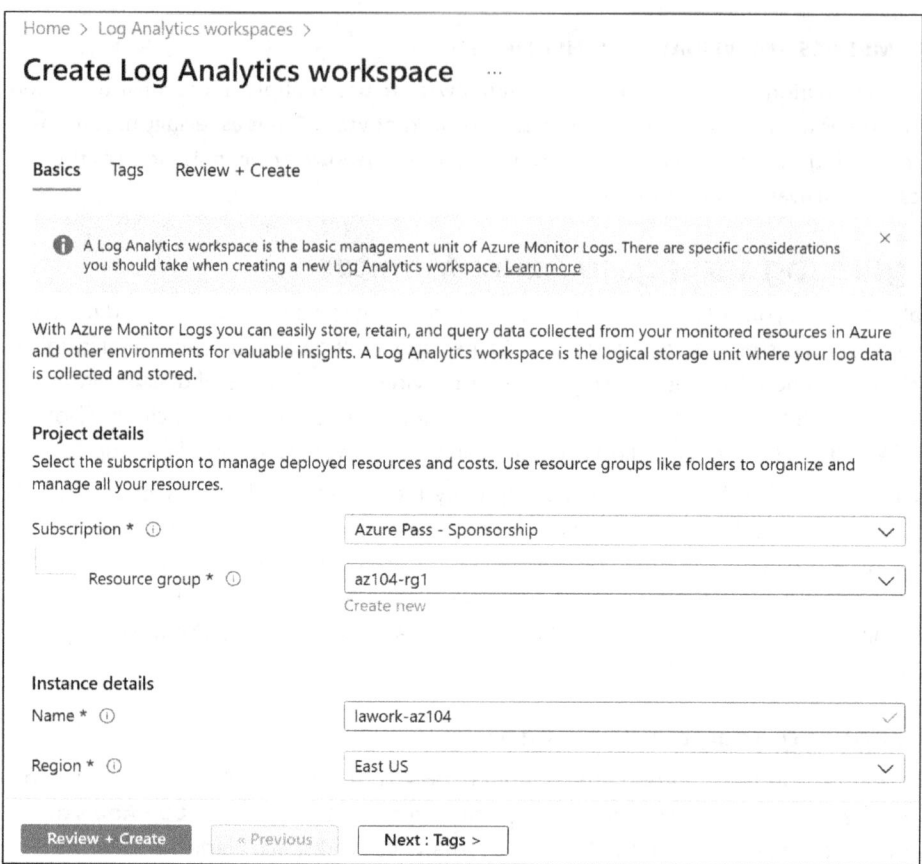

**FIGURE 5-7** Log Analytics workspace configuration

Note that Log Analytics is not available in all regions. To select an appropriate region, see the Azure Products by Region documentation at *https://azure.microsoft.com/global-infrastructure/services/.*

To select the appropriate pricing tier, review the pricing documentation at *https://azure.microsoft.com/pricing/details/monitor/.* A new workspace will default to the Free tier, which includes 5 GB of log storage per month (31 days) with per-GB pricing and per-GB charges for additional storage and retention.

After a workspace has been provisioned, you must enable data collection and configure both resource and tenant logs to store their logs within the service.

To collect event and performance data from Windows and Linux machines, open the workspace and configure the agents and data collection rules, as shown in Figure 5-8. On the Agents blade, you can obtain the workspace ID, primary key, and secondary key for associating machines with the service through the monitoring agent. You can use this information when manually onboarding clients to the workspace.

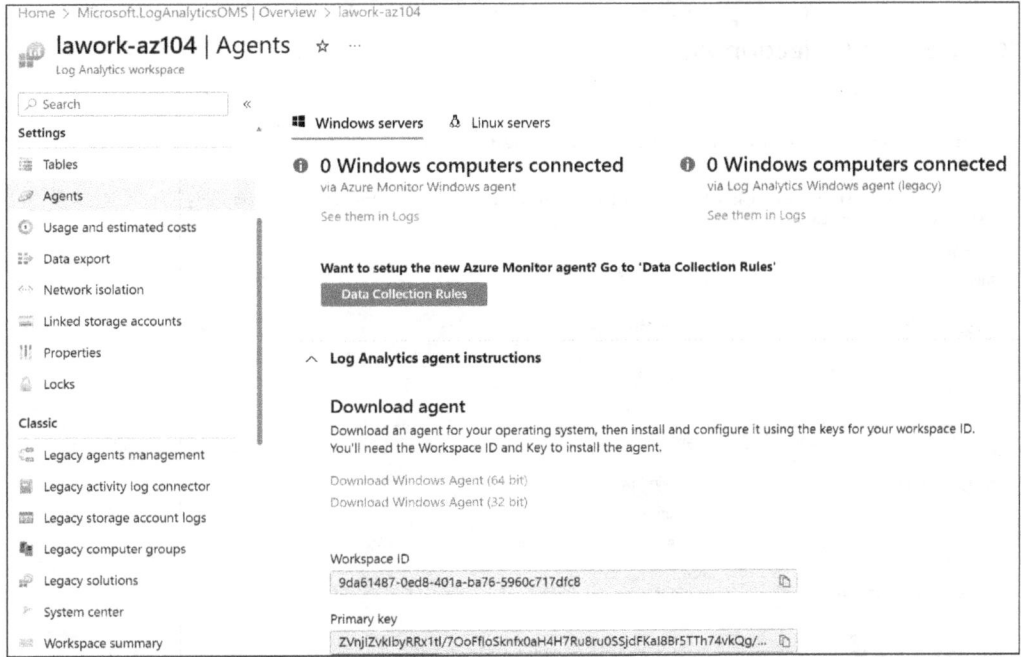

lawork-az104 | Agents ☆ ⋯
Log Analytics workspace

🔍 Search «

**Settings**

▤ Tables

𝄃 Agents

◔ Usage and estimated costs

⇉ Data export

↢ Network isolation

▤ Linked storage accounts

⦙⦙ Properties

🔒 Locks

**Classic**

⟲ Legacy agents management

▦ Legacy activity log connector

▥ Legacy storage account logs

▦ Legacy computer groups

▥ Legacy solutions

⚲ System center

▤ Workspace summary

⊞ Windows servers   △ Linux servers

ⓘ **0 Windows computers connected**
via Azure Monitor Windows agent

See them in Logs

ⓘ **0 Windows computers connected**
via Log Analytics Windows agent (legacy)

See them in Logs

**Want to setup the new Azure Monitor agent? Go to 'Data Collection Rules'**

[ Data Collection Rules ]

∧ **Log Analytics agent instructions**

**Download agent**
Download an agent for your operating system, then install and configure it using the keys for your workspace ID.
You'll need the Workspace ID and Key to install the agent.

Download Windows Agent (64 bit)
Download Windows Agent (32 bit)

Workspace ID
9da61487-0ed8-401a-ba76-5960c717dfc8 ▯

Primary key
ZVnjIZvkIbyRRx1tl/7OoFfIoSknfx0aH4H7Ru8ru0SSjdFKaI8Br5TTh74vkQg/... ▯

**FIGURE 5-8**   Log Analytics workspace agents

Click Data Collection Rules to configure the Windows event logs, Windows performance counters, Linux performance counters, Syslog, IIS Logs, custom fields, and custom logs. To create a new data collection rule, click Create. The Basics tab of the Create Data Collection Rule blade includes the following fields, as shown in Figure 5-9:

- **Rule Name**   The display name of the rule
- **Subscription**   The Azure subscription to create the rule in
- **Resource Group**   The logical group for the resource
- **Region**   The region that the rule is created in
- **Platform Type**   Whether the rule supports Windows, Linux, or all platforms

Click Next: Resources to move to the Resources tab. This tab is where you define the source virtual machines to collect logs and metrics from. For Azure VMs, scale sets, and Arc-enabled machines that might be running elsewhere, the Azure Monitor Agent will be installed on the VM. For nonconnected VMs or client machines, the Azure Monitor Agent client will need to be installed on the target VMs.

On the Resources tab, click Add Resources. Place a checkmark next to the VMs that you want to enable data collection for, then click Apply, as shown in Figure 5-10.

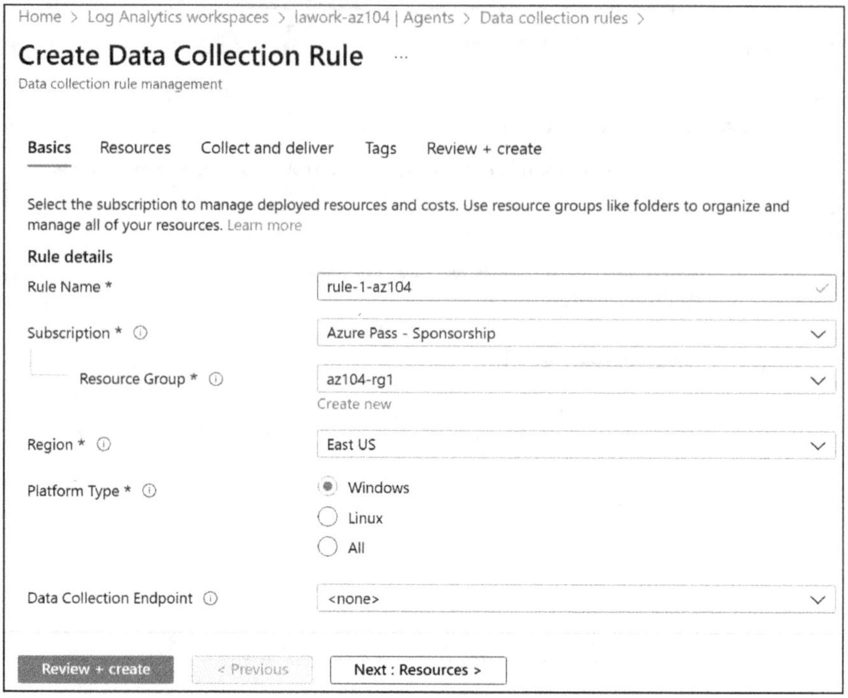

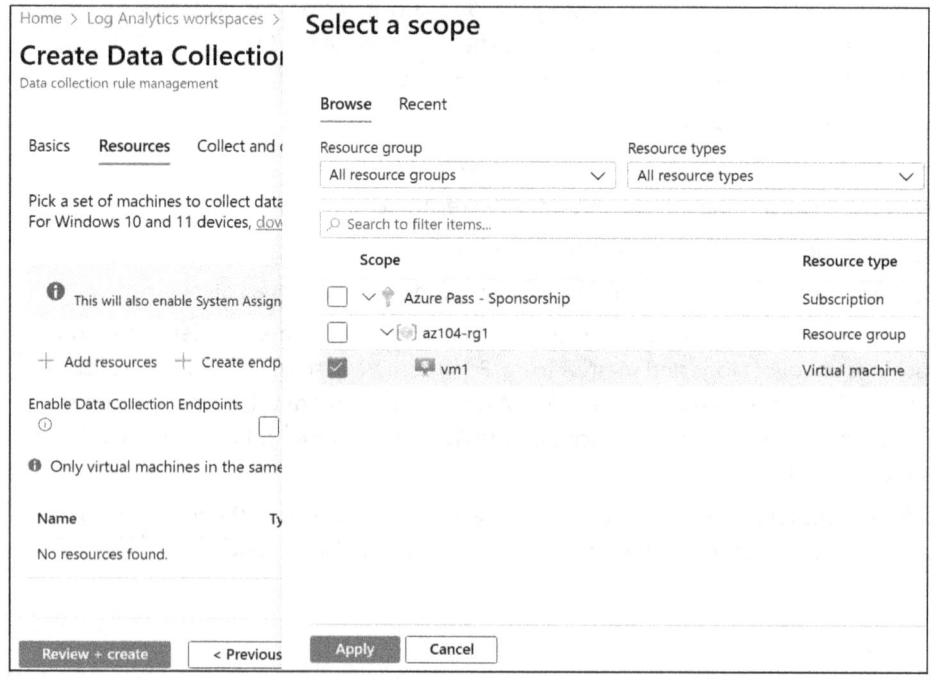

**FIGURE 5-9**  Basics tab on the Create Data Collection Rule blade

**FIGURE 5-10**  Add resources

Click Next: Collect And Deliver to move to the Collect And Deliver tab. On this tab, configure the data sources to collect. The available options here depend on which platform type was selected on the Basics tab. Click Add Data Source to add a source and destination for the rule, as shown in Figure 5-11.

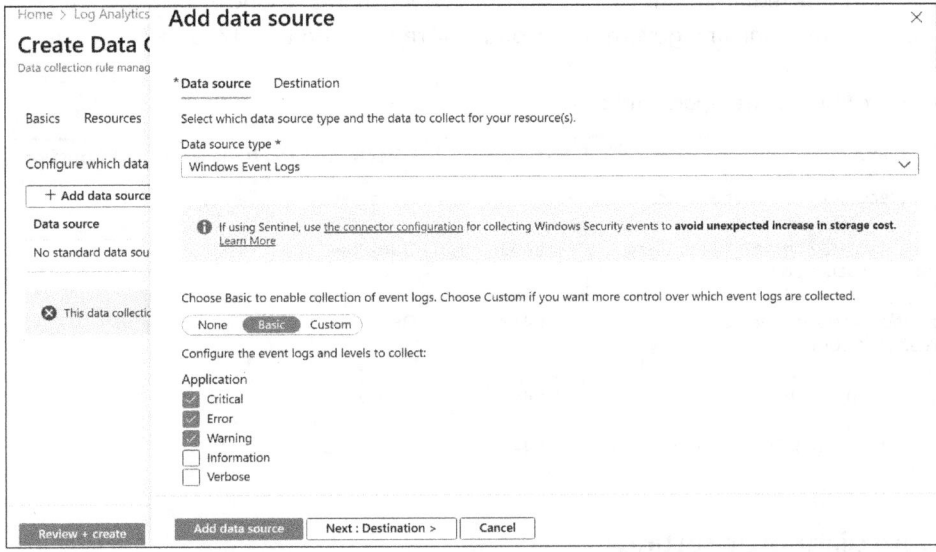

**FIGURE 5-11**   Add a data source

Click the Destination tab. By default, the selected metrics or logs will be sent to Azure Monitor Logs. Verify that the data is being sent to the Log Analytics workspace you created, as shown in Figure 5-12.

**FIGURE 5-12**   Add a data source destination

Click Add Data Source, then complete the Create Data Collection Rule blade to create the rule.

For the agent to send telemetry, you must also ensure that the required ports are available and the required URIs are whitelisted. The agent utilizes port 443 for all outbound communication. The required URIs for typical Azure deployments are shown in Table 5-1. Deployments in sovereign and government clouds will require additional changes.

**TABLE 5-1** Azure Monitor agent ports and protocols

| Agent Resource | Ports | Direction | Bypass HTTPS Inspection |
|---|---|---|---|
| global.handler.control.monitor.azure.com | Port 443 | Outbound | Yes |
| *<virtual-machine-region-name>*.handler.control.monitor.azure.com | Port 443 | Outbound | Yes |
| *<log-analytics-workspace-id>*.ods.opinsights.azure.com | Port 443 | Outbound | Yes |
| management.azure.com | Port 443 | Outbound | Yes |
| *<virtual-machine-region-name>*.monitoring.azure.com | Port 443 | Outbound | Yes |

## Configure diagnostic settings

While the resources you deploy in Azure create metrics automatically, many of them also offer richer diagnostics logs, which can be configured to send their log data to another location, such as a storage account or a Log Analytics workspace. In addition to resource logs, there are also tenant-level services, such as Microsoft Entra ID, which exist outside a subscription from which you might need to collect log data.

Diagnostics logs are one type of log data. There is also log data within the Azure activity log, and there is log data that can be obtained from virtual machines with the use of diagnostics agents that is separate from diagnostic logs associated with a tenant-level service or an Azure resource. It is important to understand the differences between the types of log data that are available and where that log data can be stored.

> **IMPORTANT RESOURCE AND TENANT LOGS ARE DIAGNOSTIC LOGS**
> Both resource logs and tenant logs are considered diagnostics logs. Diagnostics logs that you configure for a tenant service or a resource are separate from the Azure activity log and guest telemetry obtained with diagnostics agents.

The Azure activity log surfaces data at the subscription level and can be useful for understanding actions that occur within your environment against the ARM APIs. For example, when a new deployment is submitted, the events associated with that deployment—such as the time it was submitted, the resources that were created, and the user that submitted the request—are all tracked within the activity log. However, at the subscription level, you are missing any resource-level logs. For example, the activity log can show when a network security group (or NSG) was created, but it cannot show when an NSG rule was applied to traffic that was subject to the NSG, such as when a port or protocol is blocked. Diagnostic logs provide this functionality.

> **NOTE    EVENTS RETAINED FOR 90 DAYS**
>
> Events in the activity log are retained for 90 days. You can retain the data for a longer period by sending the logs to Azure Storage and/or a Log Analytics workspace.

Diagnostic logs will need to be enabled for each resource from which you want to collect additional telemetry. Note that metrics are resource-specific and captured automatically, so you only need to enable diagnostic logs to capture log data or to send metrics to another service.

> **IMPORTANT    SUPPORT FOR DIAGNOSTIC LOGS**
>
> Not all Azure resource types support diagnostic logs. A full list of services that support logs and their service-specific log schemas can be found at *https://learn.microsoft.com/azure/ azure-monitor/essentials/resource-logs-schema*.

To enable diagnostic logs through the Azure portal, you can browse to the resource itself to create the settings. The alternative and recommended method is to browse to the Azure Monitor Diagnostic Settings blade. From this blade, you can view all the resource types eligible for diagnostic logs and view the status (enabled or disabled) for log collection on each resource. Also, you can filter by subscription, resource group, resource type, and resource. An example is shown in Figure 5-13.

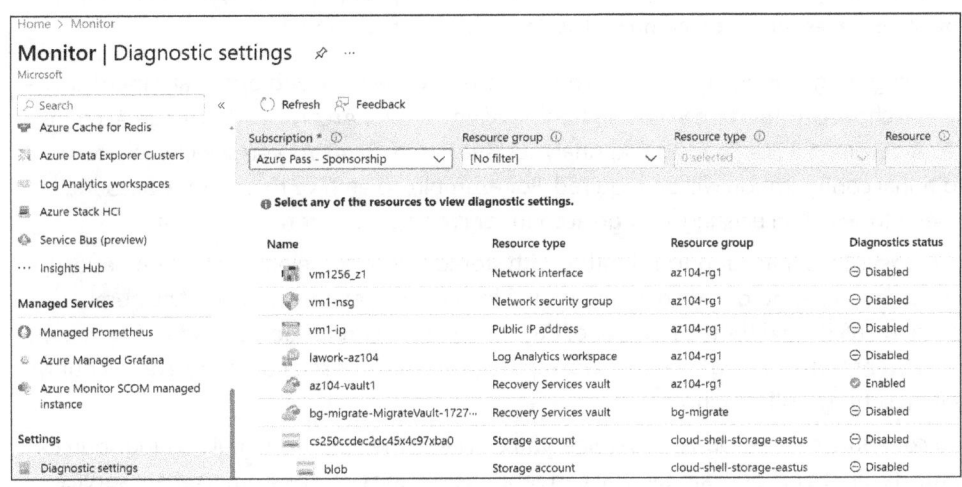

**FIGURE 5-13**   Azure Monitor Diagnostic settings

To enable diagnostic settings, click a resource with a status of Disabled. On the Diagnostic Settings blade, click Add Diagnostic Setting. Specify the Diagnostic Setting Name and select the required logs, as shown in Figure 5-14.

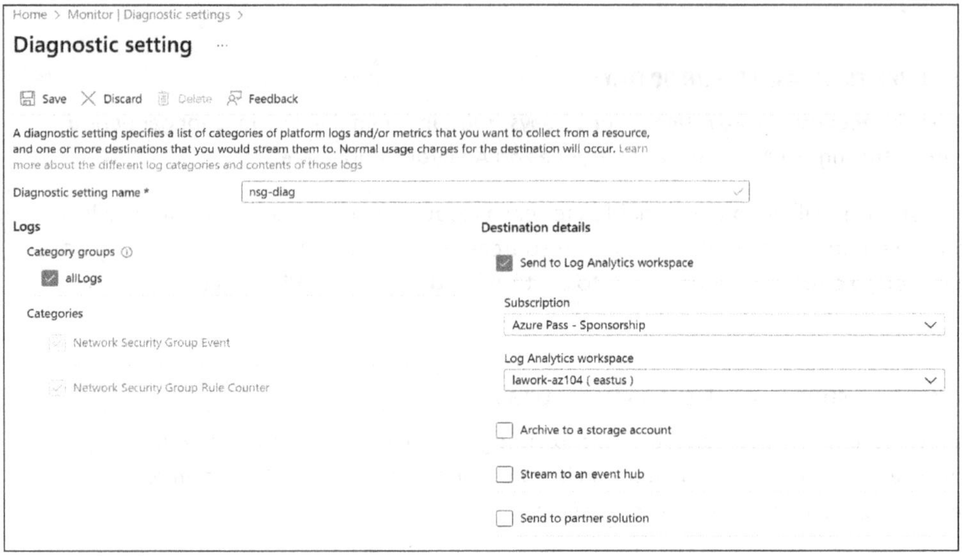

**FIGURE 5-14** Azure Monitor diagnostic settings for a resource

> **NOTE  DIAGNOSTIC LOGS**
>
> Each resource or tenant service for which you enable diagnostic logs has varying controls (or settings). For example, not all resources support a retention policy in the diagnostic settings, and not all resources support sending metric data to another location.

When configuring diagnostic settings, you select where the logs (and optionally metrics) are sent. You can choose from these valid locations to send data to: Archive To A Storage Account, Stream To An Event Hub, or Send To Log Analytics (see Figure 5-14). As you select each location, additional configuration will be required. For example, to archive to a storage account, you will need to select an existing storage account or create a new storage account.

For diagnostic logs that support retention with storage, you can select a retention period in days. A retention period of zero days means the logs will be retained forever. Any number between 1 and 365 is valid for the number of days. If you set the retention period and have only selected an Event Hub or a Log Analytics workspace (but have not selected a storage account), the retention settings will be ignored.

As you configure each resource or service, you can send the data from multiple log sources to the same destination. For example, you can send the diagnostic logs from a tenant service like Microsoft Entra ID to a Log Analytics workspace, and you can send the diagnostic logs from a resource like a Network Security Group to the same Log Analytics workspace.

It can take several moments for the setting to appear in the list of settings for the resource. Note that even though the setting has been configured, diagnostic data will not be collected until a new event is generated.

All these settings can be configured through the Azure portal, Azure PowerShell, the Azure CLI, or through the Azure Monitor REST API.

**EXAM TIP**

The Azure Diagnostics extension can also be configured through resource manager templates and the command-line tools by specifying a configuration file. For the exam, you should be aware of the schema of this configuration and how to apply it using automated tools. You can learn more about the Azure Diagnostics schema at *https://learn.microsoft. com/azure/azure-monitor/agents/diagnostics-extension-versions*.

## Query and analyze logs in Azure Monitor

As mentioned earlier, Azure Monitor stores and surfaces two types of data: metrics and logs. Metrics are numerical values such as performance counters, whereas logs can be either numerical data or text. For instance, the full text of an exception that is raised in an application or even the text of an application log from a Windows or Linux server is one example.

### Create a query

After the workspace has been configured, and tenant logs, resource logs, and machines have been onboarded, you can begin to analyze and visualize data. To interact with the data in Log Analytics, you use log queries, which are used to

- Perform interactive analysis of log data through the Azure portal in Azure Monitor and a Log Analytics workspace
- Build custom alert rules based on the logs in a workspace
- Generate visualizations that can be shared through Azure dashboards
- Export custom data sets to Excel or Power BI
- Perform automation based on log data with PowerShell or the Azure CLI

**NEED MORE REVIEW?**   **LOG QUERY USAGE**

To learn more about all the ways that log queries can be used, refer to the documentation at *https://learn.microsoft.com/azure/azure-monitor/logs/log-query-overview#where-log-queries-are-used*.

The query language used by Log Analytics is called Kusto Query Language (KQL). KQL queries are used to generate read-only requests to process data and return results. This means that the logs stored in Log Analytics are immutable and are only removed from a workspace based on the retention configuration. Queries are authored in plain text, and the schema used by Log Analytics is like that used by SQL, with databases and tables composed of columns and rows. In each table, data is organized in columns with different data types, as indicated by icons next to the column name. Column data types include text, numbers, and datetime.

Authored queries in Log Analytics can take many forms, from basic queries to very advanced queries with multiple aggregates and summarizations. Queries can be used to search terms, identify trends, analyze patterns, and provide many other insights. Queries search tables; they can start with either a table name or a search command that defines scope. The pipe (|) character separates commands, and you can add as many commands as required.

In the following example, the Heartbeat table is queried to summarize the count of computers (by IP) and by a time value (TimeGenerated) to render a chart that tracks the number of computers reporting a workspace each hour.

```
// Chart the number of reporting computers each hour
Heartbeat
| summarize dcount(ComputerIP) by bin(TimeGenerated, 1h)
| render timechart
```

To run this query, browse to Azure Monitor and click Logs to open the query interface. This query will not return data if you do not have any virtual machines deployed and running. Those machines must also be associated with the Log Analytics workspace you are querying.

The preceding query is a table-based query. Queries always begin with a scope—either a table or search-based query. Kusto queries are case-sensitive. Typically, language keywords are written in lowercase. When using the names of tables and columns in queries, ensure you are using the correct case. Table-based queries target a single table in a Log Analytics workspace (or database), while search-based queries target all tables by default.

Table-based queries start by scoping the query, and therefore tend to be very efficient and generally faster than search queries. Search queries are less structured by nature, which makes them the better choice when searching for a specific value across columns or tables. In other words, a search can scan all columns in one table or in all tables across an entire workspace for the defined value.

The amount of data being processed by a query could be enormous, which is why these queries can take longer to complete and might return large result sets that are limited by the Log Analytics service to 10,000 results.

To author queries in the Azure portal, browse to Azure Monitor, and open the Logs blade. From this blade, you can access all the subscriptions and workspaces you have rights to read from. Azure Monitor offers many sample queries for heartbeats, performance, and usage across your machines and services tracked in Log Analytics (see Figure 5-15).

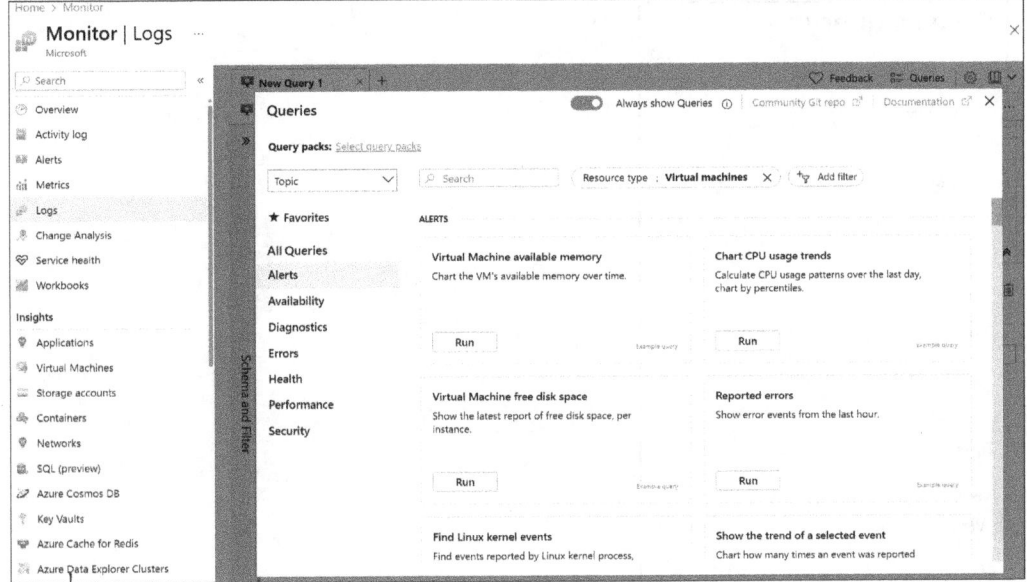

**FIGURE 5-15**   Azure Monitor logs

Select a query and click Load To Editor to open an editor with query preview, as shown in Figure 5-16.

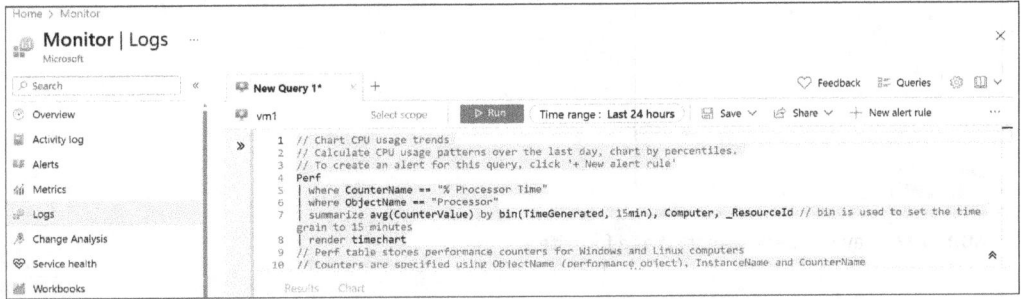

**FIGURE 5-16**   Query editor with sample query

## Save a query to the dashboard

In addition to sample queries, you can browse the schema for the currently selected work-space. This is useful for determining the proper case for table and column names because KQL is a case-sensitive query language. You can save authored queries for later or mark them as favorites so they can be retrieved later using the query explorer (see Figure 5-17).

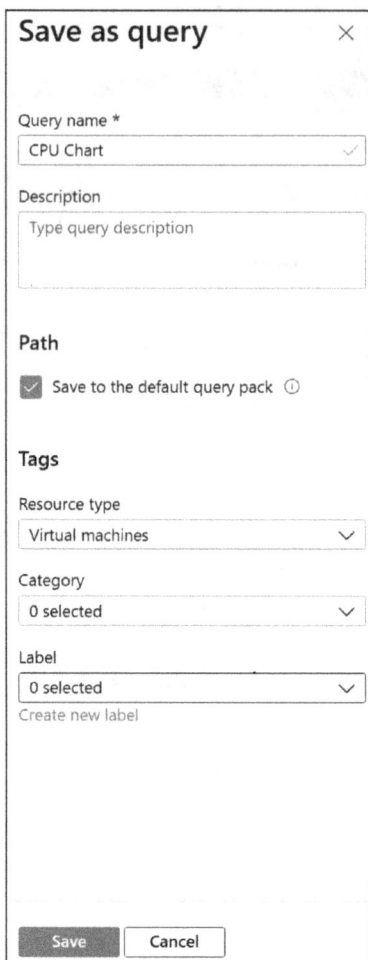

**FIGURE 5-17**   Save a query or mark it as a favorite

## Interpret graphs

In the query explorer, you can also generate charts and graphs based on the log queries. In the output section, click Chart to see the graphical representation of query results. You can choose a display option from various categories (from column, bar chart, line, pie, or area). For one of the sample queries, the stacked bar chart is shown in Figure 5-18.

You can also adjust the query and the chart type. For example, a doughnut pie chart is shown in Figure 5-19.

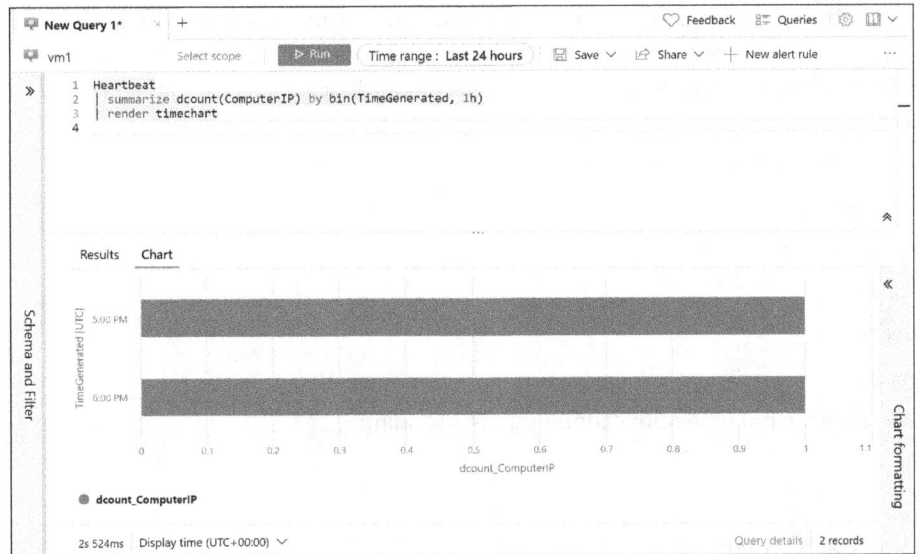

**FIGURE 5-18** Stacked bar chart

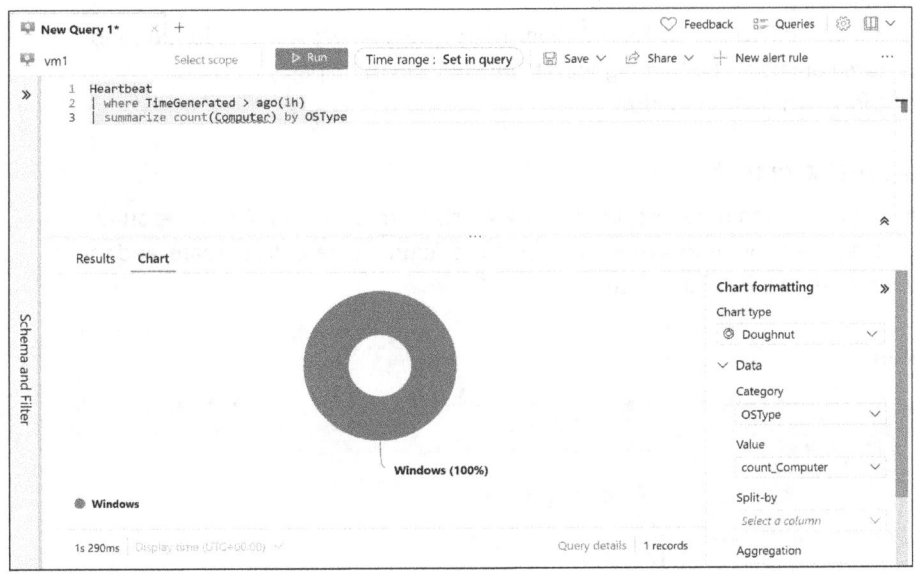

**FIGURE 5-19** Doughnut pie chart

# Set up alert rules, action groups, and alert processing rules in Azure Monitor

Alerts proactively notify you when important conditions are found in your monitoring data so you can identify and address issues before the users of your system notice them.

Azure Monitor brings a unified alerting experience to Azure, with a single pane of glass for interacting with metrics, the activity log, Log Analytics, service and resource health, and service-specific insights that provide out-of-the-box dashboards with visualizations and queries for

- Custom applications with Application Insights
- Virtual machines
- Storage accounts
- Containers
- Networks
- Key vaults

You can choose from multiple notification options, including

- Email
- SMS
- Push notifications to the Azure mobile app
- Voice
- Integration with automation services

Alerts that are generated within Azure Monitor can invoke Azure Automation runbooks, Logic Apps, Azure Functions, and even generate incidents in third-party IT service management tools, such as ServiceNow.

## Create and test alerts

To create an alert rule, open the Alerts blade (click Alerts from within the Azure Resource Configuration blade or browse to Azure Monitor in the Azure portal), click Create, and then click Alert Rules, as shown in Figure 5-20.

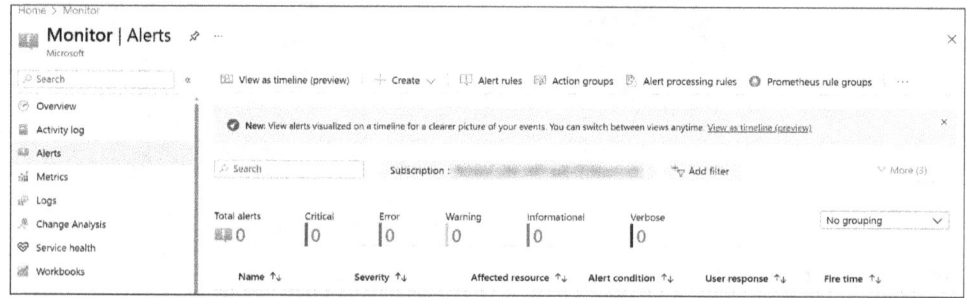

**FIGURE 5-20** The Alerts blade within Azure Monitor

Alerts in Azure Monitor are centered on alert rules. Alert rules contain the following components:

- A target resource (or resource type)
- Conditional logic for the alert with criteria based on the available signals for the target resource

- An action group, or what should happen when the alert rule condition is met
- A name and description for the alert rule

> **NOTE  AZURE MONITOR ALERT RULES**
>
> Alert rules in Azure Monitor are not the same as alerts. They are the criteria used to evaluate when an alert should be generated. An alert is generated based on the rule, and then the alerts themselves are acted upon separately, even maintaining their own state (such as New or Closed).

Click Select Scope to pick the target for the alert, which determines the available signals, as shown in Figure 5-21.

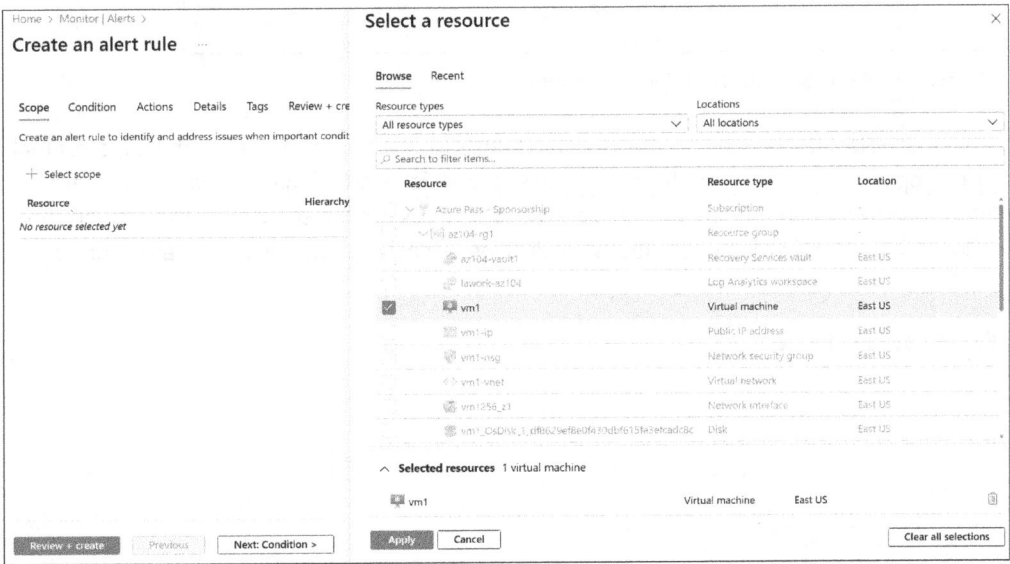

**FIGURE 5-21**  Create an alert rule

The *target resource* defines the scope and signals available for the alert. A target resource is an Azure resource that generates signals (such as metrics or the activity log) such as a virtual machine or storage account. The signal types available for monitoring vary based on the selected target (or targets, as you can select more than one target). The available signal types are as follows:

- Metrics
- Log search queries
- Activity logs

The next step is to configure the alert criteria by selecting the signal from the drop-down menu, as shown in Figure 5-22.

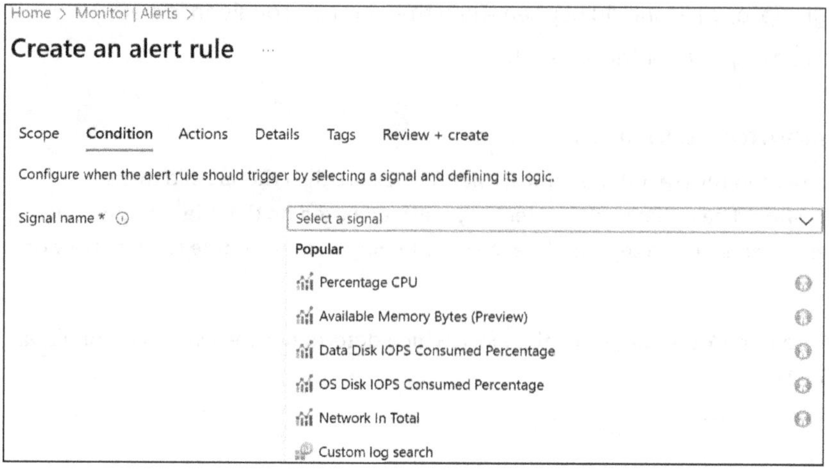

# Create an alert rule   ...

Scope   **Condition**   Actions   Details   Tags   Review + create

Configure when the alert rule should trigger by selecting a signal and defining its logic.

Signal name *  ⓘ     | Select a signal                                    ⌄ |

**Popular**

📊 Percentage CPU                                              ⓘ

📊 Available Memory Bytes (Preview)                            ⓘ

📊 Data Disk IOPS Consumed Percentage                         ⓘ

📊 OS Disk IOPS Consumed Percentage                           ⓘ

📊 Network In Total                                            ⓘ

📄 Custom log search

**FIGURE 5-22**   Azure Monitor alert rules conditions

You can select the signal from the available signals for the target and define the logic test that will be applied to the data from the signal. For example, for a virtual machine, you can use the Percentage CPU metric to generate an alert based on a custom threshold for CPU usage, as shown in Figure 5-23. The alert logic conditions are different for activity log signals or metric signals.

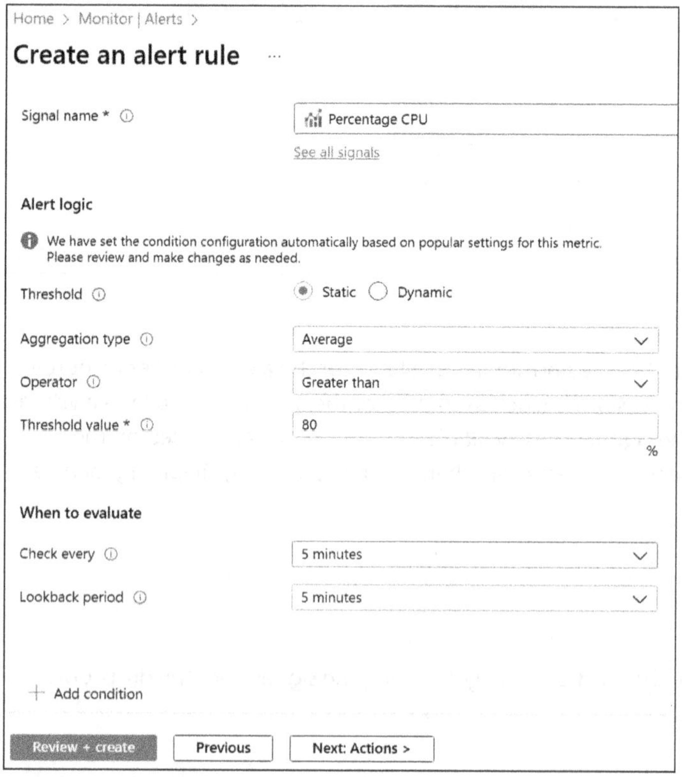

# Create an alert rule   ...

Signal name *  ⓘ     | 📊 Percentage CPU                                    |

See all signals

**Alert logic**

ⓘ We have set the condition configuration automatically based on popular settings for this metric. Please review and make changes as needed.

Threshold  ⓘ          ● Static   ○ Dynamic

Aggregation type  ⓘ    | Average                                          ⌄ |

Operator  ⓘ            | Greater than                                     ⌄ |

Threshold value *  ⓘ   | 80                                                 |
                                                                      %

**When to evaluate**

Check every  ⓘ         | 5 minutes                                        ⌄ |

Lookback period  ⓘ     | 5 minutes                                        ⌄ |

+ Add condition

| Review + create |   | Previous |   | Next: Actions > |

**FIGURE 5-23**   Azure Monitor alert condition

Configure one or more conditions for the alert rule. After the conditions are defined, click Select Action Group, as shown in Figure 5-24. An action group is a collection of actions that should occur in response to an alert being triggered.

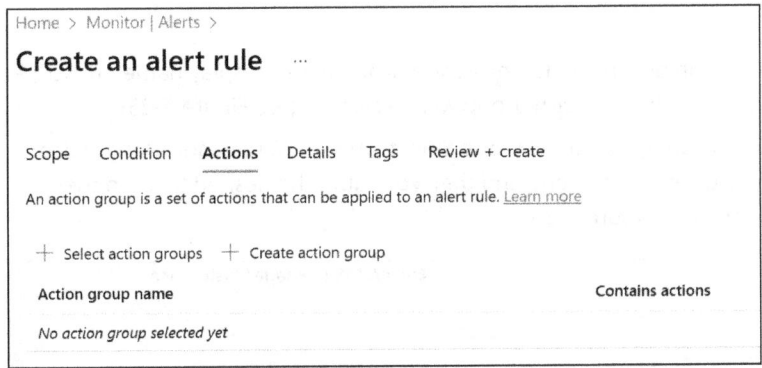

**FIGURE 5-24**  Azure Monitor action groups

Select the existing action group if you already have one. Otherwise, click Create Action Group to create a new action group, as shown in Figure 5-25.

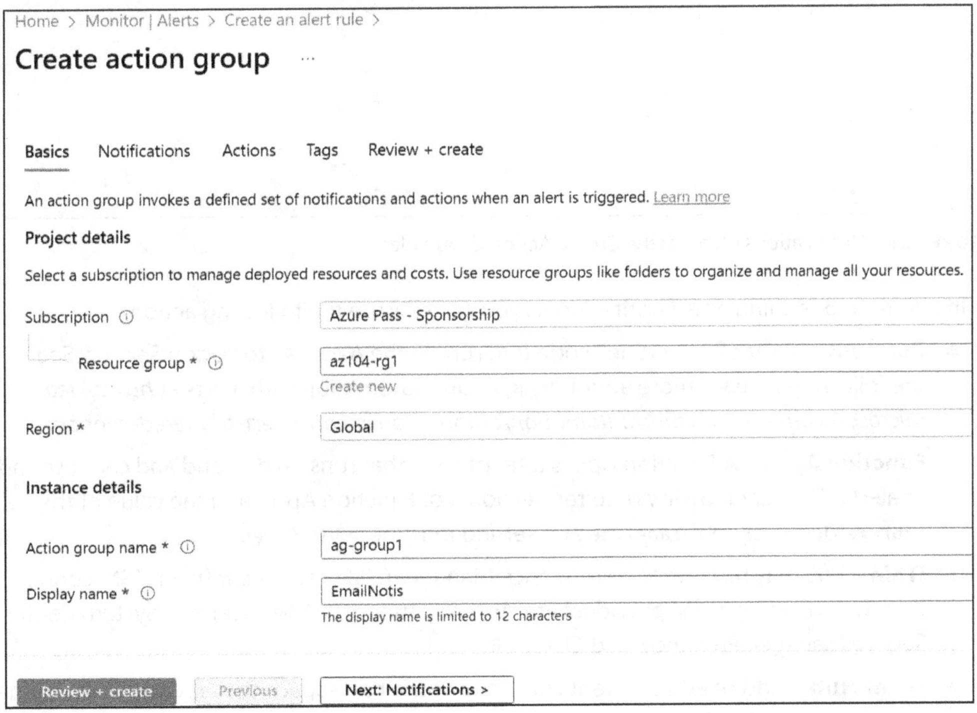

Home > Monitor | Alerts > Create an alert rule >

**Create action group**  ...

Basics   Notifications   Actions   Tags   Review + create

An action group invokes a defined set of notifications and actions when an alert is triggered. Learn more

**Project details**

Select a subscription to manage deployed resources and costs. Use resource groups like folders to organize and manage all your resources.

Subscription ⓘ              Azure Pass - Sponsorship

Resource group * ⓘ          az104-rg1
                            Create new

Region *                    Global

**Instance details**

Action group name * ⓘ       ag-group1

Display name * ⓘ            EmailNotis
                            The display name is limited to 12 characters

[ Review + create ]   [ Previous ]   [ Next: Notifications > ]

**FIGURE 5-25**  Create an action group

Action groups are separate resources and are independent of the alert rule. This means that the same action group can be used across multiple alert rules.

When creating a new action group, define the action group name, display name, subscription, and resource group in which the action group will be created (see Figure 5-25).

On the next tab, you can configure notifications. Select Email/SMS message/Push/Voice from the Notification Type drop-down menu and then configure the desired fields in the notification settings, as shown in Figure 5-26.

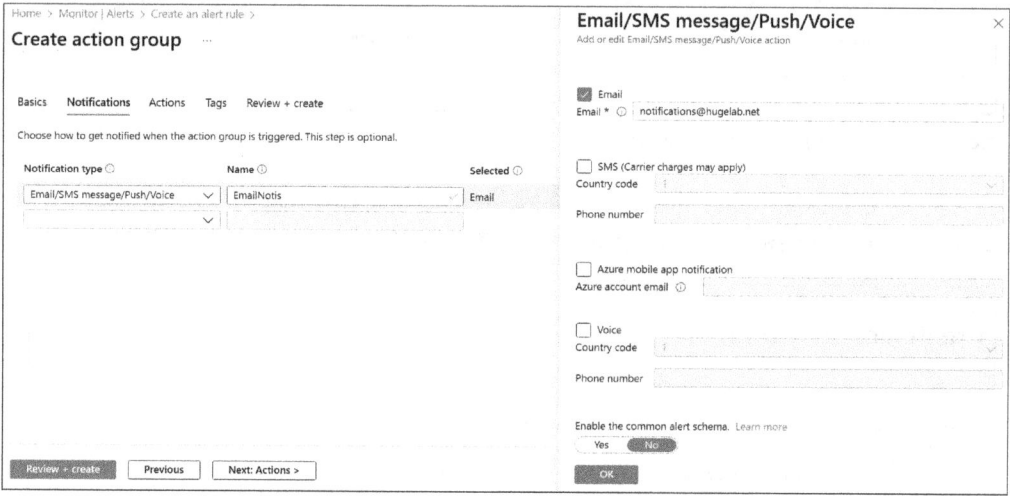

**FIGURE 5-26**  Notifications blade of the Create Action Group blade

In addition to sending email notifications, you can execute the following actions:

- **Runbook**  A set of PowerShell code that runs in the Azure Automation Service. See the following to learn more about using Azure Automation with alerts at *https://learn. microsoft.com/en-us/azure/automation/automation-create-alert-triggered-runbook*.

- **Function Apps**  A Function App is a set of code that runs on demand and can respond to alerts. This functionality requires Version 2 of Function Apps, and the value of the AzureWebJobsSecretStorageType app setting must be set to files.

- **ITSM**  You may have up to 10 IT Service Manager (ITSM) actions with an ITSM connection. The following ITSM providers are currently supported: ServiceNow, System Center Service Manager, Provance, and Cherwell.

- **Event Hub**  Add or edit an Event Hub action for a namespace that already exists in one of your Azure subscriptions.

- **Logic Apps**  A Logic App provides a visual designer to model and automate your process as a series of steps known as a workflow. There are many connectors across the cloud and on-premises to quickly integrate across services and protocols. When an alert is triggered, the Logic App can take the notification data and use it with any of the connectors to remediate the alert or start other services.

- **Webhook** Route an Azure alert notification to other systems for post-processing or custom actions. For example, you can use a webhook on an alert to route it to services that send text messages, log bugs, notify a team via chat/messaging services, or do any number of other actions.

- **Secure webhook** Uses Microsoft Entra ID to authenticate the webhook connection.

You can configure the above actions for the action group on the next tab. Select from the options available in the Action Type drop-down menu, as shown in Figure 5-27.

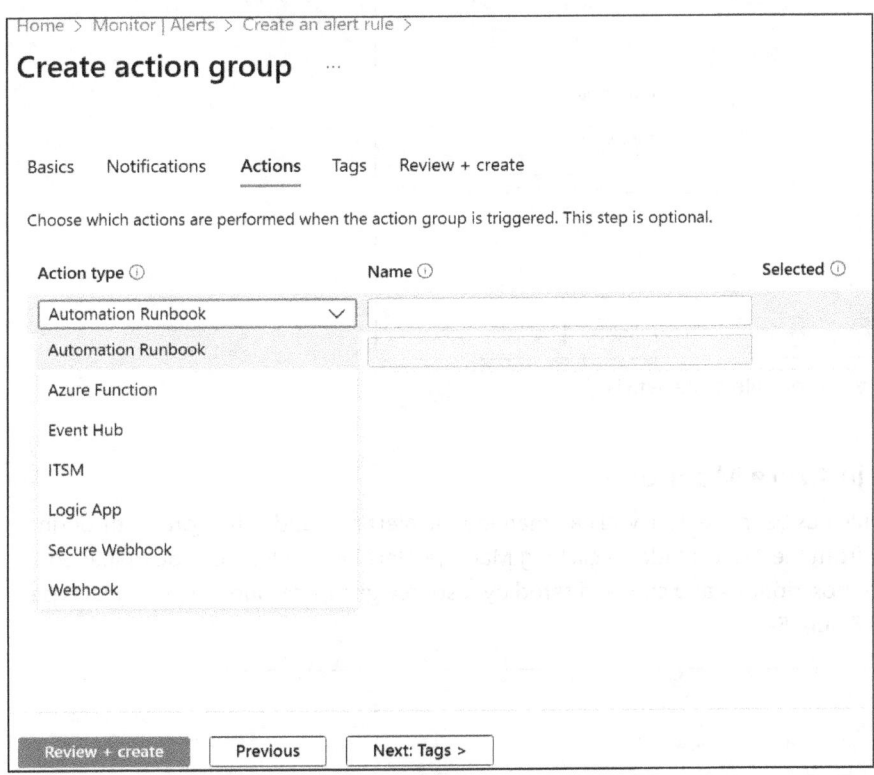

**FIGURE 5-27** Actions tab of the Create Action Group blade

Once the action group is created, specify remaining alert rule details such as the alert rule name, description, resource group to save the alert, severity, and whether to enable the alert upon creation (see Figure 5-28).

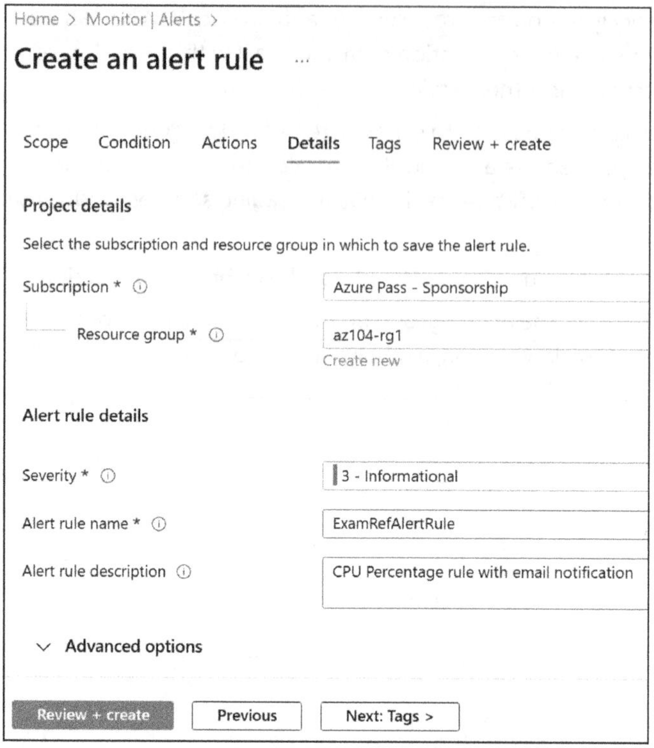

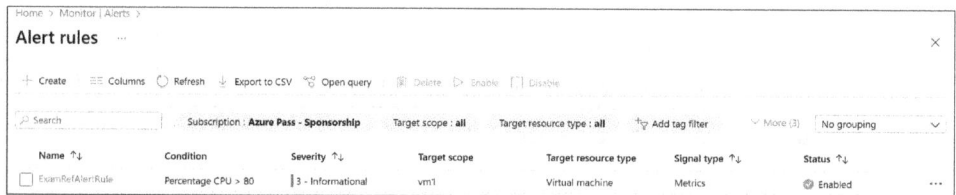

FIGURE 5-28  Action group alert rule details

## View alerts in Azure Monitor

After an alert rule has been created, you can manage the alert rule and action group through Azure Monitor from the Alerts blade by clicking Manage Alert Rules. Alerts can be managed across multiple subscriptions and can be filtered by resource group, resource type, signal type, and status (see Figure 5-29).

FIGURE 5-29  Azure Monitor new action alert rule details

Alert rules do not generate alerts immediately, and metric alerts can take up to 10 minutes. When alerts are generated, they will be distributed based on the actions defined in the action group. For example, when an email is sent, the defined users will receive a message with the alert details and a link to view the alert in the Azure portal, as shown in Figure 5-30.

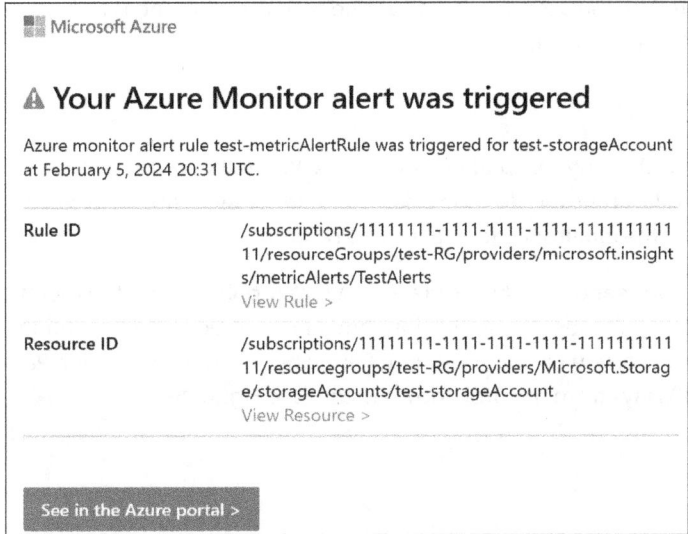

**FIGURE 5-30** Azure Monitor alert notification email

When an alert is resolved by the state of the monitor condition and changed to Resolved, notifications are sent as well.

## Analyze alerts across subscriptions

When an alert rule is created, the alert rule targets resources in a single subscription, and the alerts that are generated based on the alert rules are associated with the subscription from which they are generated. Azure operators are not limited to viewing alerts from only a single subscription through Azure Monitor, which again, provides a single pane of glass for not only managing alert rules across multiple subscriptions, but also for managing the generated alerts.

Recall that alert rules and action groups are separate entities. The alerts that are generated based on the conditional logic of an alert rule are separate entities as well. This means that they are managed independently of alert rules and maintain their own state.

Alerts can have one of three states:

- **New**   The alert is new and has not been reviewed.
- **Acknowledged**   An administrator is taking action on the issue that generated the alert.
- **Closed**   The issue that generated the alert has been resolved, and the alert has been marked as closed.

The state of an alert is updated by the user who is interacting with the alert and is not updated automatically by the Azure platform.

> **NOTE  ALERT STATE**
>
> The alert state is not the same as the monitor condition of an alert. When the Azure platform generates an alert based on an alert rule, the alert's monitor condition is set to fired and when the underlying condition clears, the monitor condition is set to resolved.

As alerts are generated, they appear on the Alerts blade in Azure Monitor. From the Alerts blade, you can view alerts for all subscriptions, and drill into one or more specific subscriptions, resource groups, and resources. Also, you can filter by Time Range by choosing Past Hour, Past 24 Hours, Past 7 Days, or Past 30 Days from the drop-down menu (see Figure 5-31).

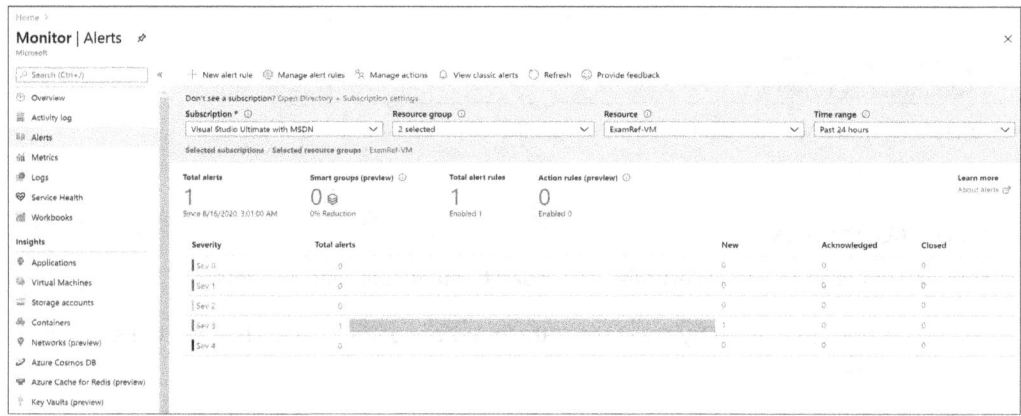

**FIGURE 5-31**   Azure Monitor Alerts dashboard

The view on this page can be filtered through the drop-down menus on the page. You can also filter, sort, and edit the columns that are displayed with the following limitations:

- When you filter by subscription, you are limited to selecting a maximum of five subscriptions.
- When filtering by resource group, you can only select one resource group at a time.
- The Resource Type filter is dynamic and is based on the selection of the resource group. You will not be able to select resource types that are not deployed to the selected resource group you are filtering with.
- The Time Range filter shows only alerts fired within the selected time window. Supported values are the past hour, the past 24 hours, the past 7 days, and the past 30 days.

Selecting an alert will open the alert details (see Figure 5-32). From this blade, you can view alert history, including any changes to monitor condition state. This is also where you can modify the alert state to New, Acknowledged, or Closed. If the state of an alert is changed, that change is included in the alert history for audit purposes.

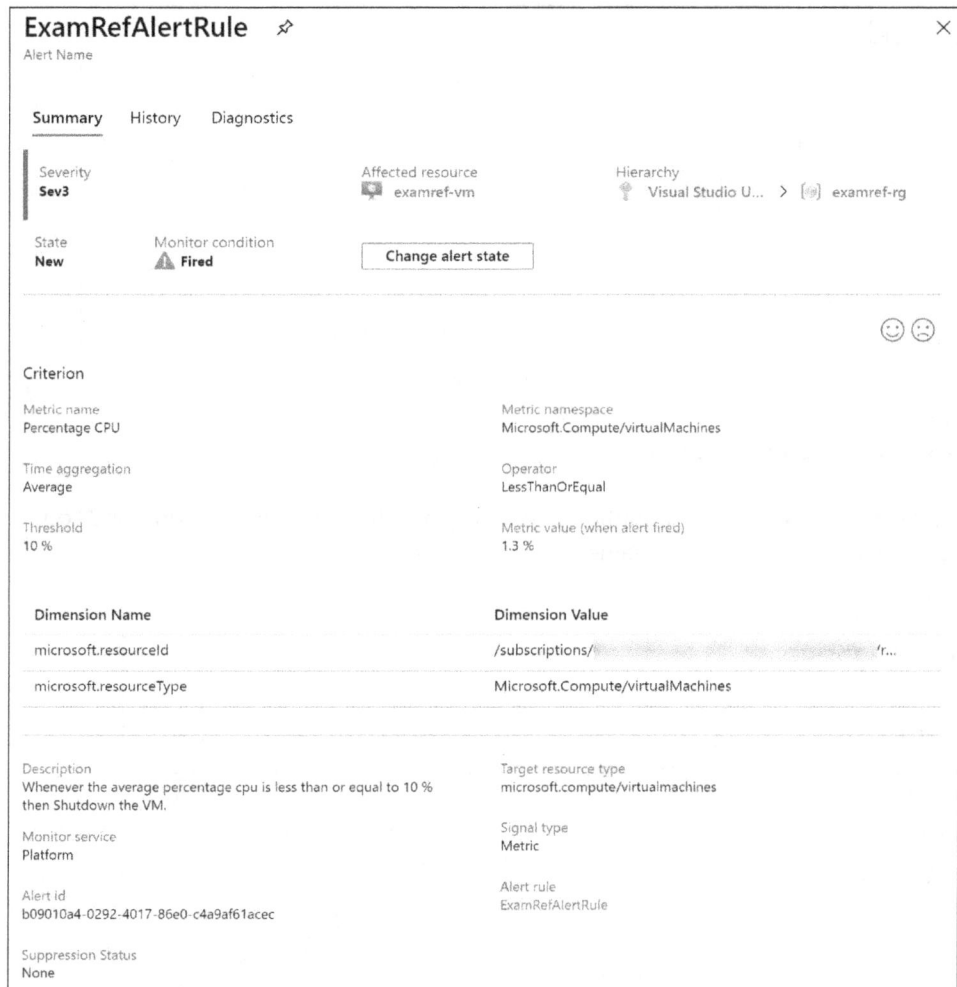

FIGURE 5-32   Azure Monitor alert details

# Configure Application Insights

Application Insights is used for development and as a production monitoring solution. It works by installing a package into your app, which can provide a more internal view of what's going on with your code. Its data includes response times of dependencies, exception traces, debugging snapshots, and execution profiles. It provides powerful smart tools for analyzing all this telemetry both to help you debug an app and to help you understand what users are doing with it. You can tell whether a spike in response times is caused by something in an app or an external resourcing issue. Application Insights provides significantly more value when your application is instrumented to emit custom events and exception information.

To create an Application Insights resource, open Azure Monitor, click Applications on the left (under Insights), and then click Create Application Insight Apps, as shown in Figure 5-33.

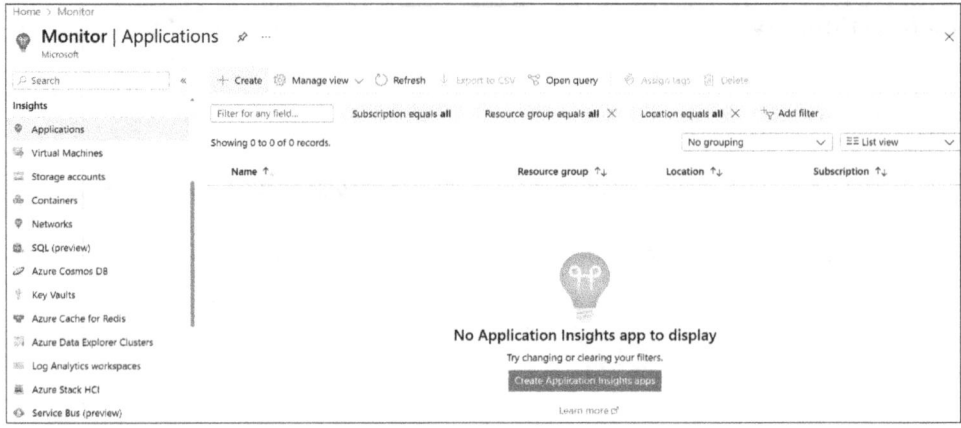

**FIGURE 5-33** Create Application Insights apps

On the Basics tab, select the subscription, resource group, region, resource mode, and Log Analytics workspace and specify the name (see Figure 5-34).

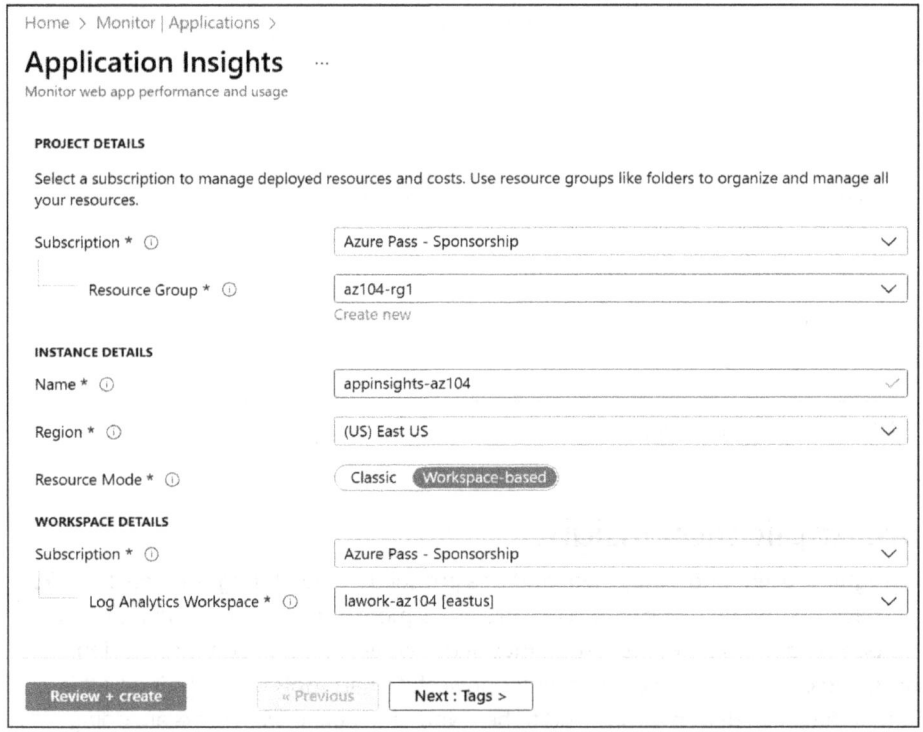

**FIGURE 5-34** Basics tab of the Application Insights blade

Application Insights provides an extensive dashboard depicting all the aspects of your application workload, as shown in Figure 5-35. The dashboard displays application performance, usage, diagnostic, and other app data. The dashboard can be customized based on your preferences.

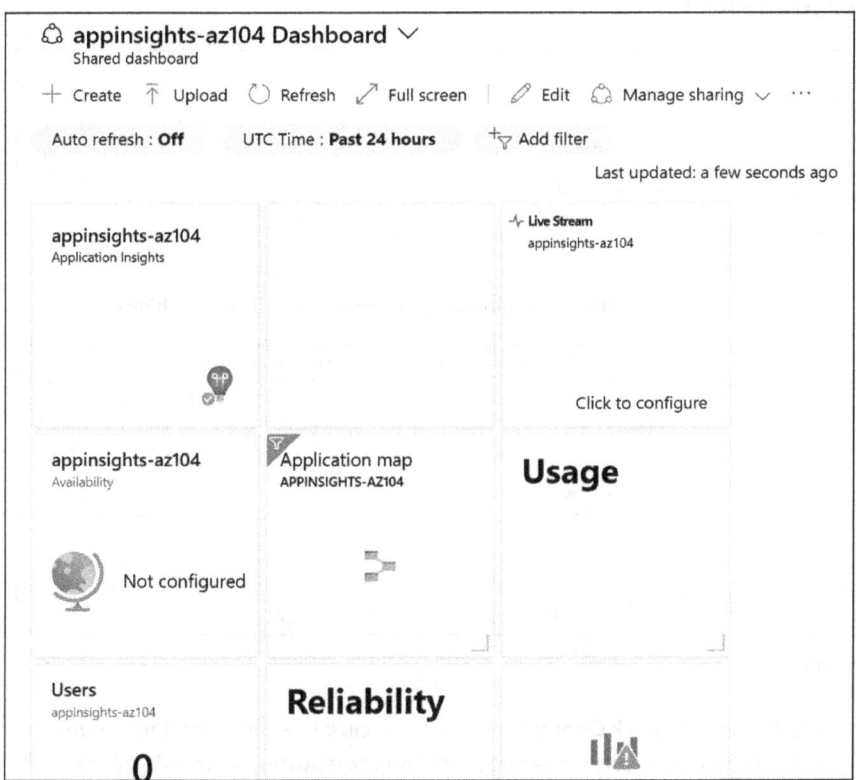

**FIGURE 5-35** Application Insights dashboard

**NEED MORE REVIEW?** **APPLICATION INSIGHTS**

You can learn more about Application Insights, including samples for emitting custom telemetry, at *https://learn.microsoft.com/azure/azure-monitor/app/app-insights-overview.*

# Configure and interpret monitoring of VMs, storage accounts, and networks using Azure Monitor Insights

The suite of Azure Monitor features includes Insights. As of this writing, there are more than 20 service types that offer different insights, including compute, storage, networking, databases, and more.

## VM insights

VM insights helps to monitor the performance and health of your VMs and VM scale sets. The insights and data gathered can then help you troubleshoot issues, see trends across logs and metrics, and create alerts. To configure VM insights from the Azure portal, search for **Monitor**. On the Monitor blade, under Insights, click Virtual Machines. The VM insights dashboard is displayed in Figure 5-36.

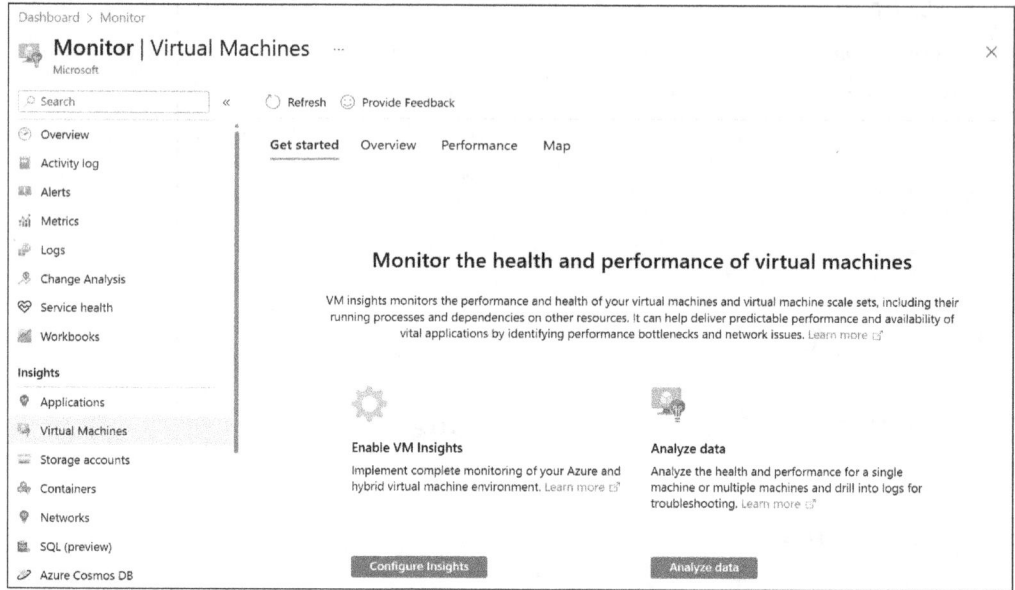

**FIGURE 5-36** VM insights

On the VM Insights dashboard, click Configure Insights, or click the Overview tab. If you have previously enabled Data Collection Rules or other Monitor features, your VMs might already be monitored. If not, they will appear on the Not Monitored tab. Figure 3-37 displays the VM insights overview on the Monitored tab.

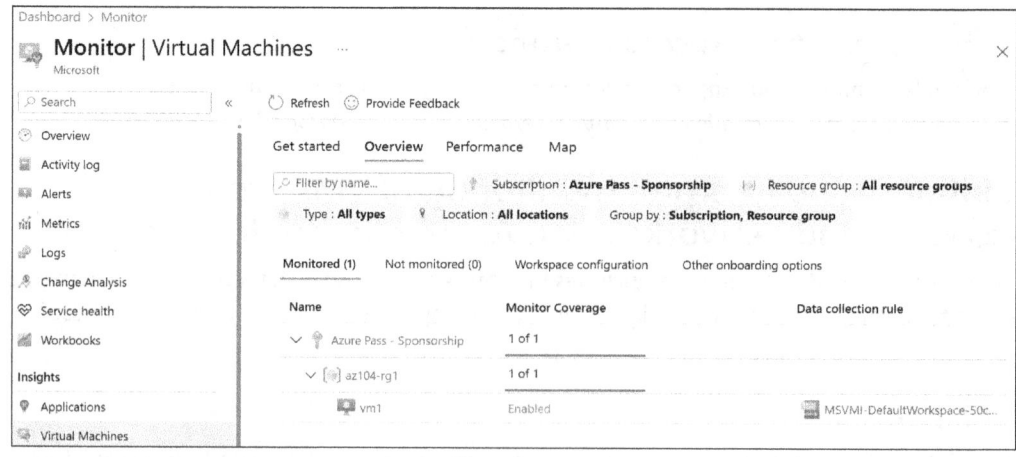

**FIGURE 5-37** VM insights overview

The Performance tab of VM insights displays some of the top charts for the VMs. This includes CPU utilization, available memory, bytes sent and received, and logical disk space used percentage. Figure 5-38 displays a portion of the Performance tab.

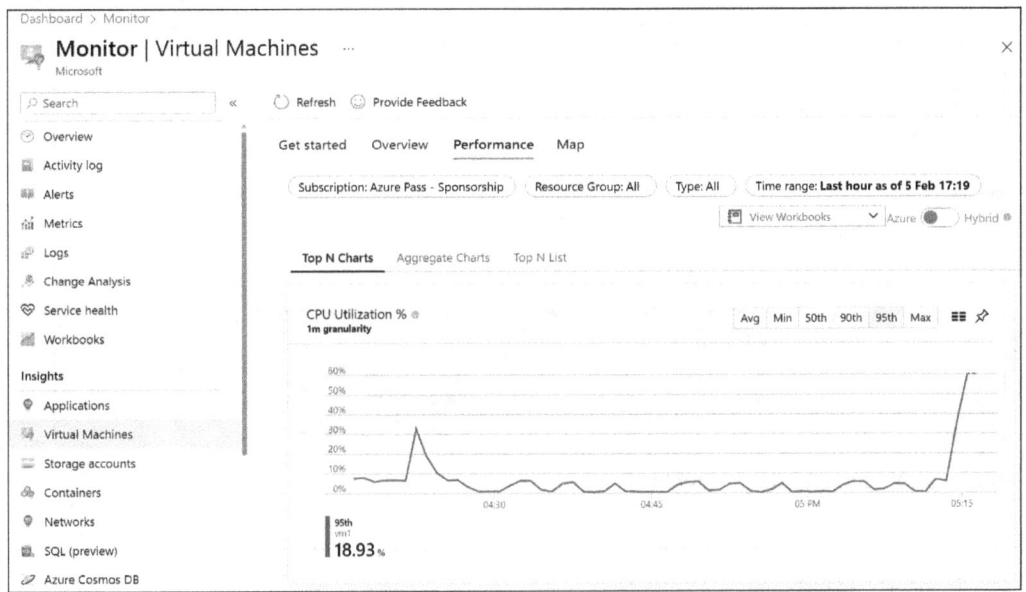

**FIGURE 5-38**   VM insights performance

## Storage account insights

Storage account insights uses a Monitor workbook to display some general metrics for the storage accounts in the subscription. These metrics include the number of transactions on the account, a timeline of the transactions, the end-to-end latency, server latency, and client errors. Figure 5-39 displays the storage account overview.

## Network insights

Network insights summarizes the network health, connectivity, and traffic for the environment. This includes the health of network interfaces, network security groups, public IP addresses, and virtual networks. Figure 5-40 displays the Network health tab of the Networks blade in Azure Monitor.

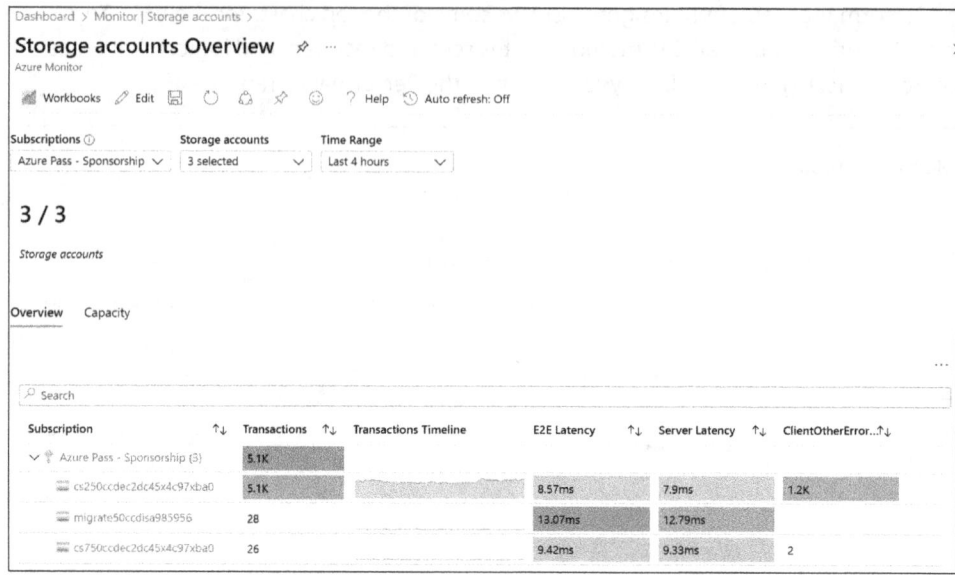

**FIGURE 5-39** Storage account insights

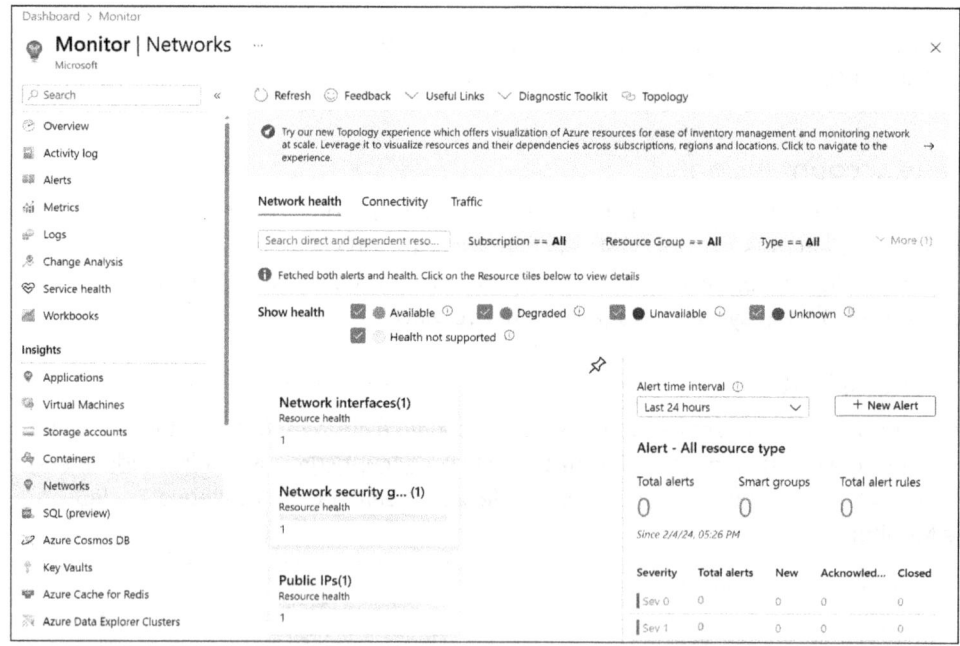

**FIGURE 5-40** Network insights

# Use Azure Network Watcher and Connection Monitor

Network Watcher provides a central hub for a wide range of network monitoring and diagnostic tools. These tools are valuable across a wide range of network troubleshooting scenarios, and also provide access to other tools listed in this skill section, such as the Network Performance Monitor and Connection Monitor.

> **NOTE  CONNECTION MONITOR**
>
> The objective list for the AZ-104 exam includes Connection Monitor with Network Watcher. However, Connection Monitor and Connection Troubleshoot were covered in Chapter 4, Skill 4.1 and will be omitted here.

## Deploy Network Watcher

Network Watcher is enabled as a single instance per Azure region. It is not deployed like a conventional Azure resource, although it does appear as a resource in a resource group.

Any subscription containing a virtual network resource will automatically have Network Watcher enabled. Otherwise, it can be enabled via the Azure portal, by choosing All Services, Network Watcher. You can also see the Network Watcher status per region. Network Watcher can also be deployed via the command line (using the `New-AzNetworkWatcher` cmdlet or the `az network watcher configure` commands), which unlike the Azure portal, provides control over the resource group used.

Some of the Network Watcher tools require the Network Watcher VM extension be installed on the VM being monitored. This extension is available for both Windows and Linux VMs. It is installed automatically when using Network Watcher via the Azure portal.

## IP Flow Verify

The IP Flow Verify tool provides a quick and easy way to test whether a given network flow will be allowed into or out of an Azure virtual machine. It will report whether the requested traffic is allowed or blocked, and in the latter case, which NSG rule is blocking the flow. It is a useful tool for verifying that NSGs are correctly configured.

It works by simulating the requested packet flow through the NSGs applied to the VM. For this reason, the VM must be in a running state.

To use IP Flow Verify via the Azure portal, open Network Watcher, and click IP Flow Verify. Select the VM and NIC to verify, and specify the protocol, direction, and remote and local IP addresses and ports, as shown in Figure 5-41.

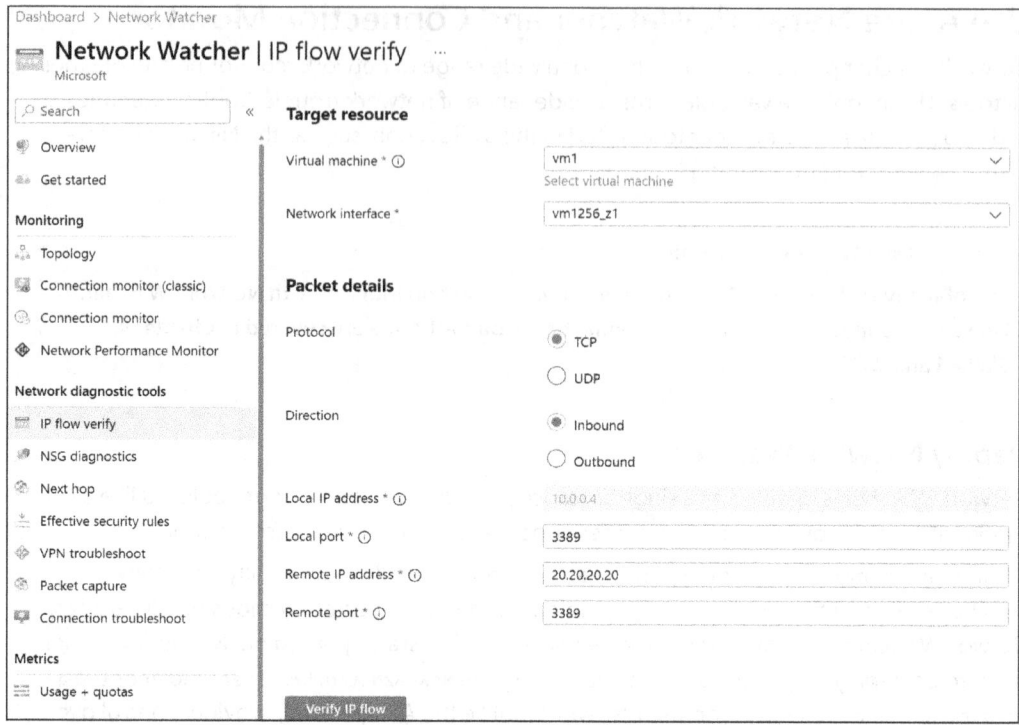

**FIGURE 5-41** Network Watcher IP Flow Verify

IP Flow Verify can also be used from PowerShell, using the `Test-AzNetworkWatcherIPFlow` cmdlet, or the Azure CLI, using the `az network watcher test-ip-flow` command.

## Next Hop

The Next Hop tool provides a useful way to understand how a VM's outbound traffic is being directed. For a given outbound flow, it shows the next hop IP address and type as well as the route table ID of any user-defined route in effect. Possible next hop types are

- Internet
- VirtualAppliance
- VirtualNetworkGateway
- VirtualNetwork
- VirtualNetworkPeering
- VirtualNetworkServiceEndpoint
- None (this is used for user-defined routes)

To use Next Hop via the Azure portal, open Network Watcher and click Next Hop. Select the source VM, NIC, IP address, and the destination address, as shown in Figure 5-42. The destination can be any IP address, either on the internal network or the internet.

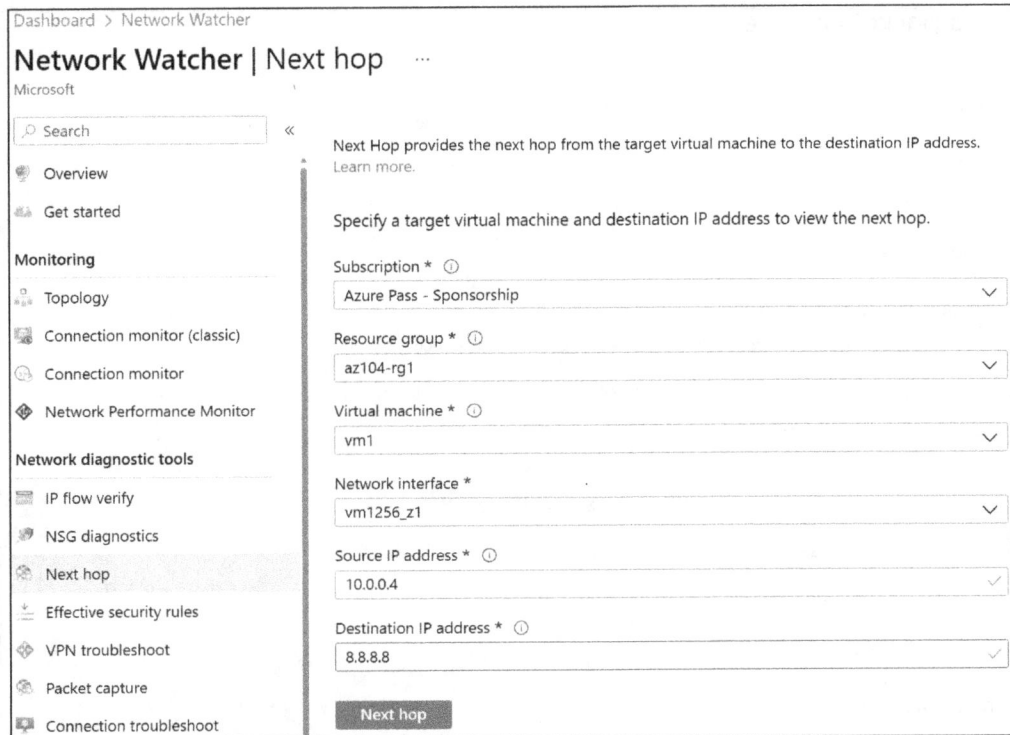

**FIGURE 5-42**  Network Watcher Next Hop

Next Hop can also be used from PowerShell using the `Get-AzNetworkWatcherNextHop` cmdlet, or the Azure CLI using the `az network watcher show-next-hop` command.

## Packet Capture

The Packet Capture tool captures network packets entering or leaving your virtual machines. It is a powerful tool for deep network diagnostics.

You can capture all packets, or a filtered subset based on the protocol and local and remote IP addresses and ports. You can also specify the maximum packet and overall capture size, and a time limit (captures start almost immediately once configured).

Packet captures are stored as a file on the VM or in an Azure storage account, in which case NSGs must allow access from the VM to Azure Storage. These captures are in a standard format and can be analyzed off line using common tools such as WireShark or Microsoft Message Analyzer.

To use the Packet Capture tool, open Network Watcher and click Packet Capture, Add. Select the VM, give the capture a name, and specify the destination, packet and total size, time limit, and filters. An example is shown in Figure 5-43.

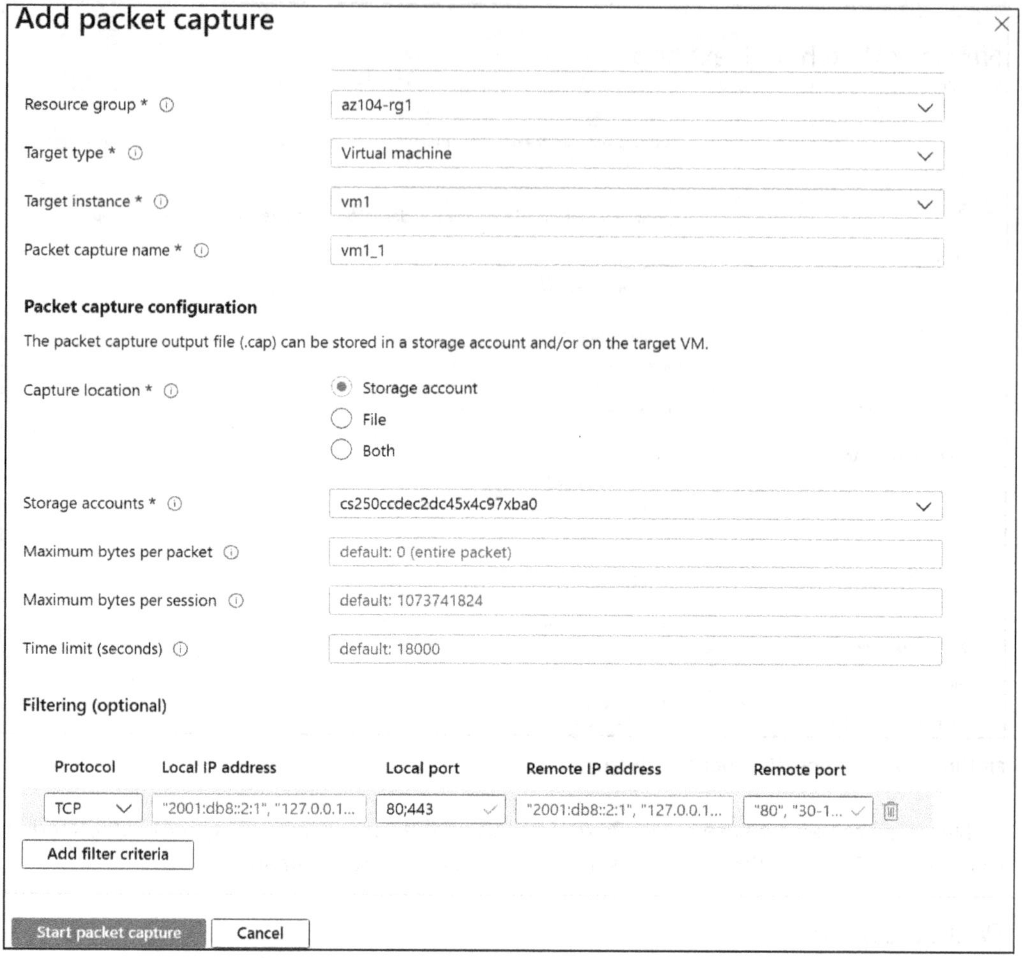

## Add packet capture

| | | |
|---|---|---|
| Resource group * ⓘ | az104-rg1 | ⌄ |
| Target type * ⓘ | Virtual machine | ⌄ |
| Target instance * ⓘ | vm1 | ⌄ |
| Packet capture name * ⓘ | vm1_1 | |

**Packet capture configuration**

The packet capture output file (.cap) can be stored in a storage account and/or on the target VM.

Capture location * ⓘ
- ⦿ Storage account
- ◯ File
- ◯ Both

| | | |
|---|---|---|
| Storage accounts * ⓘ | cs250ccdec2dc45x4c97xba0 | ⌄ |
| Maximum bytes per packet ⓘ | default: 0 (entire packet) | |
| Maximum bytes per session ⓘ | default: 1073741824 | |
| Time limit (seconds) ⓘ | default: 18000 | |

**Filtering (optional)**

| Protocol | Local IP address | Local port | Remote IP address | Remote port | |
|---|---|---|---|---|---|
| TCP ⌄ | "2001:db8::2:1", "127.0.0.1... | 80;443 ⌄ | "2001:db8::2:1", "127.0.0.1... | "80", "30-1... ⌄ | 🗑 |

Add filter criteria

Start packet capture    Cancel

**FIGURE 5-43**   Network Watcher Packet Capture

## Network topology

The network topology view in Network Watcher provides a diagrammatic view of the resources in your virtual network. It is not a diagnostic or alerting tool. It is a quick and easy way to review your network resources and manually check for misconfiguration.

A limitation of the tool is that it only shows the topology within a single virtual network. All common network resource types are supported, although for application gateways, only the backend pool connected to the network interface is shown.

To view the network topology via the Azure portal, open Network Watcher and click Topology. Select the resource group and virtual network, and the topology will be shown.

An example topology is given in Figure 5-44.

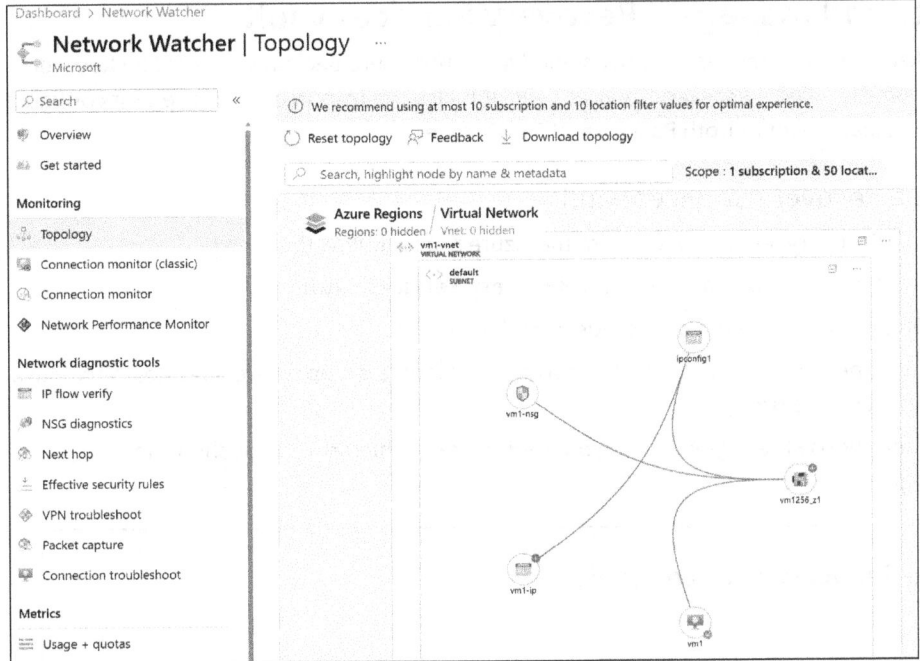

**Network Watcher | Topology** ...
Microsoft

🔍 Search    «

ⓘ We recommend using at most 10 subscription and 10 location filter values for optimal experience.

🔗 Overview

🔗 Get started

**Monitoring**

Topology

Connection monitor (classic)

Connection monitor

Network Performance Monitor

**Network diagnostic tools**

IP flow verify

NSG diagnostics

Next hop

Effective security rules

VPN troubleshoot

Packet capture

Connection troubleshoot

**Metrics**

Usage + quotas

🔄 Reset topology    🗨 Feedback    ⬇ Download topology

🔍 Search, highlight node by name & metadata        Scope : 1 subscription & 50 locat...

**Azure Regions** / **Virtual Network**
Regions: 0 hidden / Vnet: 0 hidden

vm1-vnet
VIRTUAL NETWORK

default
SUBNET

ipconfig1

vm1-nsg

vm1256_z1

vm1-ip

vm1

**FIGURE 5-44** Viewing network topology in Network Watcher

The underlying topology data can be downloaded in JSON format via Azure PowerShell or the Azure CLI, using the `Get-AzNetworkWatcherTopology` cmdlet or the `az network watcher show-topology` command, respectively.

# Skill 5.2: Implement backup and recovery

Azure Backup is a service that allows you to back up on-premises servers, cloud-based virtual machines, and virtualized workloads such as SQL Server and SharePoint to Microsoft Azure. It also supports backup of Azure Storage file shares.

Azure Site Recovery is a business continuity/disaster recovery tool that helps replicate resources. The source resources can be on-premises, in another cloud, or in Azure. You create a replication policy to define how you want the resources replicated.

**This skill covers how to:**

- Create and manage a Recovery Services vault
- Configure Azure Site Recovery
- Create an Azure Backup vault
- Create and configure backup policy
- Configure and review backup reports

# Create and manage a Recovery Services vault

Within Azure, a single resource is provisioned for either Azure Backup or Azure Site Recovery. This resource is called a *Recovery Services vault*. It is also the resource that is used for configuration and management of both Backup and Site Recovery.

## Create a Recovery Services vault

To create a Recovery Services vault from the Azure portal, follow these steps:

1. From the Azure portal, search for **Recovery Services vaults**.

2. On the Site Recovery Vaults blade, click Create.

3. Enter the name of the vault and choose the resource group where it resides, or create a new resource group.

4. Then choose the region where you want to create the resource, as shown in Figure 5-45.

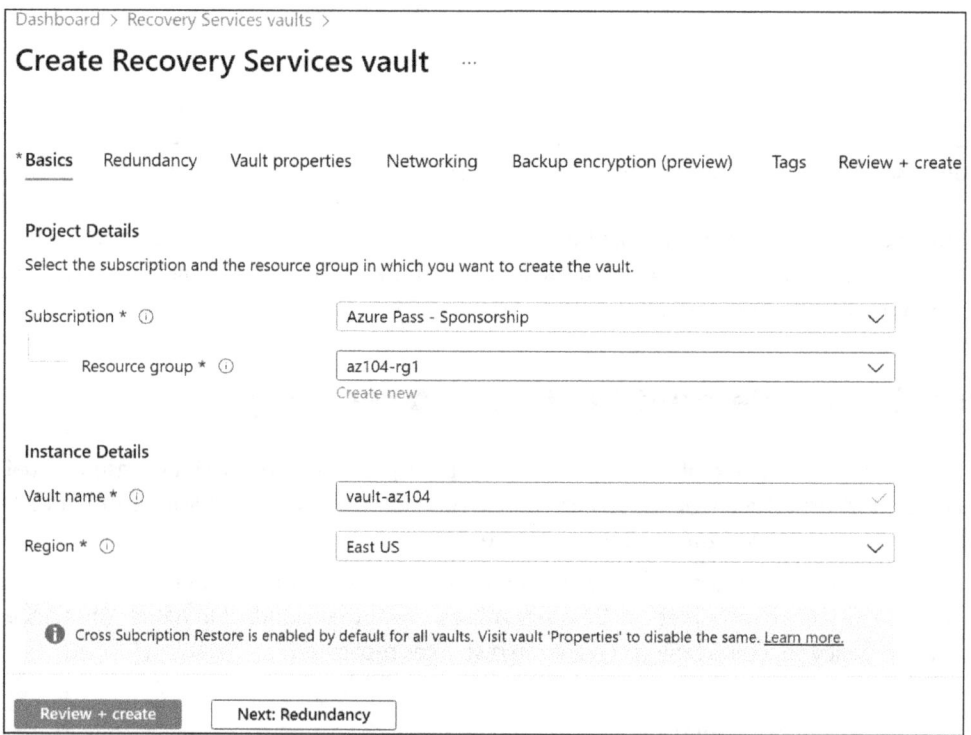

FIGURE 5-45   Recovery Services vault basics

5. Click Next: Redundancy. You can configure two options on the Redundancy tab, as shown in Figure 5-46:

- **Backup Storage Redundancy**  Choose whether the data stored in the vault is replicated locally redundant, zone-redundant, or geo-redundant.
- **Cross Region Restore**  Select whether restores can occur across regions.

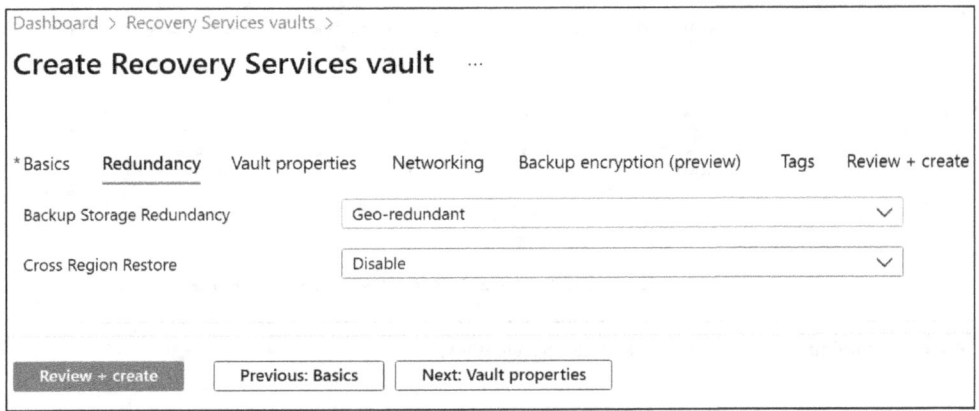

Dashboard > Recovery Services vaults >

## Create Recovery Services vault  …

| * Basics | Redundancy | Vault properties | Networking | Backup encryption (preview) | Tags | Review + create |

Backup Storage Redundancy     Geo-redundant ⌄

Cross Region Restore          Disable ⌄

Review + create     Previous: Basics     Next: Vault properties

**FIGURE 5-46**  Recovery services vault redundancy

6. Click Next: Vault Properties. Then specify whether immutability is enabled for the vault. Vault immutability ensures that restore points of the vault are created to prevent items from being deleted early.

7. Click Next: Networking Choose whether to make the vault accessible from the internet from all networks or to use a private endpoint and deny public access.

8. Click Next: Backup Encryption. Choose whether to use the default Microsoft-managed key to encrypt the backup items, or to use a customer-managed key from an Azure key vault.

9. Click Review + Create, and then click Create.

## Use soft delete to recover Azure VMs

The default behavior when you delete a backup is that the backup is deleted and lost forever. When soft delete is enabled, you can save and recover your data when backup data are deleted even in the event of an overwrite. This feature must be enabled in the Recovery Services vault. Choose Properties, Security Settings to see soft delete options (see Figure 5-47). When you use soft delete, backup data is retained for 14 days after deletion.

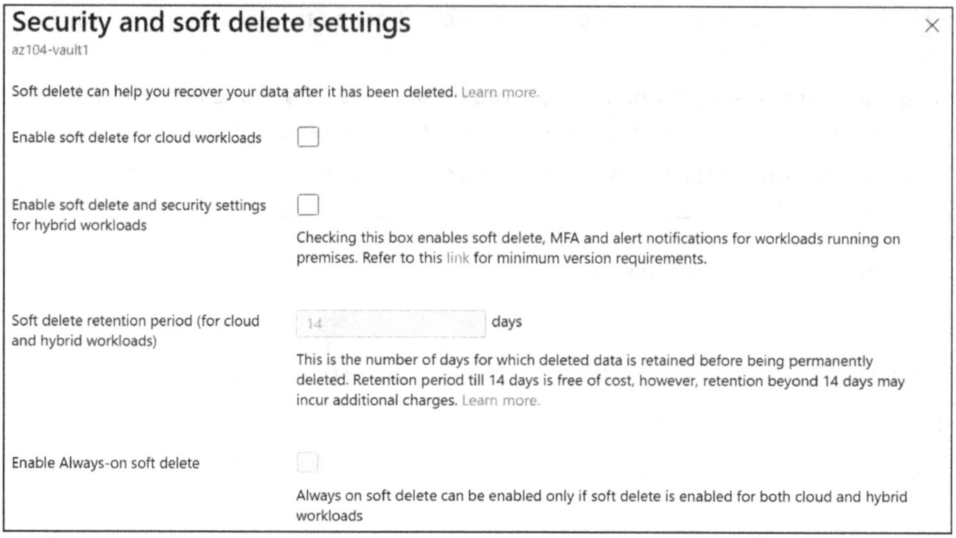
Security and soft delete settings
az104-vault1
Soft delete can help you recover your data after it has been deleted. Learn more.
Enable soft delete for cloud workloads
Enable soft delete and security settings for hybrid workloads
Checking this box enables soft delete, MFA and alert notifications for workloads running on premises. Refer to this link for minimum version requirements.
Soft delete retention period (for cloud and hybrid workloads)
14
days
This is the number of days for which deleted data is retained before being permanently deleted. Retention period till 14 days is free of cost, however, retention beyond 14 days may incur additional charges. Learn more.
Enable Always-on soft delete
Always on soft delete can be enabled only if soft delete is enabled for both cloud and hybrid workloads

**FIGURE 5-47** Enabling soft delete for a Recovery Services vault

> ***NEED MORE REVIEW?*** **SOFT DELETE FOR AZURE VM BACKUP**
>
> **You can learn more about using soft delete with Azure VM Backup at *https://learn.microsoft.com/azure/backup/backup-azure-security-feature-cloud?tabs=azure-portal#soft-delete.***

If the soft delete option is enabled, you can delete the backup data by clicking Stop Backup and then clicking Delete Backup Data. You will be prompted to provide a reason for deleting backup data that will be stored with the activity log of the deletion. Once deleted, your soft-deleted backup item will appear, as shown in Figure 5-48.

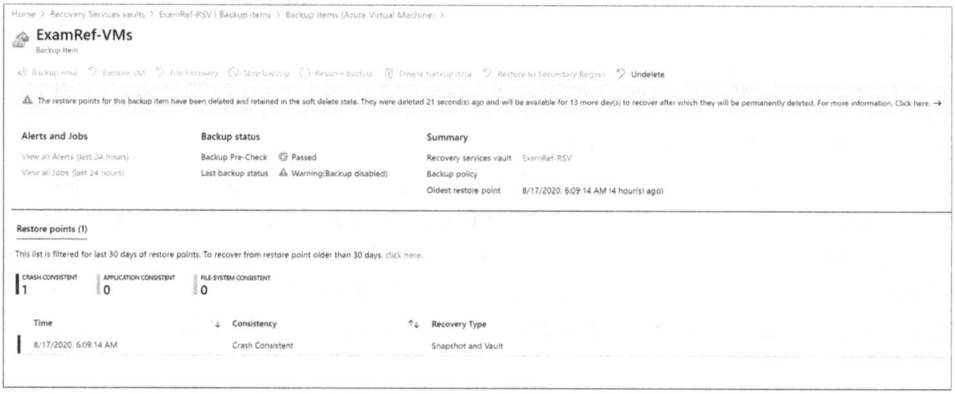
ExamRef-VMs
Undelete
Alerts and Jobs
Backup status
Summary
Backup Pre-Check
Passed
Last backup status
Warning(Backup disabled)
Recovery services vault
Backup policy
Oldest restore point
8/17/2020, 6:09:14 AM (4 hour(s) ago)
Restore points (1)
1
0
0
Time
Consistency
Recovery Type
8/17/2020, 6:09:14 AM
Crash Consistent
Snapshot and Vault

**FIGURE 5-48** Soft delete–enabled backup item after deletion

You can click Undelete anytime within 14 days of the retention period (see Figure 5-49). Once the data is restored, you can click Resume Backup again.

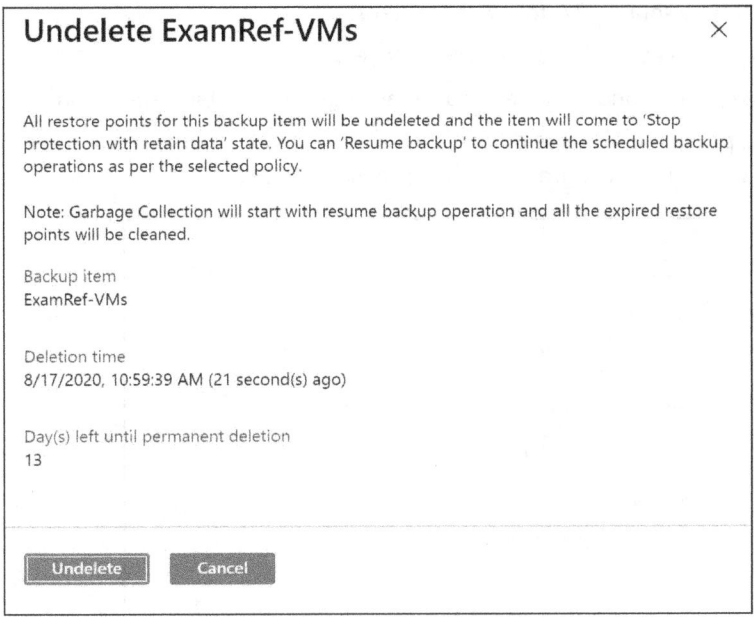

**Undelete ExamRef-VMs**                                    ✕

All restore points for this backup item will be undeleted and the item will come to 'Stop protection with retain data' state. You can 'Resume backup' to continue the scheduled backup operations as per the selected policy.

Note: Garbage Collection will start with resume backup operation and all the expired restore points will be cleaned.

Backup item
ExamRef-VMs

Deletion time
8/17/2020, 10:59:39 AM (21 second(s) ago)

Day(s) left until permanent deletion
13

[ Undelete ]  [ Cancel ]

**FIGURE 5-49**  Undelete option for soft-deleted ExamRef-VMs

## Configure Azure Site Recovery

Every organization will have its own business continuity and disaster recovery (BCDR) plans to handle unpredictable circumstances with unexpected outages that occur. Azure Site Recovery service enables you to replicate, failover, and failback virtual machines as needed. Azure Site Recovery solution addresses these major replication scenarios:

- Azure VMs from one region to another
- On-premises VMs (VMware, Hyper-V, and physical servers) to Azure
- On-premises VMs to another site

Suppose you need to replicate Azure VMs from one region to another, as an example. First, you would need to create a Recovery Services vault. As a best practice, you should always validate the target subscription readiness by checking the appropriate VM SKU and major feature availability. You also need to take into consideration the regions that you are using. For cross-region recovery, the vault *and* the resource group the vault is deployed to *must* be in a different region than the VMs that you are replicating.

For enterprise environments, you should also consider allow-listing the URLs for outbound connectivity to required Azure resources and service tag-based NSG rules. You would also need minimum Site Recovery Contributor rights for configuring the replication and Site Recovery Operator rights for executing the failover and failback operations.

To enable replication from a source VM, follow these steps:

1. Open the Recovery Services vault and click Site Recovery.

2. On the Site Recovery blade, under Azure Virtual Machines, click Enable Replication.

3. On the Source tab, provide the source details, such as the location, deployment model, subscription, and source resource group, as shown in Figure 5-50.

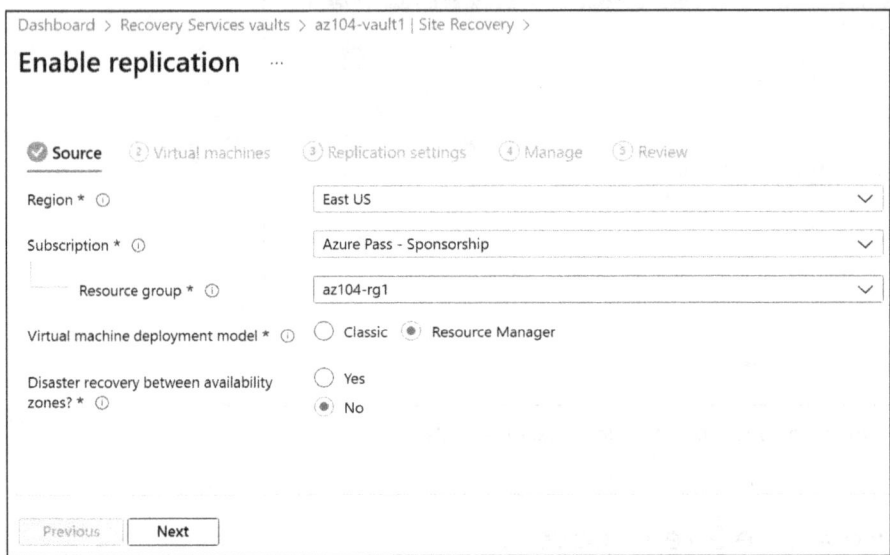

**FIGURE 5-50**   Source configuration when enabling replication

4. On the next tab, select the source VM for replication (see Figure 5-51).

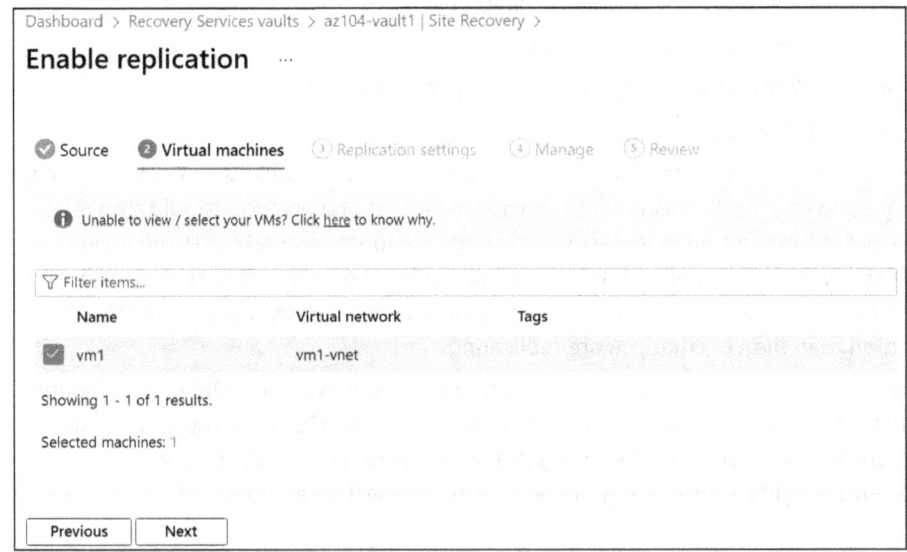

**FIGURE 5-51**   Source VM selection when enabling replication

5. Next, you must configure target environment settings, as shown in Figure 5-52. Target settings include

- **Target Location** The target Azure region, different from the source
- **Target Subscription** If you plan to recover the machines in a different subscription
- **Target Resource Group** The resource group of the VM object after recovery
- **Failover Virtual Network** The virtual network in the target region that the VM should be associated with
- **Failover Subnet** The subnet within the virtual network that the VM NIC should be associated with
- **Storage** Managed disk, disk churn, and disk cache storage options for each VM that is being replicated
- **Availability Options** Target availability set, availability zone, and/or proximity placement groups after recovery
- **Capacity Reservation** Ensure that the VM has guaranteed capacity allocation after replication

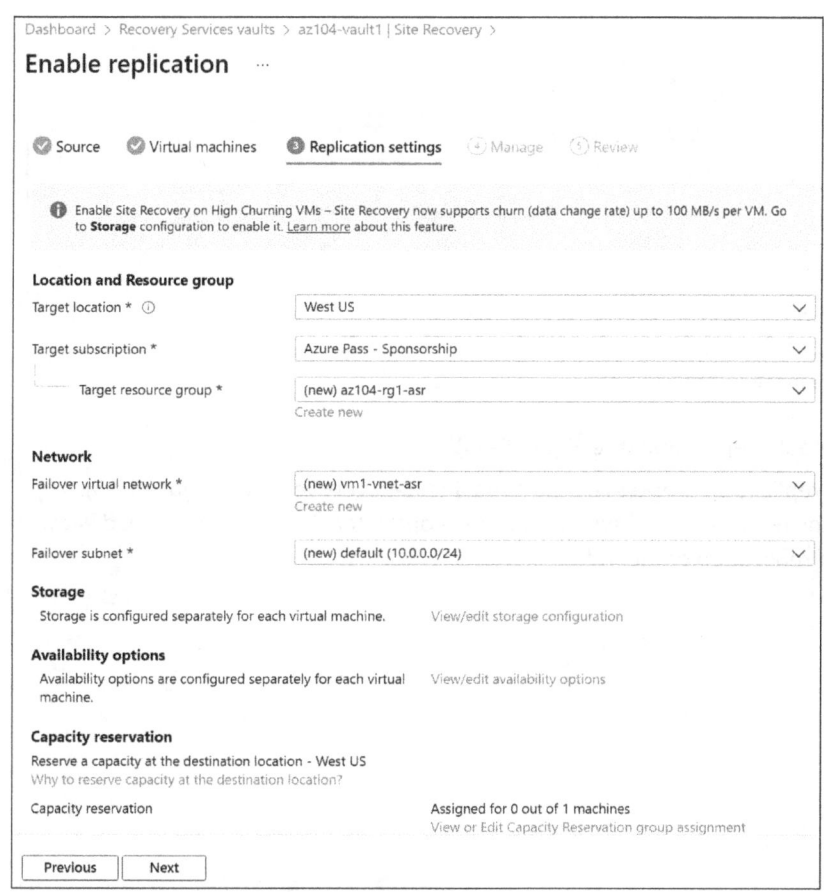

FIGURE 5-52 Target settings for the replication

6.  On the Manage tab, configure the replication policy that is used. You can also define replication groups if a cluster of machines running the same workload needs consistency across VMs, as shown in Figure 5-53.

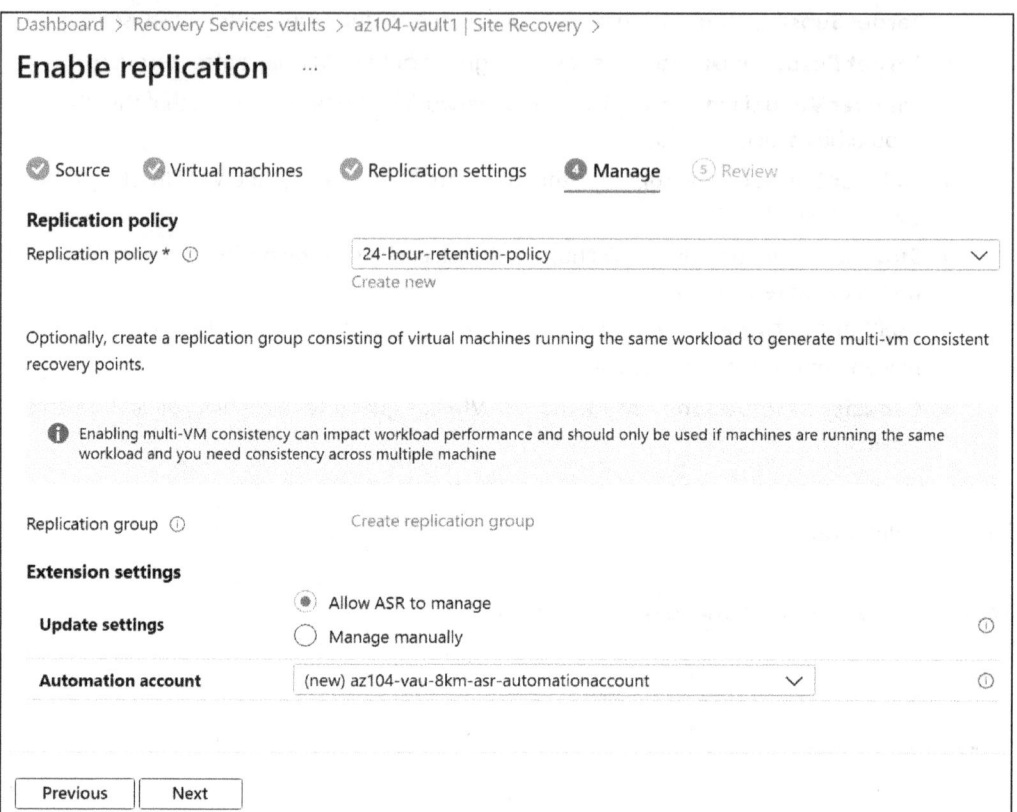

**FIGURE 5-53** Configure the replication policy

7.  Finally, click Enable Replication (see Figure 5-54).

You can track the replication progress by selecting Site Recovery Jobs (see Figure 5-55). It takes a while to get the replication and synchronization completed. You cannot proceed with further steps without replication of the VM.

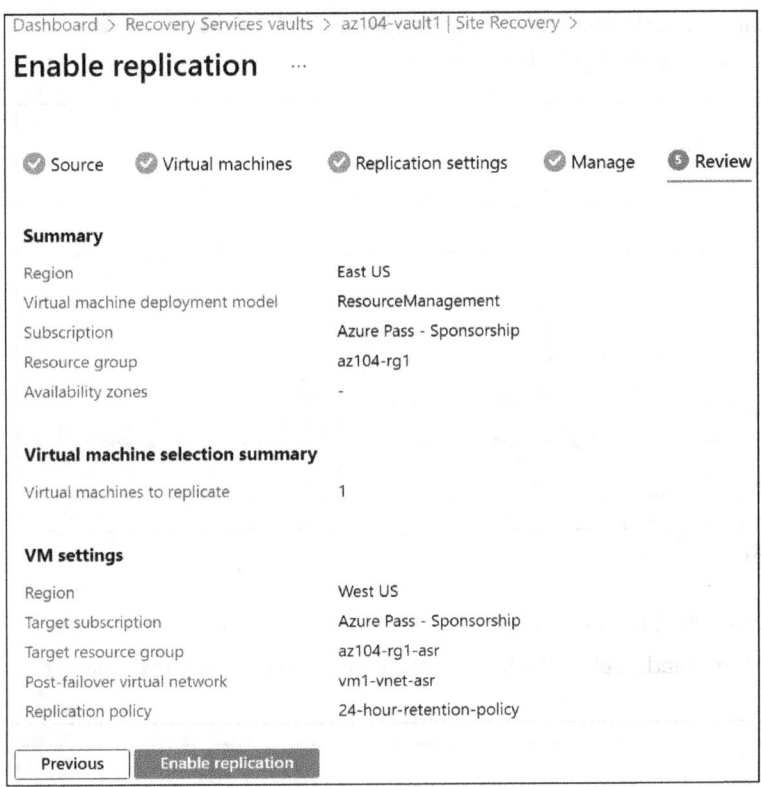

**FIGURE 5-54** Review of all the Enable Replication settings

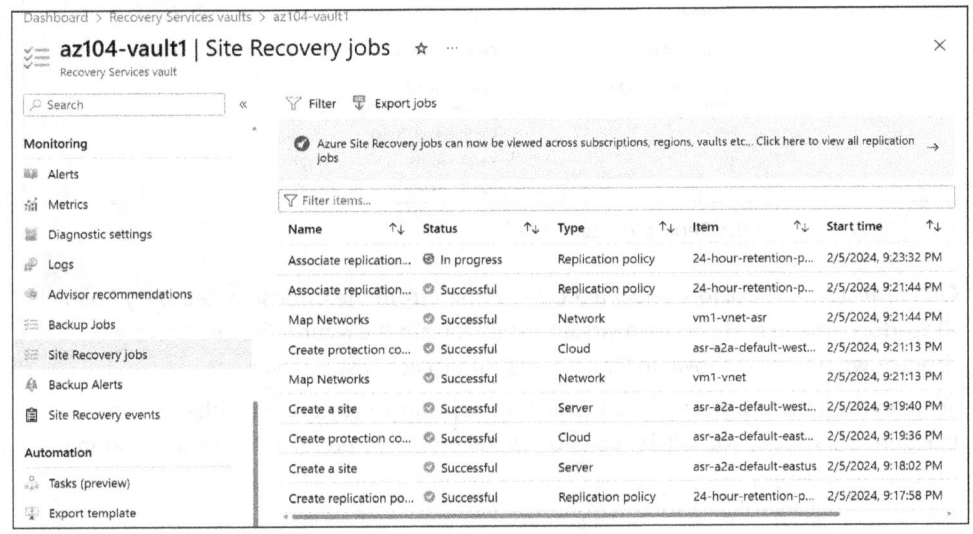

**FIGURE 5-55** Site Recovery jobs

Once replicated, you can see the Source VM listed in the Recovery Services vault under Replicated Items. The overview is shown in Figure 5-56.

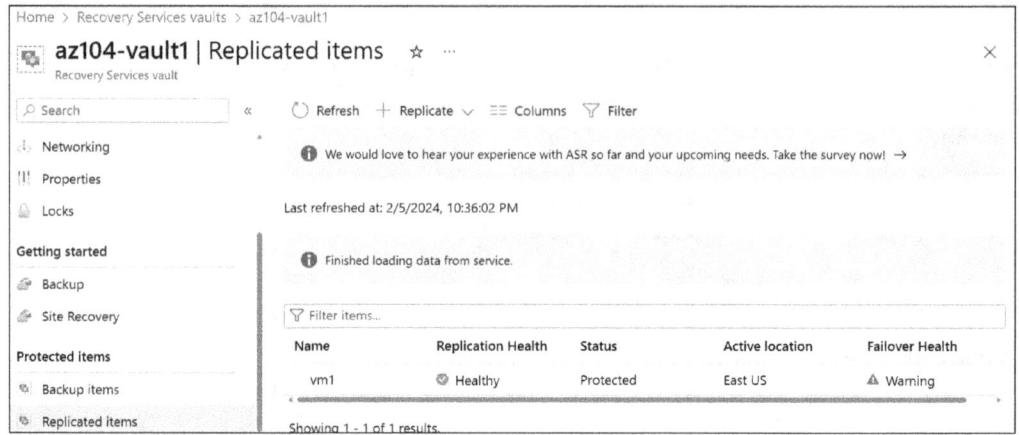

FIGURE 5-56   Replicated Items

After replication has completed, it's time to test failover:

1. On the Replicated Items blade, select the VM. On the command bar, click Test Failover (see Figure 5-57).

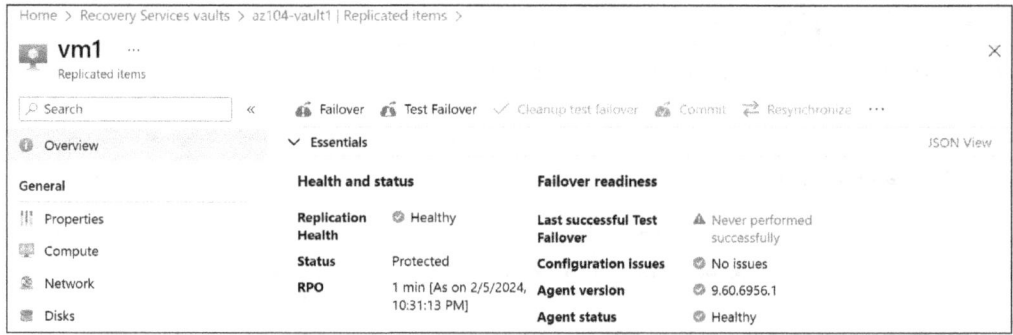

FIGURE 5-57   Test Failover on the Replicated Items blade

2. On the Test Failover blade, select a recovery point from the Choose A Recovery Point drop-down menu and choose a virtual network from the Azure Virtual Network drop-down menu, as shown in Figure 5-58. Then, click Test Failover.

3. You can track the progress of the test failover by using Site Recovery jobs, as shown in Figure 5-59. Now, you will be able to see the test VM created in the target resource group.

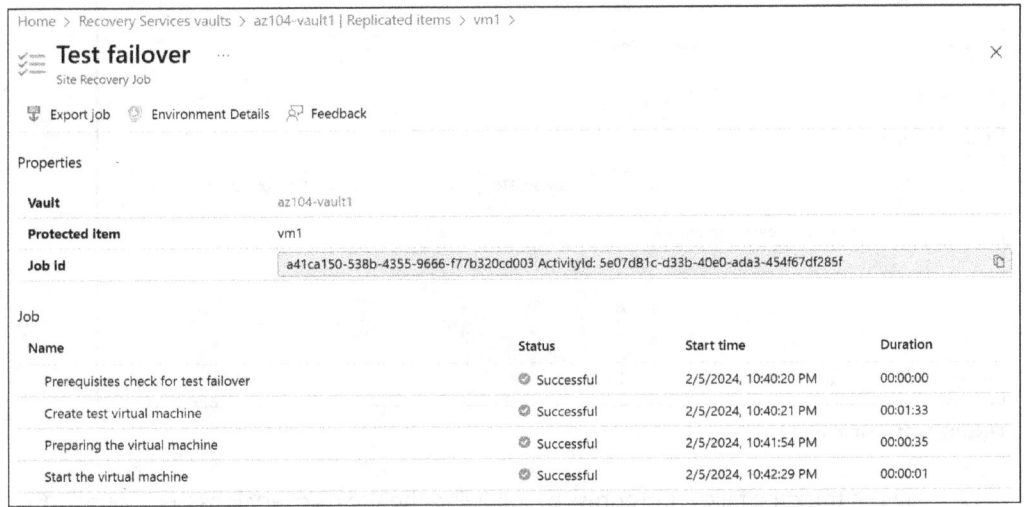

**Test failover** ···
vm1

⊘ Pre-validation successful

**Test failover direction**

Source ⓘ                                    East US (Zone 1)

Destination ⓘ                              West US

**Test failover settings**

Recovery point * ⓘ          Latest processed (lowest RTO) (2/5/2024, 10:35:00 PM) (1 out of 1 disks)   ⌄

Azure virtual network * ⓘ    vm1-vnet-asr (mapped)                                                     ⌄

⚠ It is recommended that the networks selected for test failover and failover operations are different.
Learn more about DR Drills ↗

[ Test failover ]    [ Cancel ]

**FIGURE 5-58**   Test Failover blade

---

Home > Recovery Services vaults > az104-vault1 | Replicated items > vm1 >

**Test failover** ···                                                                        ✕
Site Recovery Job

🖥 Export job    ⊙ Environment Details    ⬚ Feedback

Properties

| | |
|---|---|
| **Vault** | az104-vault1 |
| **Protected item** | vm1 |
| **Job Id** | a41ca150-538b-4355-9666-f77b320cd003 ActivityId: 5e07d81c-d33b-40e0-ada3-454f67df285f |

Job

| Name | Status | Start time | Duration |
|---|---|---|---|
| Prerequisites check for test failover | ⊘ Successful | 2/5/2024, 10:40:20 PM | 00:00:00 |
| Create test virtual machine | ⊘ Successful | 2/5/2024, 10:40:21 PM | 00:01:33 |
| Preparing the virtual machine | ⊘ Successful | 2/5/2024, 10:41:54 PM | 00:00:35 |
| Start the virtual machine | ⊘ Successful | 2/5/2024, 10:42:29 PM | 00:00:01 |

**FIGURE 5-59**   Test failover jobs

4.  You can delete the test VM after verifying the VM and network details. To delete the VM and other resources, click Cleanup Test Failover, as shown in Figure 5-60.

**FIGURE 5-60** Clean up test failover

5. You can now run an actual failover. Click Failover in the Command bar.

6. On the Failover blade, select the recovery point and verify the failover direction, as shown in Figure 5-61. Click Failover.

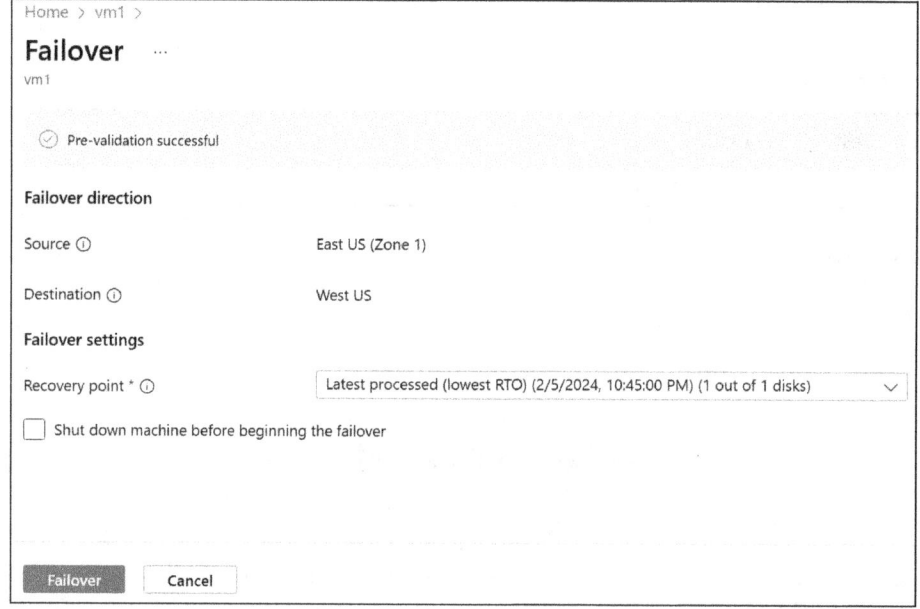

**FIGURE 5-61** Failover blade

7. You can track the failover progress by following the site recovery jobs (see Figure 5-62). A target Azure VM will be created with the same configuration and target settings provided earlier.

8. You should validate by logging into the VM.

9. Click Commit to complete the failover process.

**FIGURE 5-62** Failover completed

10. You should also consider protecting your VM again by clicking Re-protect, which will reverse the process (see Figure 5-63).

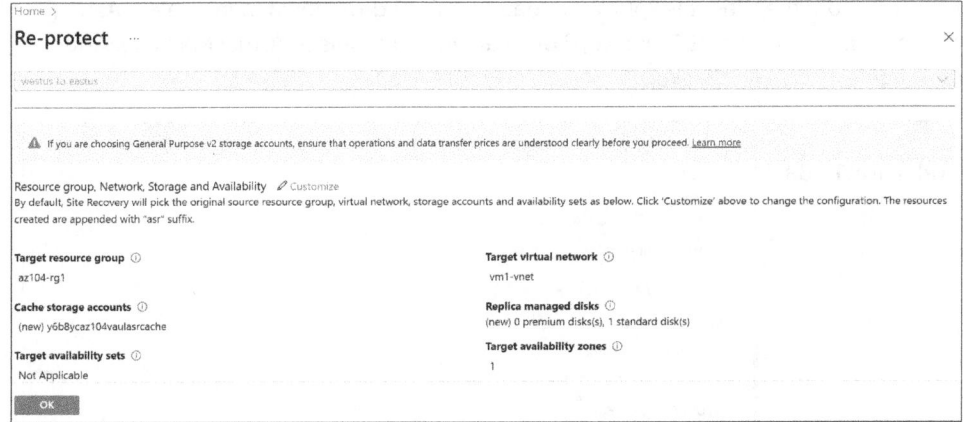

**FIGURE 5-63** Re-protect option

> **NEED MORE REVIEW?** **SITE RECOVERY SCENARIOS**
>
> Learn about the VMware site recovery to Azure at *https://learn.microsoft.com/azure/site-recovery/tutorial-prepare-azure*. Learn about Hyper-V VM site recovery to Azure at *https://learn.microsoft.com/azure/site-recovery/tutorial-prepare-azure-for-hyper*.

11. Once the VM is protected again, you can perform a failback to get to the original state. Similarly, you can use Site Recovery for other scenarios.

> **NEED MORE REVIEW?** **AZURE MIGRATE**
>
> If you want to migrate an on-premises workload to Azure, see *https://learn.microsoft.com/azure/migrate/migrate-services-overview*.

# Create an Azure Backup vault

Azure Backup service can be used to back up and restore various cloud as well as on-premises resources. Recovery Services vault is used to enable Azure Backup and to configure the backup policies.

For Azure workloads, the Azure Backup service can back up the following resources:

- Virtual machines
- SAP HANA databases running in an Azure VM
- Azure file share
- SQL Server databases running in an Azure VM

When you back up an Azure virtual machine, you can restore an entire virtual machine or you can restore individual files from the virtual machine, and it is quite easy to set up. To back up a VM in Azure with Azure Backup, open the Recovery Services vault, and click Backup under Getting Started. From the Where Is Your Workload Running? drop down menu, select Azure, and from the What Do You Want To Backup? drop down menu, select Virtual Machine. After making these selections, click Backup, as shown in Figure 5-64.

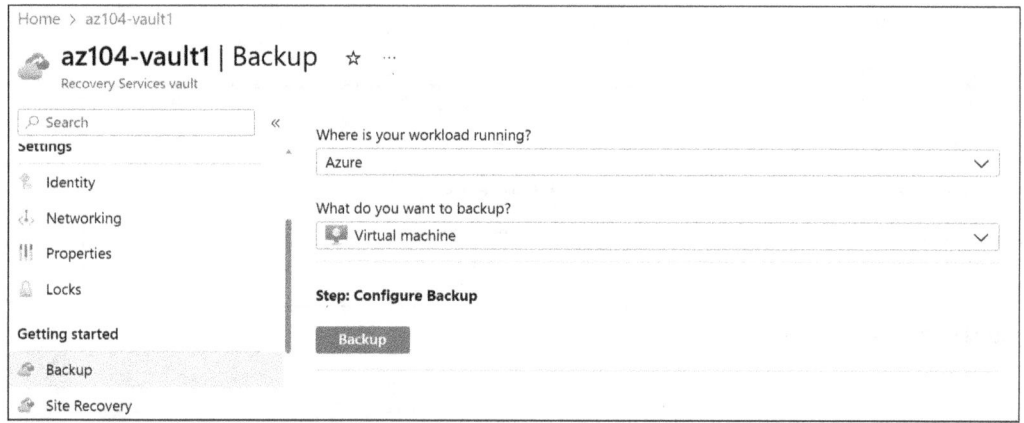

**FIGURE 5-64** Configuring Azure Backup to protect virtual machines

On the Configure Backup blade, first select whether to back up using a Standard or Enhanced policy types. Standard policies are used for simple backups of non trusted launch VMs, or VMs that aren't using ultra disks or Premium SSD v2 disks, either of which require Enhanced. After selecting the type, select the specific policy that matches that type, or choose the default policy. Finally, click Add to add the VM to the backup configuration. A completed configuration using a Standard policy for VM1 is displayed in Figure 5-65.

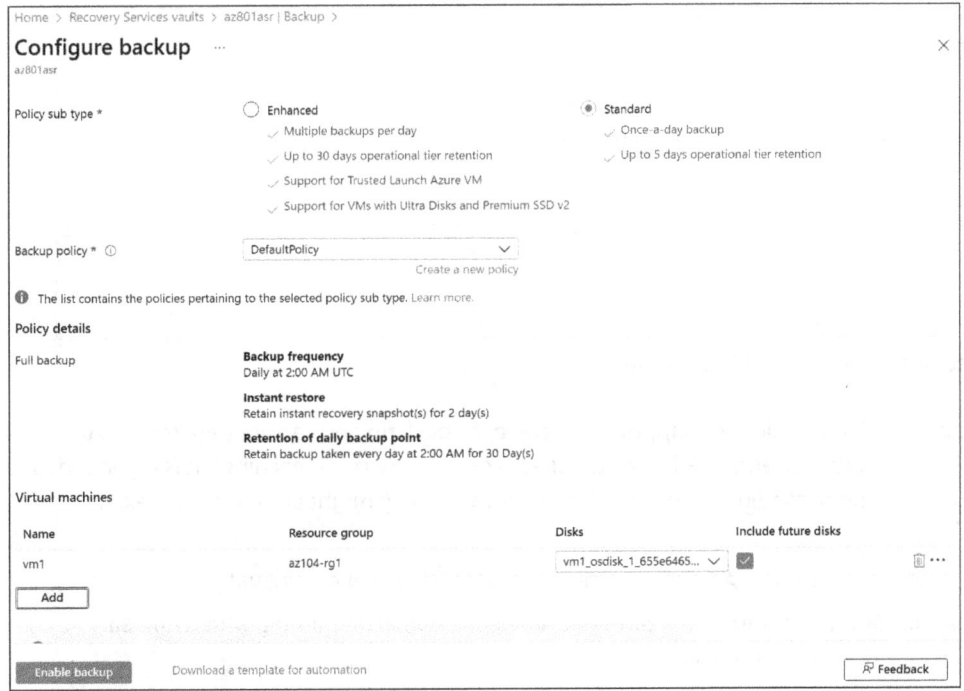

**Configure backup**  ...
az801asr

Policy sub type *
- ○ Enhanced
  - ✓ Multiple backups per day
  - ✓ Up to 30 days operational tier retention
  - ✓ Support for Trusted Launch Azure VM
  - ✓ Support for VMs with Ultra Disks and Premium SSD v2
- ● Standard
  - ✓ Once-a-day backup
  - ✓ Up to 5 days operational tier retention

Backup policy * ⓘ
DefaultPolicy
Create a new policy

ⓘ The list contains the policies pertaining to the selected policy sub type. Learn more.

**Policy details**

Full backup
**Backup frequency**
Daily at 2:00 AM UTC

**Instant restore**
Retain instant recovery snapshot(s) for 2 day(s)

**Retention of daily backup point**
Retain backup taken every day at 2:00 AM for 30 Day(s)

Virtual machines

| Name | Resource group | Disks | Include future disks |
|------|----------------|-------|----------------------|
| vm1 | az104-rg1 | vm1_osdisk_1_655e6465... | ✓  🗑 ⋯ |

[ Add ]

[ Enable backup ]    Download a template for automation    [ 🗩 Feedback ]

**FIGURE 5-65**  Azure Backup configuration

After the VMs are selected, click Enable Backup.

> **NOTE  AZURE VM PROTECTION AND VAULT STORAGE REDUNDANCY TYPE**
>
> When protecting IaaS VMs by using Azure Backup, only VMs in the same region as the Recovery Services vault are available for backup. Because of this, it is best practice to choose Geo-redundant storage or Read Access Geo-redundant storage to be associated with the Recovery Services vault. This ensures that if a regional outage affects VM access, there is a replicated copy of the backup in another region.

When you click Enable Backup, behind the scenes, the VMSnapshot (for Windows) or VMSnapshotLinux (for Linux) extension is automatically deployed by the Azure fabric controller to the VMs. This makes snapshot-based backups possible, which means a snapshot of the VM is taken first, and then this snapshot is streamed to the Azure Storage associated with the Recovery Services vault. The initial backup is not taken until the day/time configured in the backup policy, though an ad-hoc backup can be initiated at any time. To do so, navigate to the Protected Items section of the Recovery Services vault properties, click Backup Items, and click Azure Virtual Machine under Backup Management Type. The VMs that are enabled for backup are listed here. To begin an ad-hoc backup, right-click a VM and select Backup Now, as shown in Figure 5-66.

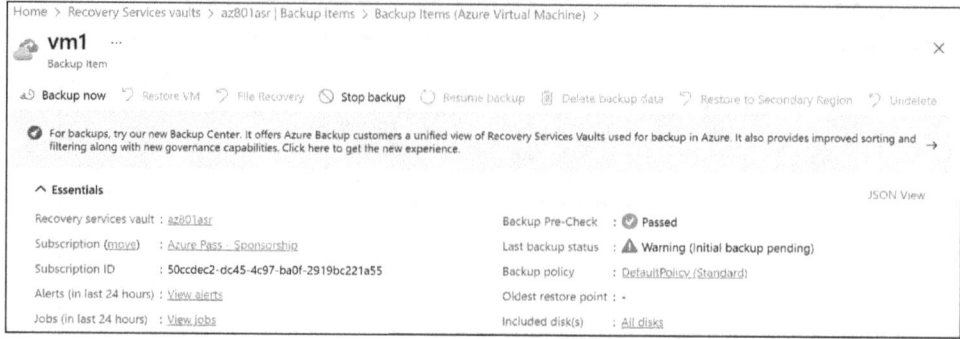

**FIGURE 5-66**    Starting an ad-hoc backup

Azure Backup also directly supports the ability to back up and restore data from Azure Files, SQL Server databases, and SAP HANA databases on Azure virtual machines. It is a good idea to have a basic understanding of the capabilities because they might appear on the exam.

> **NEED MORE REVIEW?    AZURE FILES AND SQL SERVER IN AN AZURE VM**
>
> Each specific type of storage has different capabilities when integrating with Azure Backup. Learn more about each service:
>
> - **Azure Files:** *https://learn.microsoft.com/azure/backup/backup-azure-files? tabs=backup-center*
>
> - **SQL Server on Azure virtual machines:** *https://learn.microsoft.com/azure/backup/ backup-azure-sql-database*
>
> - **SAP HANA on Azure virtual machines:** *https://learn.microsoft.com/azure/backup/ sap-hana-database-about*.

After backing up a virtual machine using Azure Backup, there are two methods to restore data: Restore VM and File Recovery.

To restore a recovery point as a new virtual machine, open the Recovery Services vault, click Backup Items, click Azure Virtual Machine, and then click the virtual machine you want to restore from the list. You'll see a list of all the restore points available for restoration, as shown in Figure 5-67.

Right-click the desired restore point and select Restore VM, or click Restore VM at the top of the page (refer to Figure 5-67).

You can then restore to a new virtual machine by selecting Create New, or you can restore over an existing virtual machine by selecting Replace Existing.

Figure 5-68 shows the Restore Virtual Machine blade with Create New selected. Here, you can specify the virtual machine name, resource group, virtual network, subnet, and storage account.

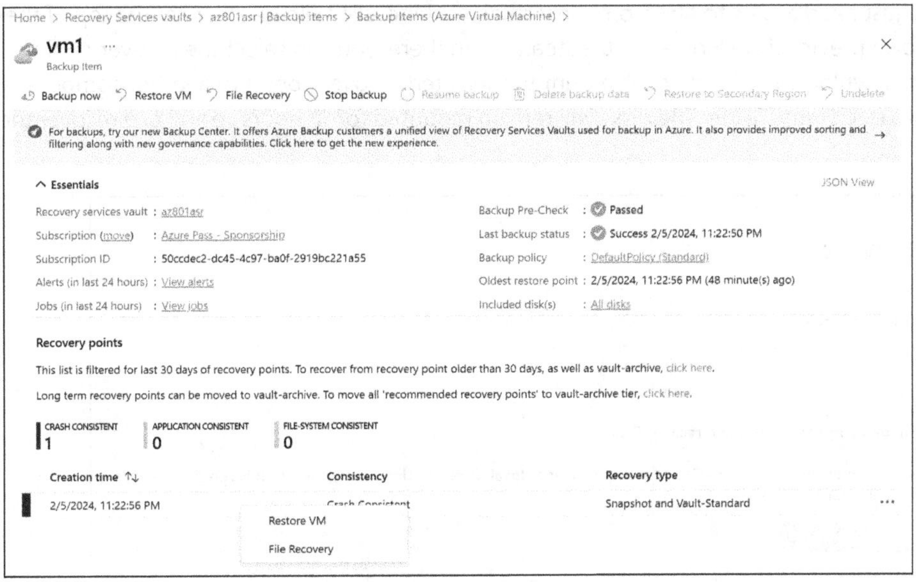

**FIGURE 5-67** Available restore points for a virtual machine.

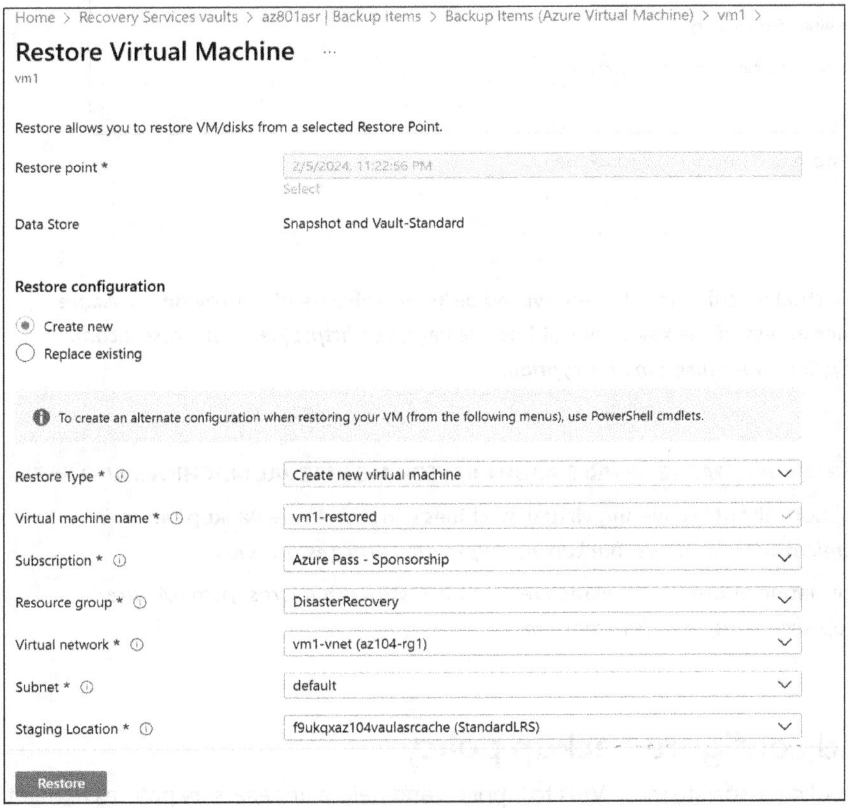

**FIGURE 5-68** Restore to a new virtual machine

If you just need access to files from the virtual machine, click File Recovery at the top of the page shown previously in Figure 5-66 instead. From there, you can select the recovery point and then download a script that will mount the selected recovery point to another computer as local disks (see Figure 5-69). The disks will remain mounted for 12 hours so you can recover the needed data.

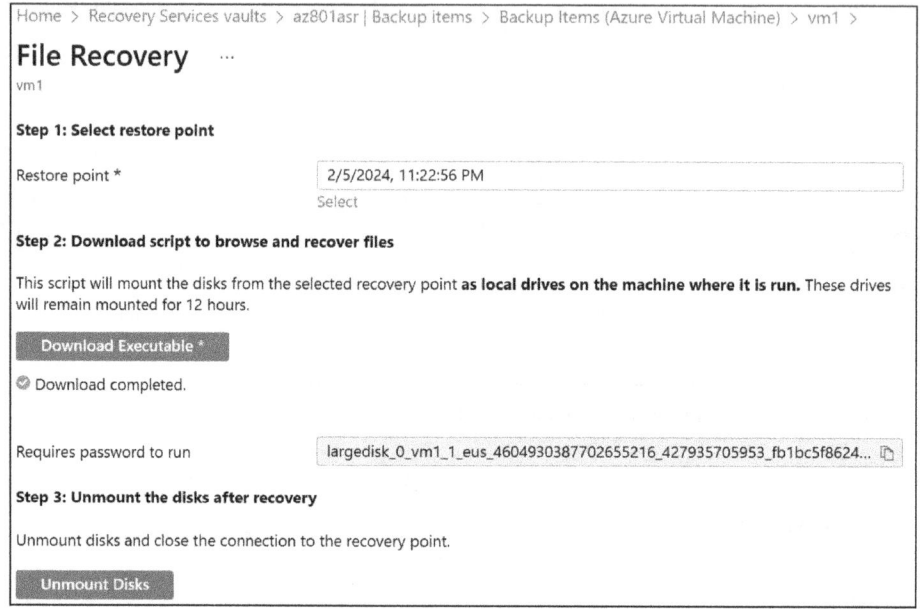

**FIGURE 5-69**   Restore to a new virtual machine

*EXAM TIP*

To restore a virtual machine that has encrypted disks, you also need to provide the Azure Backup service access to the key vault holding the keys. See *https://learn.microsoft.com/azure/backup/backup-azure-vms-encryption*.

*NEED MORE REVIEW?*   **MORE DETAILS ABOUT RESTORING VIRTUAL MACHINES AND FILES**

You can learn more about recovering virtual machines with the Azure Backup service at *https://learn.microsoft.com/azure/backup/backup-azure-arm-restore-vms*.

For more information about file-level recovery, see *https://learn.microsoft.com/azure/backup/backup-azure-restore-files-from-vm*.

## Create and configure backup policy

You can edit a policy, associate more VMs to a policy, and delete unnecessary policies to meet compliance requirements.

To view your current backup policies in the Azure portal, open the Recovery Services vault blade, and then click Backup Policies, as shown in Figure 5-70. Click an existing policy to view the policy details, or click Add to create a new policy.

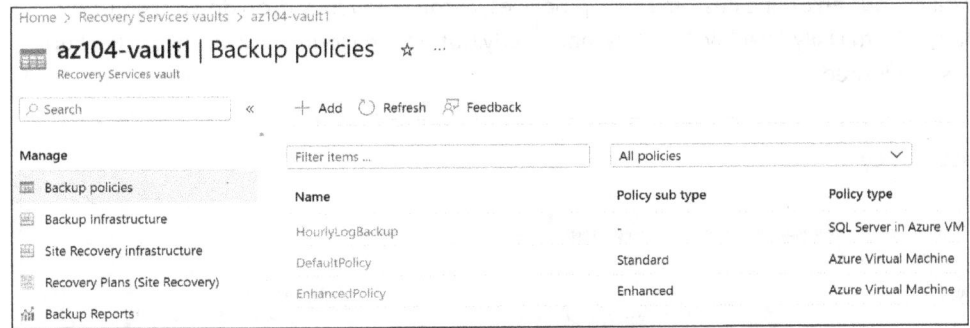

**FIGURE 5-70**  Backup policies in the Recovery Services vault

You can create five different types of policies from this view, as shown in Figure 5-71:

- **Azure Virtual Machine**   Specify the backup frequency, retention period, and the backup point on a weekly, monthly, and yearly schedule.
- **Azure File Share**   Schedule a daily backup for an Azure file share.
- **SQL Server In Azure VM**   Use SQL Server-specific backup technology, such as full, differential, and log backups, with an associated schedule for each option. Also, you can enable SQL backup compression.
- **SAP HANA In Azure VM (Database via Backint)**   Use SAP HANA—specific backup technology such as full, differential, and log backup with an associated schedule for each option.
- **SAP HANA In Azure VM (DB instance via snapshot)**   Use a snapshot to back up a SAP HANA instance.

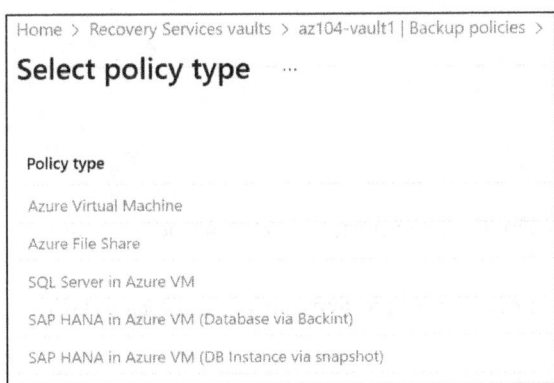

**FIGURE 5-71**  Available backup policy options in the Azure portal

## Define backup policies

An Azure Backup policy defines how often backups occur and how long the backups are retained. The default policy accomplishes a daily backup at 05:30pm UTC and retains backups for 30 days. You can define custom backup policies. In the Frequency drop down menu, you can choose from Daily, Weekly, Monthly, and Yearly options. In Figure 5-72, a custom backup policy is configured.

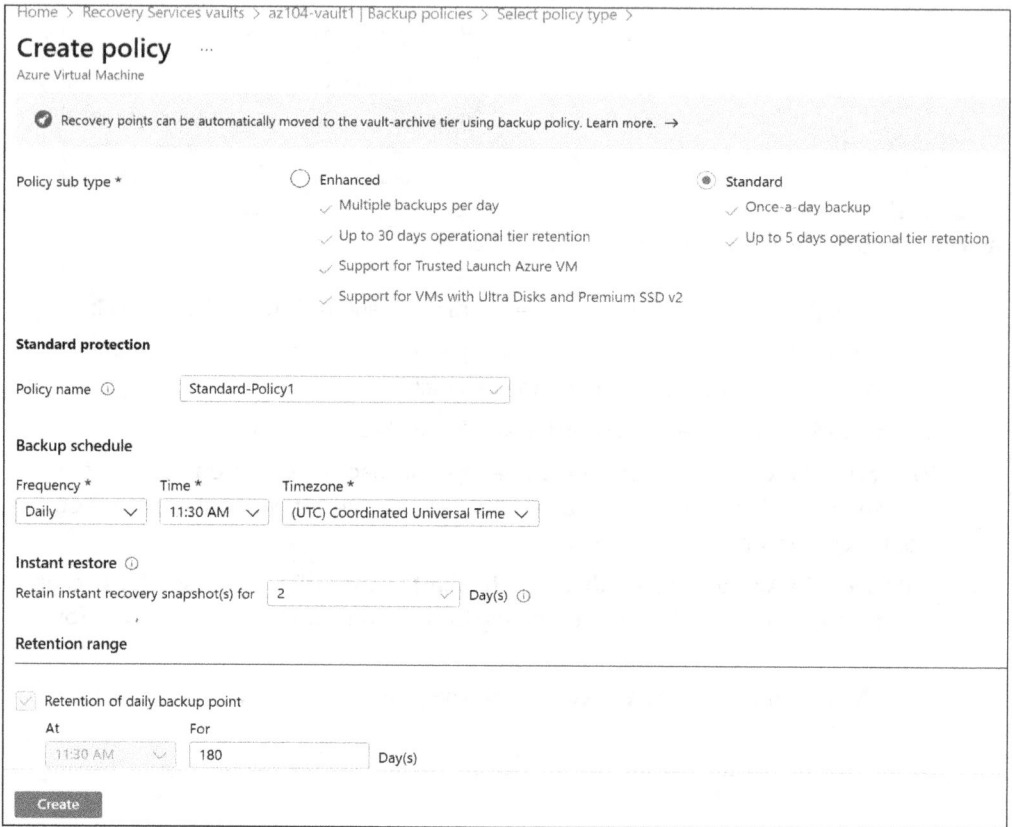

**FIGURE 5-72**   Configuring a custom backup policy

## Implement backup policies

To implement a backup policy, open the policy in the Azure portal and click Associated Items, as shown in Figure 5-73.

The Associated Items blade in Figure 5-74 shows all the resources currently associated with the policy.

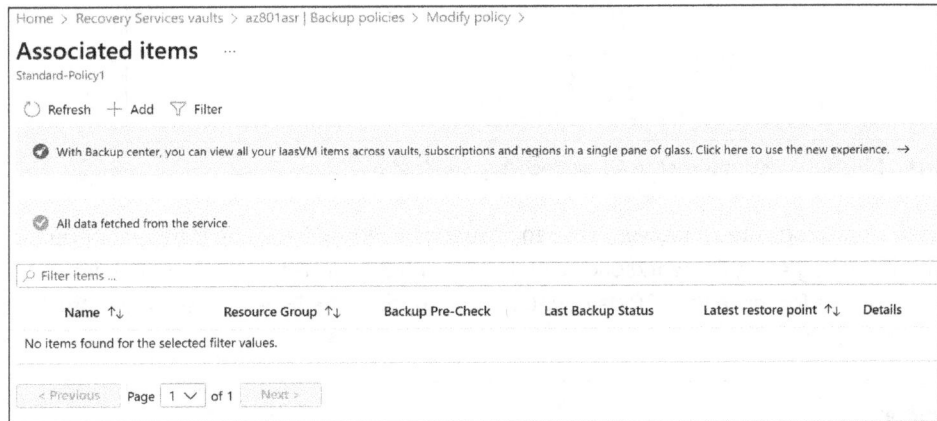

FIGURE 5-73   Associated Items in the Modify Policy blade

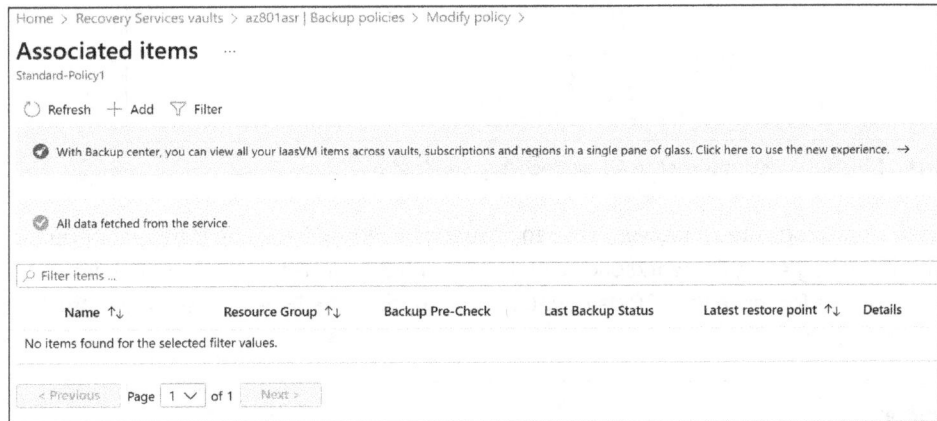

FIGURE 5-74   The Associated Items blade for the backup policy

Clicking Add will open the Backup Goal blade, where you can add other virtual machines or file shares to be backed up using the goals defined in the policy.

## Configure and review backup reports

Using Azure Backup reports, you can visualize data across your Recovery Services vaults and Azure subscriptions to provide insight into your backup activity. This reporting solution is currently widely supported for Azure virtual machine backup and file and folder backup scenarios when using the MARS (Microsoft Azure Recovery Services) agent. For other supported scenarios, see *https://learn.microsoft.com/azure/backup/configure-reports?tabs=recovery-services-vaults#supported-scenarios*.

To configure the backup reports, you need to create or use an existing Log Analytics workspace to store the backup reporting data. Also, you need a Recovery Services vault, which records all the backup operations as diagnostic data. Creating a Recovery Services vault is discussed earlier in Skill 5.2 in this chapter (see "Create a Recovery Services vault"). To configure diagnostics for the Recovery Service Vault, open the Recovery Services vault and then choose Diagnostic Settings, Add Diagnostic Setting (see Figure 5-75).

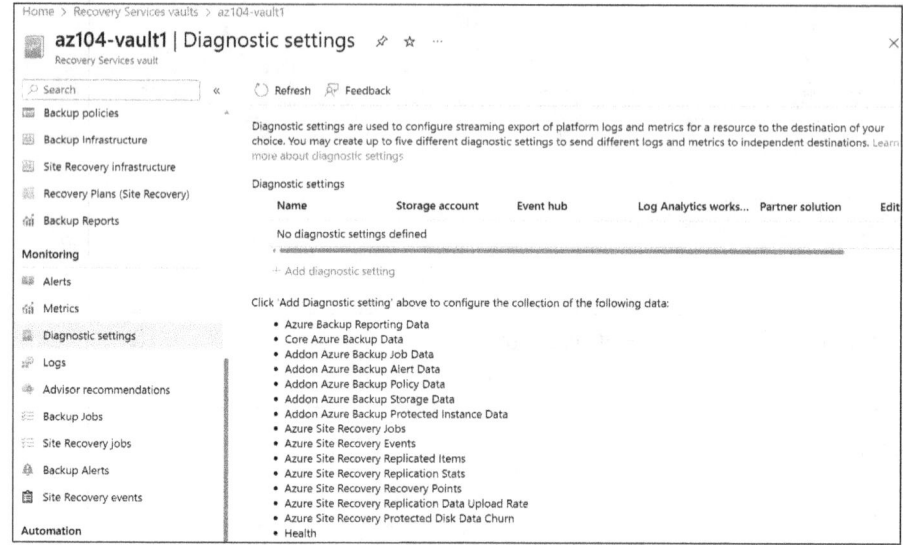

**FIGURE 5-75**   Diagnostic settings for the Azure Recovery Services vault

In this example, log categories were chosen as shown in Figure 5-76, and data is configured to be sent to the Log Analytics workspace with the default 30-day retention setting. If you want to retain data for more than 30 days, you need to update the Retention setting in the Log Analytics workspace.

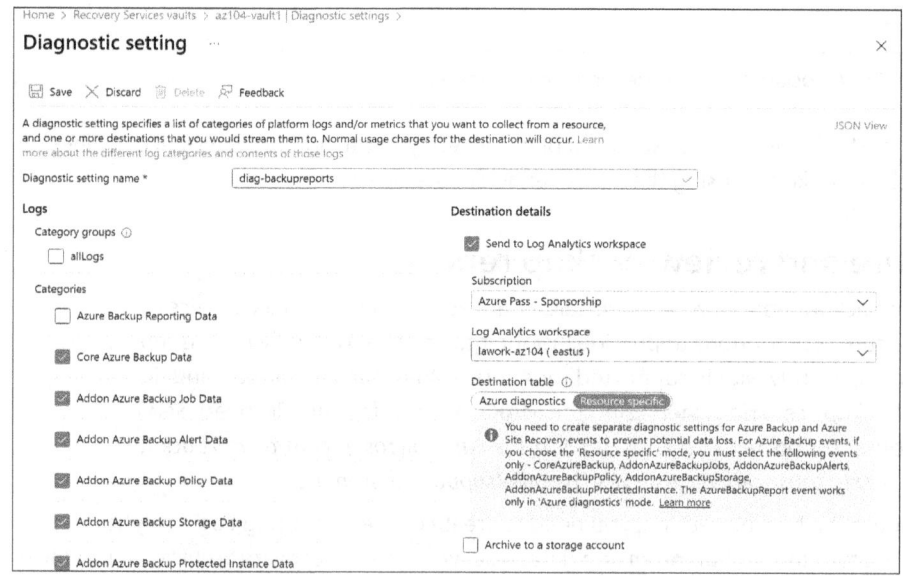

**FIGURE 5-76**   Diagnostic settings for the Azure Recovery Services vault

Once diagnostic settings are configured, you can view the backup report data in the Recovery Services vault by clicking Backup Reports under Manage, as shown in Figure 5-77.

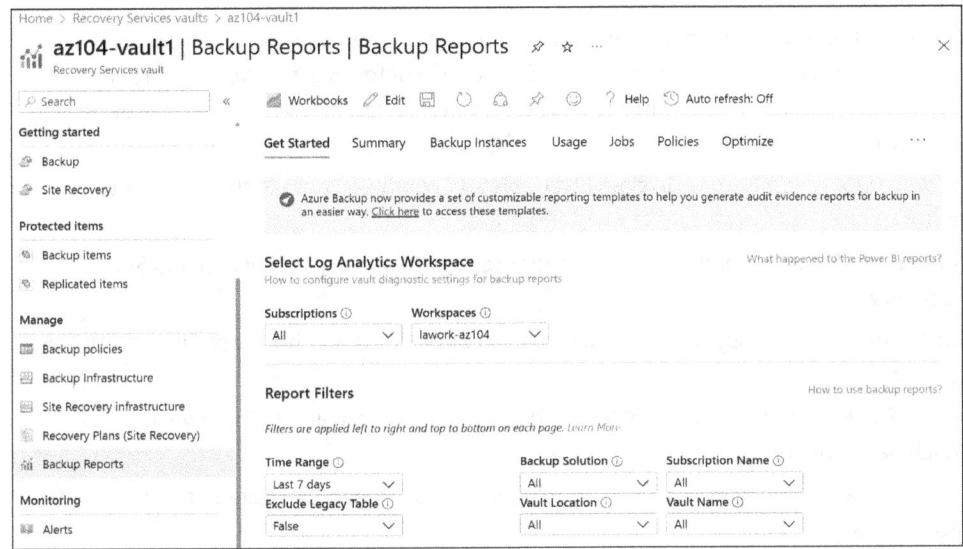

**FIGURE 5-77** Backup reports for the Azure Recovery vault

# Chapter summary

Here are some of the key takeaways from this chapter:

- Azure Monitor is a single pane of glass for accessing Azure metrics, tenant and resource diagnostic logs, Log Analytics, service health, and alerts.
- You can configure alerts based on metric alerts (captured from Azure Monitor Metrics) to activity log alerts that can provide notification by email, webhook, SMS, Logic Apps, or even an Azure Automation runbook.
- Azure Log Analytics can consolidate machine data from on-premises and cloud-based workloads and this data is indexed and categorized for quick searching. Data can be collected from both Windows and Linux machines.
- Azure Log Analytics has many management solutions that help administrators gain value out of complex machine data. These solutions contain prebuilt visualizations and queries that help surface insights quickly.
- Queries in Log Analytics can be saved for quick access and visualized and shared using Azure dashboards. To analyze data outside of Log Analytics you can export the data to Excel and Power BI.
- Network Watcher contains a suite of tools that can be used to help monitor network connections and traffic flows.

- The Azure Backup service can back up and restore an entire virtual machine. You can also use it to restore files from a recovery point without recreating the entire virtual machine.

- Azure Backup can be used to protect files and folders, applications, and IaaS virtual machines. This cloud-based data protection service helps organizations by providing offsite backups of on-premises servers and protection of VM workloads they have already moved to the cloud.

- Backup data is retained for 14 days after deletion when you enable the soft delete feature.

- A Recovery Services vault is used to configure and manage both Backup and Site Recovery.

- An Azure Backup policy defines how often backups occur and how long the backups are retained.

- Azure Site Recovery service enables us to replicate, failover, and failback virtual machines as needed.

- Recovery Services vaults can use diagnostic settings to enable detailed backup reports stored in Log Analytics.

## Thought experiment

In this thought experiment, apply what you have learned. You can find answers to these questions in the next section.

You are the administrator for Trey Research Pharmaceuticals. As a leader in the design and manufacturing of cutting-edge treatments for cancer patients, Trey Research needs to ensure that users' data within the organization is protected as they handle sensitive data and can't accommodate any data loss. Users have their own assigned VMs in Azure that are deployed in Canada Central region.

Trey Research needs to ensure the backup solution has the following features:

1. All users' data must be backed up daily at 6 PM Eastern time. The data should be retained for one year from the date it is backed up. What should be configured with Azure Backup to facilitate this requirement?

2. If any user's backup data is accidently deleted, then they should be able to restore it within two weeks. Which setting allows recovery for accidentally deleted items?

3. Users should be able to restore their VMs as well as files and folders from the backup data. Which features of Azure Backup provide this functionality?

# Thought experiment answers

This section contains the answers to the thought experiment for the chapter.

1. Create a backup policy with the schedule to execute the backup at 6 PM Eastern time with the retention of daily backup points for 365 days.

2. Enable the soft delete feature under Security Settings by visiting the Properties of the Recovery Services vault.

3. Leverage Restore VM and File Recovery options for restoring the VM and restoring files and folders respectively.

# CHAPTER 6

# Exam Ref AZ-104 Microsoft Azure Administrator exam updates

## The purpose of this chapter

In all the other chapters, the content should remain unchanged throughout this edition of the book. However, this chapter will change over time, with an updated PDF posted online so you can see the latest version of the chapter, even after you purchase this book.

Why do we need a chapter that updates over time? For three reasons.

1. To add more technical content to the book before it is time to replace the current book edition with the next edition. This chapter will include additional technology content and possibly additional PDFs containing more content.

2. To communicate detail about the next version of the exam, to tell you about our publishing plans for that edition, and to help you understand what that means to you.

3. To provide an accurate mapping of the current exam objectives to existing chapter content. While exam objectives evolve and are updated and products are renamed, much of the content in this book will remain accurate and relevant. In addition to covering any content gaps that appear through additions to the objectives, this chapter will provide explanatory notes on how the new objectives map to the current text.

After the initial publication of this book, Microsoft Press will provide supplemental updates as digital downloads for minor exam updates. If an exam has major changes or accumulates enough minor changes, we will then announce a new edition. We will do our best to provide any updates to you free of charge before we release a new edition. However, if the updates are significant enough in between editions, we may release the updates as a low-priced standalone e-book.

If we do produce a free updated version of this chapter, you can access it on the book's product page. Simply visit *MicrosoftPressStore.com/ERAZ1042e/downloads* to view and download the updated material.

## About possible exam updates

Microsoft reviews exam content periodically to ensure that it aligns with the technology and job role associated with the exam. This includes, but is not limited to, incorporating functionality and features related to technology changes, changing skills needed for success within a job role, and revisions to product names. Microsoft updates the exam details page to notify candidates when changes occur. If you have registered this book and an update occurs to this chapter, you will be notified by Microsoft Press when the updated chapter is available.

## Impact on you and your study plan

Microsoft's information helps you plan, but it also means that the exam might change before you pass the current exam. That affects you, and it affects how we deliver this book to you. This chapter gives us a way to communicate in detail about those changes as they occur. But you should watch other spaces as well.

For news, bookmark and check these sites:

- **Microsoft Learn**: Check the main source for up-to-date information: *microsoft.com/learn.* Make sure to sign up for automatic notifications on that page.
- **Microsoft Press**: Find information about products, offers, discounts, and free downloads: *microsoftpressstore.com.* Make sure to register your purchased products.

As changes arise, we will update this chapter with more detail about exam and book content. At that point, we will publish an updated version of this chapter, listing our content plans. That detail will likely include the following:

- Content removed, so if you plan to take the new exam version, you can ignore those sections when studying.
- New content planned related to new exam topics, so you know what's coming

The remainder of the chapter shows the new content that may change over time.

# News and commentary about the exam objective updates

The current official Microsoft Study Guide for the AZ-104 Microsoft Azure Administrator exam is located at *https://learn.microsoft.com/en-us/credentials/certifications/resources/study-guides/az-104.* This page has the most recent version of the exam objective domain.

This statement was last updated in October 2023, before the publication of *Exam Ref AZ-104 Microsoft Azure Administrator.*

This version of this chapter has no news to share about the next exam release.

| Exam objective | Chapter |
| --- | --- |
| Configure and manage storage accounts<br>  ■ Configure Azure storage redundancy<br>  ■ Configure object replication<br>  ■ Configure storage account encryption<br>  ■ Manage data by using Azure Storage Explorer<br>  ■ Manage data by using AzCopy | 2 |
| Configure Azure Files and Azure Blob Storage<br>  ■ Create and configure a file share in Azure storage<br>  ■ Configure Azure Blob Storage<br>  ■ Configure storage tiers<br>  ■ Configure soft delete, versioning, and snapshots<br>  ■ Configure blob lifecycle management | 2 |
| **Deploy and manage Azure compute resources** | |
| Automate deployment of resources<br>  ■ Interpret an Azure Resource Manager (ARM) template<br>  ■ Modify an existing ARM template<br>  ■ Deploy resources from a template<br>  ■ Export a deployment template<br>  ■ Interpret and modify a Bicep file | 3 |
| Create and configure virtual machines<br>  ■ Create a virtual machine<br>  ■ Configure Azure Disk Encryption<br>  ■ Move VMs from one resource group or subscription to another<br>  ■ Manage VM sizes<br>  ■ Manage VM disks<br>  ■ Deploy VMs to availability sets and zones<br>  ■ Deploy and configure Virtual Machine Scale Sets | 3 |
| Provision and manage containers<br>  ■ Create and manage an Azure Container Registry (ACR)<br>  ■ Provision a container by using Azure Container Instances (ACI)<br>  ■ Provision a container by using Azure Container Apps (ACA)<br>  ■ Manage sizing and scaling for containers | 3 |
| Create and configure Azure App Service<br>  ■ Provision an App Service plan<br>  ■ Configure scaling for an App Service plan<br>  ■ Create an App Service<br>  ■ Map an existing custom DNS name to an App Service<br>  ■ Configure certificates and TLS for an App Service<br>  ■ Configure backup for an App Service<br>  ■ Configure networking settings for an App Service<br>  ■ Configure deployment slots for an App Service | 3 |

| Exam objective | Chapter |
|---|---|
| **Configure and manage virtual networking** | |
| Configure and manage virtual networks in Azure<br>■ Create and configure virtual networks and subnets<br>■ Create and configure virtual network peering<br>■ Configure public IP addresses<br>■ Configure user-defined network routes<br>■ Troubleshoot network connectivity | 4 |
| Configure secure access to virtual networks<br>■ Create and configure network security groups and application security groups<br>■ Evaluate effective security rules<br>■ Deploy and configure Azure Bastion Service<br>■ Configure service endpoints for Azure services<br>■ Configure private endpoints for Azure services | 4 |
| Configure name resolution and load balancing<br>■ Configure Azure DNS<br>■ Configure load balancing<br>■ Troubleshoot load balancing | 4 |
| **Monitor and back up Azure resources** | |
| Monitor resources in Azure<br>■ Interpret metrics in Azure Monitor<br>■ Configure log settings in Azure Monitor<br>■ Query and analyze logs in Azure Monitor<br>■ Set up alert rules, action groups, and alert processing rules in Azure Monitor<br>■ Configure Application Insights<br>■ Configure and interpret monitoring of VMs, storage accounts, and networks by using Azure Monitor Insights<br>■ Use Azure Network Watcher and Connection Monitor | 5 |
| Implement backup and recovery<br>■ Create and manage a Recovery Services vault<br>■ Configure Azure Site Recovery<br>■ Create an Azure Backup vault<br>■ Create and configure backup policy<br>■ Configure and review backup reports | 5 |

# Index

## A

ACA (Azure Container Apps), 123
    connecting to, 184–186
    creating an instance, 178–184
    provisioning a container, 178
    scaling and sizing, 187–189
access control, 16
    blob storage, 77–78, 86–88
    role-based, 16–19
    scope, 18–19
access keys, storage account, 83–84
access tiers, Azure Blob Storage, 69–70, 110–112
accountability, organizational, 55
ACI (Azure Container Instances), 123
    connecting to, 177–178
    creating, 174–177
    scaling and sizing, 186–187
ACR (Azure Container Registry), 168
    Access Keys blade, 172
    creating an instance from the Azure portal, 170–171
    managing, 172
    tiers, 169
action groups, 316–318
activity logs, 304–305
additive model, 16–17
ADFS (Active Directory Federation Services), 1
administrator role
    permissions, 49–50
    subscription, 49
agents, Log Analytics workspace, 301
AKS (Azure Kubernetes Service), 123
alert/s, 311–312
    action groups, 316–318
    analyzing across subscriptions, 319–321
    Azure Monitor, 294

budget, 57
    creating, 312–313
    rules, 298, 313–315
    target resource, 313
    viewing, 318–319
algorithm, spreading, 168
alias record, 269–270
aligned availability set, 163
allocation, public IP address, 228–229
App Service, 189–190, 199–200
    backup, 204–205
    creating, 193–196
    deployment slots, 210–211
    managed certificates, 200–201
    mapping to a custom DNS name, 196–199
    network settings, 205–206
    private key certificates, 201–203
    public key certificate, 203
App Service plan
    creating, 190–192
    provisioning, 190
    scaling, 192–193
append blob, 66, 107
application, three-tier architecture, 246
Application Insights, configuration, 321–323. *See also* Azure Monitor, insights
applying, resource tags, 41
architecture, three-tier application, 246
archive access tier, Azure Blob Storage, 69–70
ARM (Azure Resource Manager) template, 16, 124
    for adding a public IP address, 128–129

Complete mode, 131–132
for creating a network interface, 127
for creating a virtual network, 126–127
for defining a virtual machine resource, 129
deployment, 133–135, 137
editing, 133–134
elements, 125
exporting from a deployment, 137–139
functions, resourceGroup(), 126
Incremental mode, 132
modifying an existing, 131
parameters, 136, 137
UI, 135
validation, 135
variables, syntax, 126
ASG (application security group), 246–247, 251–253
assigned group, 5
async blob copy, 100–101
authentication, 84–85, 86–88
availability, 161
    set
    storage account replication mode, 89–91
    zone, 159–160
az deployment group create command, 142
AZ-104 Microsoft Azure Administrator exam, 358–359
    updates, 358–359
azcopy command, 99–100
AzNetworkWatcherNextHop cmdlet, 329